# Sex, Smoke, and Spirits: The Role of Chemistry

ACS SYMPOSIUM SERIES **1321**

# Sex, Smoke, and Spirits: The Role of Chemistry

**Brian Guthrie**, Editor
*Cargill, Incorporated*
*Plymouth, Minnesota, United States*

**Jonathan D. Beauchamp**, Editor
*Fraunhofer Institute for Process Engineering and Packaging IVV*
*Freising, Germany*

**Andrea Buettner**, Editor
*Department of Chemistry and Pharmacy, Friedrich-Alexander-Universität Erlangen-Nürnberg*
*Erlangen, Germany*
*Fraunhofer Institute for Process Engineering and Packaging IVV*
*Freising, Germany*

**Stephen Toth**, Editor
*International Flavors & Fragrances*
*Union Beach, New Jersey, United States*

**Michael C. Qian**, Editor
*Department of Food Science and Technology, Oregon State University*
*Corvallis, Oregon, United States*

**Sponsored by the**
**ACS Division of Agricultural and Food Chemistry, Inc.**

American Chemical Society, Washington, DC

**Library of Congress Cataloging-in-Publication Data**

Names: Guthrie, Brian (Chemist), editor. | Beauchamp, Jonathan, editor. | Buettner, Andrea, editor. | Toth, Stephen, editor. | Qian, Michael, editor.

Title: Sex, smoke, and spirits : the role of chemistry / Brian Guthrie, Jonathan D. Beauchamp, Andrea Buettner, Stephen Toth, Michael C. Qian, editors.

Description: Washington, DC : American Chemical Society, [2019] | Series: ACS symposium series ; 1321 | Includes bibliographical references and index.

Identifiers: LCCN 2019030422 (print) | LCCN 2019030423 (ebook) | ISBN 9780841234673 (hardcover) | ISBN 9780841234635 (ebook other)

Subjects: LCSH: Smell. | Sex (Biology) | Smoke--Physiological effect. | Alcoholic beverages--Flavor and odor. | Psychophysiology. | Chemistry, Technical.

Classification: LCC QP458 .S49 2019 (print) | LCC QP458 (ebook) | DDC 612.8/6--dc23

LC record available at https://lccn.loc.gov/2019030422

LC ebook record available at https://lccn.loc.gov/2019030423

# Foreword

The purpose of the series is to publish timely, comprehensive books developed from the ACS sponsored symposia based on current scientific research. Occasionally, books are developed from symposia sponsored by other organizations when the topic is of keen interest to the chemistry audience.

Before a book proposal is accepted, the proposed table of contents is reviewed for appropriate and comprehensive coverage and for interest to the audience. Some papers may be excluded to better focus the book; others may be added to provide comprehensiveness. When appropriate, overview or introductory chapters are added. Drafts of chapters are peer-reviewed prior to final acceptance or rejection.

As a rule, only original research papers and original review papers are included in the volumes. Verbatim reproductions of previous published papers are not accepted.

**ACS Books Department**

# Contents

## Indexes

# Preface

*Sex*, *smoke*, and *spirits*: these three nouns are associated with acts of pleasure and indulgence but can equally invoke feelings of impropriety or abstention in some people and certain cultures. In terms of human evolution, however, each word signifies an essential element in the survival and procreation of our species during its conspecific and environmental interactions. The importance of sex from a biological perspective need not be elaborated on here, but the associated pleasurable and emotional aspects played critical roles in the creation of societal structures, cohabitation, and the evolution of human relationships. Smoke from wildfires would elicit fear and cause flight, but in a different context it was an intrinsic part of essential fire used for cooking, warmth and thereby survival. And spirits (alcohol) was encountered naturally as a byproduct in fermenting fruits but was also actively exploited to preserve foods and later purposefully produced for consumption; distillation has even recently been posited as the primary driver for early nomadic peoples to become settlers (1). Sex, smoke, and spirits are similarly integral aspects of many cultural celebrations and rites of passage, both religious and secular, and all have been invoked in some way or other in ritualistic or spiritual observances to connect with a creator or for mental transcendence.

Our evolutionary interactions with these three primeval phenomena have laid the foundations for their societal importance today, with their influences seemingly ingrained in our collective psyche, both positively and negatively. Contemporarily, sex, smoke, and spirits are at the forefront of consumerism and marketing. Sex sells. The smoke (tobacco) and spirits (alcohol) industries are behemoths of consumer goods, primarily geared toward pleasure and indulgence. So, how does chemistry play a role? If you have picked up this book with the hope of reading a romantic thriller, then prepare to be disappointed; but if applied chemistry turns your knobs, then you will be rewarded with a compendium of cutting-edge scientific research centering on the topics of smoke, sex, and spirits.

The contributions to this book are drawn from three symposia organized under the auspices of the Division of Agriculture and Food Chemistry (AGFD) of the American Chemical Society (ACS), held at the 255th ACS National Meeting that took place from March 18 - 22, 2018 in New Orleans, LA, USA. These symposia were:

- "Chemistry of Sex", organized by Terry E. Acree, John W. Finley, Stephen J. Toth, Michael H. Tunick, Kathryn. D. Deibler, and Alyson E. Mitchell;
- "Up in Smoke: Chemistry of Smoky Odors in Food & the Environment", organized by Jonathan D. Beauchamp, Brian D. Guthrie, and Andrea Buettner; and
- "Chemistry of Spirits", organized by Keith R. Cadwallader, Michael Granvogl, and Michael. C. Qian.

While seemingly disparate topics, discussions between the session organizers during the conference identified many commonalities: the chemistries across these areas represent significant challenges that require the development of new methodologies applied in creative ways. Thus, besides the cultural and societal links between sex, smoke, and spirits, these emerging methods, approaches, and applications are the unifying elements underpinning this book.

Sex and relationships play an implicit role in human well-being, and consumer products are often associated or marketed with connotations toward sex and arousal. The first section of this book contains chapters on topics concerning the chemical origins of attraction, behavior, and pleasure. Fragrance and certain foods, such as chocolate, are well known to drive attraction, emotion, and craving. While psychologists have studied these in great detail, the complete picture of their chemical nature and psychological drivers are still emerging. These are of great importance, especially for fragrance providers, since these are critical determinants of product value. The individual chapters of the sex section include: a report on the chemical basis of perceived fragrance pleasantness and the related molecular structures, stereochemistry, chirality, and other molecular descriptors; a presentation of novel data exploring interpersonal relationships using experimental design theory and statistical models; and a review of the chemistry of chocolate – often considered an aphrodisiac – and its relationship with the consumer in terms of its constituent aroma compounds.

Over the course of history, humans have had significant positive and negative interactions with fire and smoke. Smoke and fire find their way into many human rituals and cuisines. The middle section of this book contains chapters that focus on the complex chemistries and sensory perceptions associated with our interactions with smoke and smoky foods. The smoke chapters include: a review of the chemical structure-to-smoky odor relationships in the perception of smoky odors; an exploration of the specific chemical reactions that occur during wood pyrolysis and impart the characteristic flavors on charcoal-grilled foods; a report on a novel zeolite filter that removes potentially harmful compounds while retaining the smoky character during the smoking process of foods; a presentation of the development and characterization of a lab-scale smoke generator for smoking foods; and finally three chapters dealing with smoky alcoholic beverages, namely Bavarian wheat-beer and Scotch whiskies from the island of Islay, and how the associated odor-active compounds elicit the characteristic flavor, as well as a report on a novel method for quantifying their amounts.

Smoky beers and whiskies offer a transition to the final section of the book, which deals with other distilled spirits. Spirits are well known for their characteristic flavor subtleties and regional variations. The analysis of the flavor-active volatiles of distilled spirits can be challenging due to high levels of ethanol that can make volatile isolation difficult. The chapters in the spirits section of this book encompass reports on complex flavor chemistries of distilled spirits, which often show a high degree of flavor diversity that requires precise, quantitative analysis of volatiles. The spirits chapters include: a review of the diverse nature of spirits in terms of their production and flavor characteristics; a discourse summarizing traditional Chinese baijiu distilled spirit in terms of volatile composition; a report on sesame flavor in baijiu; an exploration of the aroma composition of barley-based Tibetan Qingke liquor in comparison to other baijiu spirits; a similar treatment of Wuliangye liquor; empirical data on the aroma composition of the *Folium isatidis* leaf, a raw material used for Chinese liquor; an exploration of how adding water to your spirit beverage impacts its flavor; a report on using the "sensomics" approach to unravel the relationship between chemical content and sensory impact in rum; and finally a report on the use of a stir-bar sorptive extraction (SBSE) in the analysis of volatile compounds in Tequila.

Overall, this book presents a snapshot of the latest research covering emerging challenges in chemistry in relation to sexual attraction, smoky flavors, and alcoholic spirits. We hope that researchers in the field will benefit from the range of studies presented in this book and their notable commonalities.

We are grateful to the authors for contributing their chapters and to the numerous reviewers who carefully read and commented on the chapters of this book to help improve their quality. We would like to express thanks for the wonderful help and support – and patience! – from the staff at the ACS Books Editorial Office, especially Chris Moffitt and Amanda Koenig.

## References

1.   Rogers, A. *Proof: The Science of Booze*; Mariner Books: New York, 2014.

**Brian Guthrie**
Cargill Corporate Fellow
Cargill, Incorporated
14800 28th Avenue N.
Plymouth, Minnesota 55447, United States

**Jonathan D. Beauchamp**
Department of Sensory Analytics
Fraunhofer Institute for Process Engineering and Packaging IVV
Giggenhauser Str. 35
85354 Freising, Germany

**Andrea Buettner**
Chair of Aroma and Smell Research
Department of Chemistry and Pharmacy
Friedrich-Alexander-Universität Erlangen-Nürnberg Henkestr. 9
91054 Erlangen - Germany
Fraunhofer Institute for Process Engineering and Packaging IVV
Giggenhauser Str. 35
85354 Freising, Germany

**Stephen Toth**
International Flavors & Fragrances
1515 State Highway #36
Union Beach, New Jersey 07735, United States

**Michael C. Qian**
Department of Food Science and Technology
Oregon State University,
Corvallis, Oregon 97330, United States

## Chapter 1

# Fragrance and Attraction

**Anubhav P. S. Narula***

**International Flavors & Fragrances, 1515 Highway 36,
Union Beach, New Jersey 07735, United States**
***E-mail:anubhav.narula@iff.com.**

The fragrance of flowers, fruits, and fauna of natural scents have long been highly popular. What makes this attraction possible? What is the chemistry of fragrance and flavors? This chapter is based on the plenary lecture given at the "Chemistry of Sex" symposium held during the 2018 Spring ACS meeting in New Orleans, Louisiana. This chapter will focus on the chemical basis of fragrance attraction and presents a rich tapestry of chemistry and the vibrant fragrance industry behind our emotional attraction to scents and perfumes. Perfumery examples delineating how minor changes in the structure, stereochemistry, shape, and chirality of a fragrance ingredient or a scent molecule influence its smell, pleasantness, and usefulness are highlighted. Thanks to the technological advances in the art of organic synthesis, perfumes, once the luxury of kings and queens, have now become available for everyone to enjoy. Brief glimpses of the power of synthetic chemistry and biotechnology in a quest for new scents are presented.

## Introduction

Fragrance and attraction are intertwined through a multi-sensory communication of pheromones and neurotransmitters. Why are bees attracted to flowers? Why does a humming bird typically feed on red flowers? Besides pollination, it has been speculated that flowers developed fragrances, spicy smelling natural scents not only to attract insects but to also act as a defense mechanism to ward off plant eaters.

People are naturally attracted to smells that are pleasing, soothing, and calming. As a result, there has been a significant rise in the use of homecare fragrances, incense sticks, and candles. Fragrances have been known to evoke childhood memories and can touch our hearts, which has been used for beneficial purposes like aromatherapy. Fragrances also impart feeling like naturalness, cleanliness, softness, and pleasantness; as a result, they are widely used in personal hygiene and homecare products.

## Flavor and Fragrance Business

Because of the enormous utility of fragrances, the attraction to fragrances and flavor has become a huge global business worth $26.3 billion in 2017. The top five flavors and fragrance companies (*1*) control 61.5% of the total share of the market.

Hundreds of new fragrances are launched every year in an attempt to please and attract consumers with exhilarating scents that impart relaxing, soothing, energizing and joyful feelings. In spite of that, the bestselling fragrance for women in 2017 in the USA was Chanel No 5. Chanel No 5 was created in 1921 and has remained one of the most admired and beloved perfumes. It has stood the test of time, and appealed to every generation of users over the last 97 years. In addition, it is worth pointing out that one of the oldest continuously produced fragrance in the world is Jicky by Guerlain (*2*) that was launched in 1889.

## Language of Fragrances

Let us now consider common descriptors that are used by perfumers to describe fragrance scents. There is an incredible odor diversity that exists in fragrance and olfaction today. These notes belong to the following five main odor categories:

- Fresh Notes
- Floral Notes
- Oriental Notes
- Woody Notes
- Animal Notes

Each of these notes are further comprised of many other widely used notes including: citrus, grapefruit, marine, green, galbanum, aromatic, fruity notes of apple, peach, pear, strawberry, nectarine, plum, and pineapple.

Floral notes comprise all of the most appreciated fragrant flowers like rose, jasmine, tuberose, gardenia, orange blossoms, violet, and lavender. Oriental notes include amber, incense, resins, and woody notes that are an essential part of every perfume. Woody notes primarily include sandal, cedar, patchouli, and vetiver class. Oudh and agar wood have gained popularity lately. Other classes of appreciated notes include spicy, leather, and mossy notes along with animal notes such as musk, ambergris, civet, and others. Sweet and gourmand notes are also well liked by consumers and perfumers alike.

## Fragrance Design

Conceptually, fragrance creation or design uses an olfactive pyramid of various notes that are layered on top of each other like a piece of fine art. A fragrance can be comprised of up to hundreds of ingredients that are harmoniously blended like a musical symphony. These notes are divided into top, middle, and bottom notes.

### Top Notes

Aroma chemicals having citrus, fruity, green, aldehydic, marine and ozone odor impart, freshness, juiciness, friskiness, and impact to the fragrance. A select group of top notes are shown in Figure 1.

## Center Notes

The next part of the perfumery triangle are center notes which represent the heart and body of the fragrance. This group mainly includes floral notes such as rose, jasmine, violet, lily of the valley, muguet, tuberose, orange blossoms, iris, and others along with some long-lasting fruity, spicy, and herbaceous notes.

## Base Notes

The last part of the perfumery triangle is known as the base or bottom notes that use fragrance ingredients belonging to sweet, powdery, musk, and woody (sandal, patchouli, cedar, vetiver, and amber) scent molecules. These ingredients impart longevity and substantivity to the fragrance.

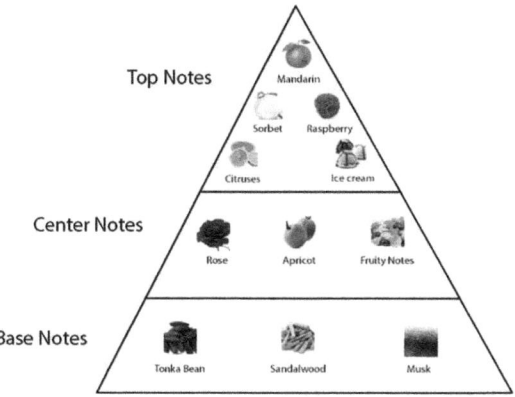

*Figure 1. Perfumery Triangle.*

## Precious Essential Oils Used in Fragrances

As described earlier, a fragrance is a symphony of hundreds of aroma chemicals (3). These aromas are derived from various essential oils (4); absolutes made from select flowers, fruits, citrus, precious woods, taste enhancing spices, sweet and creamy absolutes, and amber and musk odorants; and aromatherapy materials. Modern fragrances contain both natural and synthetic fragrance ingredients that chemists have discovered in order to complement and enhance creativity, innovation, and differentiation. A compilation of diverse essential oils, taste enhancing spices, creamy and floral absolutes, and other natural notes used in perfumery include:

- Jasmine Absolute Indian;
- Rose Bulgarian, Turkish rose, tuberose, and gardenia;
- Orange blossom, violet, hyacinth, geranium, and lavender;
- Peony, iris, and orchids;
- Eucalyptus, citronella, and turpentine essential oils.
- Spice oils such as cinnamon, ginger, cardamom, clove, and nutmeg;
- Lemon, bergamot, citrus, grapefruit, and orange essential oils;
- Fruits like peach, pear, apple, nectarine, strawberry, and melon;
- Sandalwood essential oil;

- Cedar wood essential oil;
- Patchouli essential oil;
- Vertiver essential oil;
- Oudh wood or agar wood;
- Ambergris, which is now replaced with mostly synthetic and natural amber molecules;
- Musk tincture, which is also replaced with mostly synthetic and natural musk molecules;
- Natural vanilla, synthetic vanillin, coumarin, lactonic, sweet, and gourmand molecules; and
- Aromatherapy essential oils.

## Analysis of Essential Oils

Fragrance chemists use instrumental analysis techniques, such gas chromatography-mass spectrometry (GC-MS), NMR, IR, and UV, to analyze and identify various components that are present in the essential oils (5). GC-olfactometry (6) is an important technique that is used to identify key components of a given essential oil, fruit, or spice. In addition, synthesis can be used to corroborate the structure of an unknown aroma chemical present in these natural fragrance ingredients. Figure 2 describes some of the key components that have been identified in Italian lemon oil.

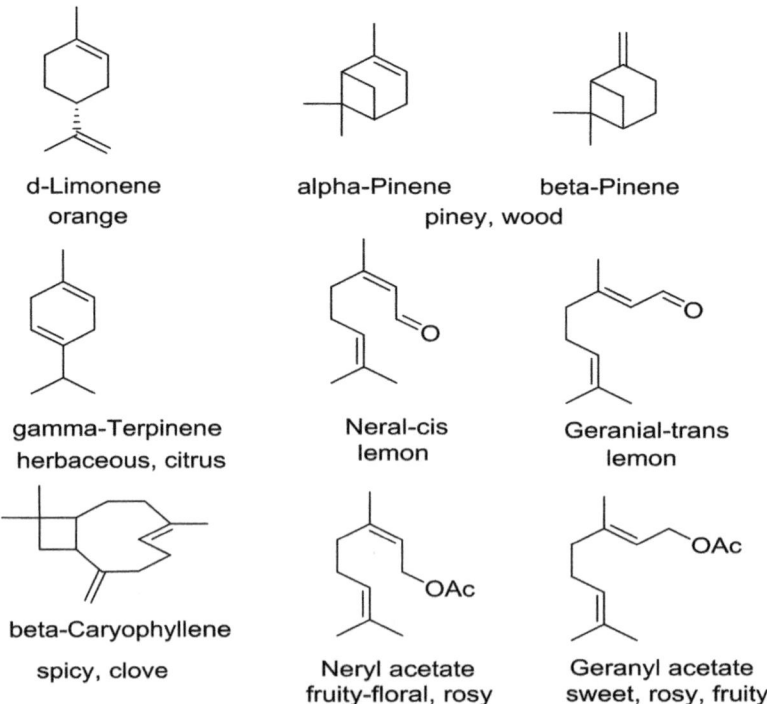

Figure 2. Key components of Italian lemon oil.

## Search for New Molecules

There are multiple benefits of discovering new molecules including:

- Ensuring a consistent supply of a key odor that contributes to the fragrance ingredient of an essential oil;
- Providing a cost effective alternate for very expensive naturally occurring essential oils that are in short supply or for a source that has been labeled as endangered or unsustainable;
- Having quality assurance while also providing molecules for safety testing, as well as testing for stability in functional applications including biodegradability, human health, and toxicity compliance concerns;
- Proprietary status of new molecules enhances creativity and offers competitive advantage for the inventing company or individual; and
- New fragrances launched are difficult to copy.

As a result, several flavor and fragrance companies have invested huge resources in developing hundreds of new fragrance ingredients (7) that are superior in performance and odor to many naturally occurring aroma chemicals.

## Synthetic Technologies Used for the Discovery of New Molecules

Over the last century, chemists have discovered new chemical reactions (8) and technologies (9) to create carbon-carbon bonds that are at the heart of synthesizing organic molecules. These endeavors have led to the synthesis of novel, unique, and complex structures. Many scientists have been awarded Nobel Prizes for their groundbreaking discoveries, which have been utilitized in advancing the art of organic synthesis. A few such technologies that have wide utility in flavor and fragrance are cited as follows:

- Name Reactions: Diels-Alder reaction (Robert Diels and Kurt Alder shared the 1950 Nobel Prize in Chemistry), pericyclic, free radical, biomimetic reactions, Mannich reaction, Wittig reaction, and Prins reaction (discovered in 1912 by Dr. Prins who also worked at International Flavors & Fragrances [IFF]);
- Condensation: Aldol, Dieckman, Knoevenagel, Michael, Stobbe, acyloin, hydroformylation, and metathesis cyclization (Robert Grubbs, Richard Schrock, and Yves Chauvin shared the 2005 Nobel Prize in Chemistry);
- Rearrangements: Claisen, Carrol, Favorskii, Cope, and oxy-Cope;
- Organometallic: Grignard, organolithium and cuprates, Reformatsky, Simmons-Smith cyclopropanation, and ethynylation among others;
- Oxidation: per acids, $CrO_3$, $NaIO_4$, $KMnO_4$, $H_2O_2$, TEMPO, Ozone, and $RuO_4$ among others;
- Reductions: using vitride, diborane (Herbert C. Brown was awarded the Nobel Prize in 1979), $LiAlH_4$, and $NaBH_4$;
- Asymmetric: hydrogenation or reduction and chiral epoxidation (Barry Sharpless, Ryoji Noyori, and Williams Knowles shared the 2001 Nobel Prize in Chemistry);
- Functional group transformations: aldehyde, ketone, alcohol, ether, acetal and ketal, ester, lactone, epoxide, and nitriles among others; and
- Hetero atoms incorporation: N (pyridine, pyrazine, pyrimidine), O (furan, pyran), and S (thio analogs).

## Discovery of IFF Classic Fragrance Molecules

Over the last 70 years, IFF scientists have discovered numerous differentiating aroma chemicals and developed commercial processes for fragrance ingredients like Galaxolide, Iso E Super, Cashmeran, Vertofix, Bacdanol, Lyral, Phenoxanol, phenyl ethyl alcohol, Helional, Canthoxal, Triplal, Trimofix, δ-Damascone, and Kharismal that have greatly influenced the fragrance industry. Many of these chemicals have become indispensable ingredients in perfumery.

The abundant availability of these unique and cost effective fragrance ingredients have allowed perfumers across the globe to create classic fragrances that have delighted and attracted consumers of both genders and all generations. Using these hedonically superior ingredients, IFF perfumers have created some of the world's most popular fragrances, which are presented in Figure 3.

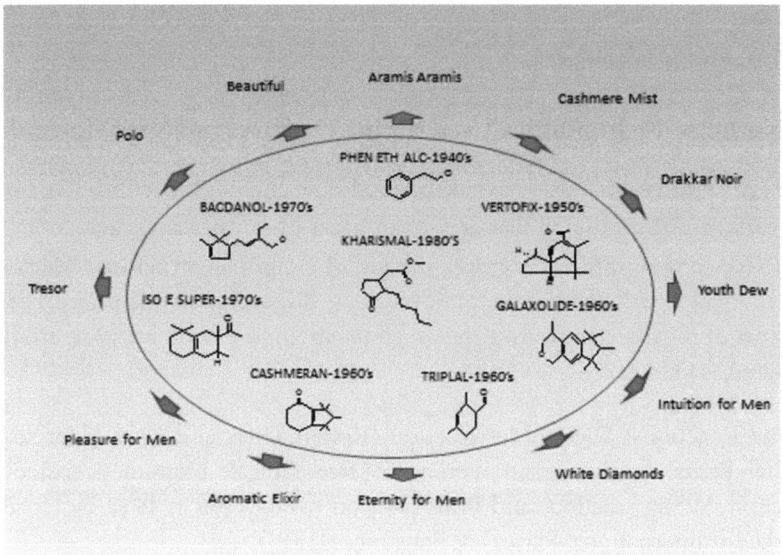

*Figure 3. Iconic fragrances created using IFF classic fragrance ingredients.*

## Technologies Used To Synthesize Fragrance Molecules for Consumer Fragrances

Over the next few sections, examples of the technologies used in the development of recent aroma chemicals for creating fine and consumer fragrances are provided. Since consumer fragrances employ very harsh bases (with pH varying from 2–13), fragrances for detergents, shampoos, hair conditioners, softeners, soaps, shower gels, homecare, and fabcare must use ingredients with good chemical stability within that specific medium needed for functional application.

## Examples of Diels-Alder Technology

Diels-Alder technology has been used extensively in the synthesis of numerous flavor and fragrance molecules that are used in consumer and fine fragrances. The odor description and structures of a few key ingredients like Lyral, aldehyde AA, Isocyclocitral, Iso E Super, Isoprecyclemone B, Melafleur, Myrac aldehyde, Camek, Oriniff, and δ-damascone are shown in Figure 4.

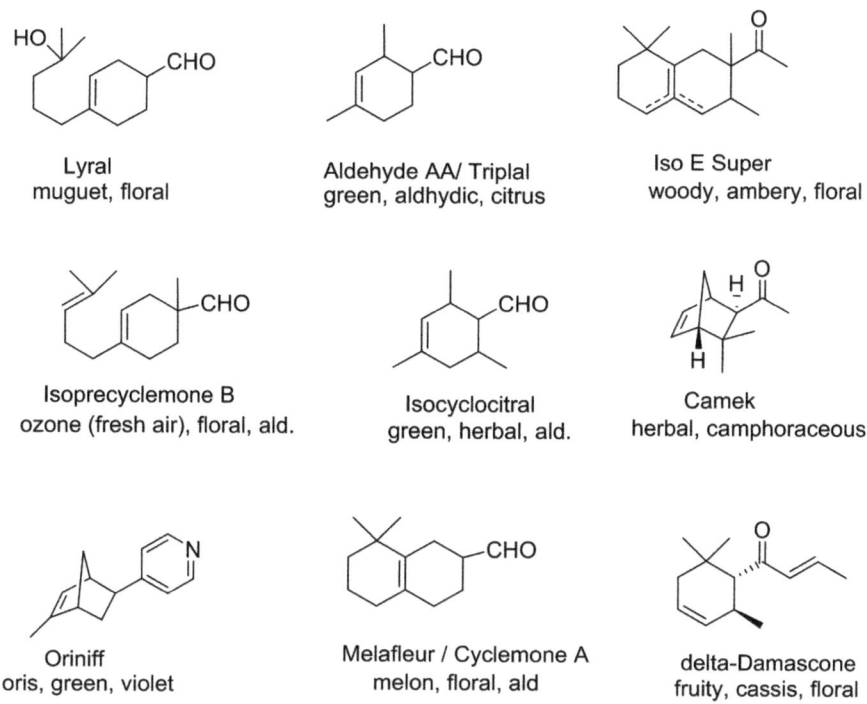

Figure 4. Fragrance ingredients based on Diels-Alder technology.

## Use of Nitriles (*10*) in Perfumery by Functional Group Transformations

As previously stated, consumer fragrances in soaps, detergents, conditioners, and softeners employ very stringent conditions and harsh bases; because of this, perfumers need to use fragrance ingredients that would be stable in those environments. Fragrance ingredients with aldehyde groups play a large role in the design of functional perfumes, but unfortunately, aldehydes undergo many side reactions in the basic media such as aldol condensations and polymerizations.

It has been observed that a simple transformation of an aldehyde group into a nitrile group present in a fragrance ingredient not only enhances its base stability, but also retains its odor. Figure 5 highlights a select group of aromatic nitrile fragrance ingredients such as Fleuranil, Salicynalva, Khusinil, and a few acyclic nitrile fragrance ingredients such as Azuril, Peonile, Citralva, Citronalva, and Lemonalva that perform well in basic applications.

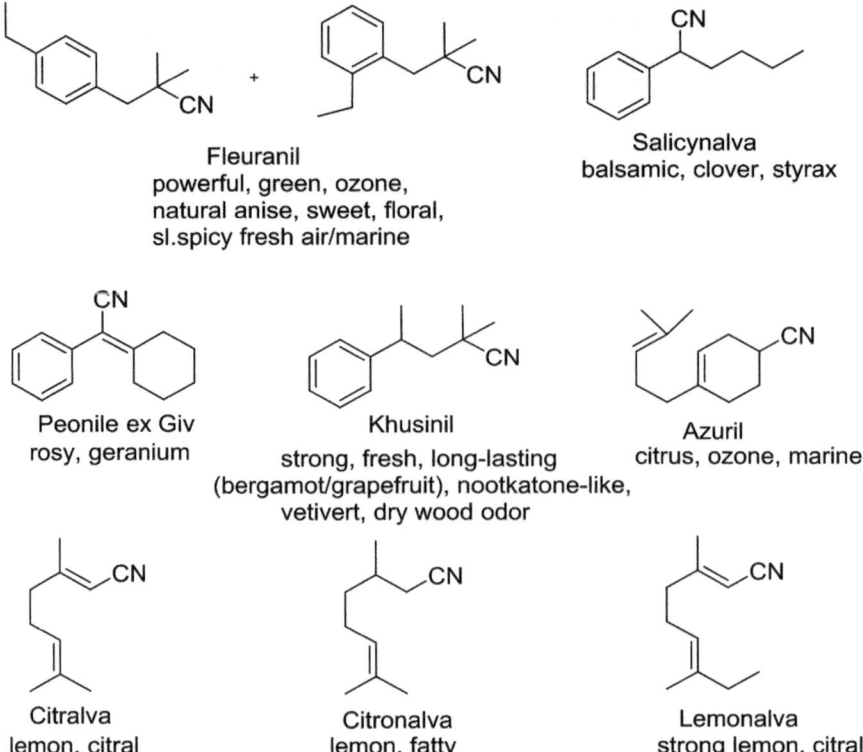

**Fleuranil**
powerful, green, ozone,
natural anise, sweet, floral,
sl.spicy fresh air/marine

**Salicynalva**
balsamic, clover, styrax

**Peonile ex Giv**
rosy, geranium

**Khusinil**
strong, fresh, long-lasting
(bergamot/grapefruit), nootkatone-like,
vetivert, dry wood odor

**Azuril**
citrus, ozone, marine

**Citralva**
lemon, citral

**Citronalva**
lemon, fatty

**Lemonalva**
strong lemon, citral

*Figure 5. Examples of nitriles used in perfumery.*

## More Examples of Diverse Technologies Employed in the Discovery of New Molecules for Creation of Consumer Fragrances

Cassiffix (*11*) was the first long-lasting cassis note discovered that did not contain a sulfur atom. It was prepared in four steps starting from campholenic aldehyde, which was converted to α-methylene campholenic aldehyde using a Mannich reaction, followed by its Diels-Alder reaction with Isoprene to produce a mixture of aldehydes. These aldehydes ~~in the~~ after chemical reduction, followed by cyclization with acid, produced Cassiffix.

Prismantol, a spicy, ginger, woody note, was prepared using two technologies: hydroformylation of R-(+)-Limonene and an Ene reaction (*12*). Both Montaverdi and Vivaldie were derived from cis-3-Hexenol and are powerful green (*13*) odorants. Arctical and Ozofleur are functional fragrance ingredients that provide benefits to consumer fragrances due to their stability in high pH. Arctical (*14*) was developed as a base stable note in place of n-Decanal with its fresh, citrus, aldehydic notes. Figure 6 describes the structure and odor of new recently introduced ingredients.

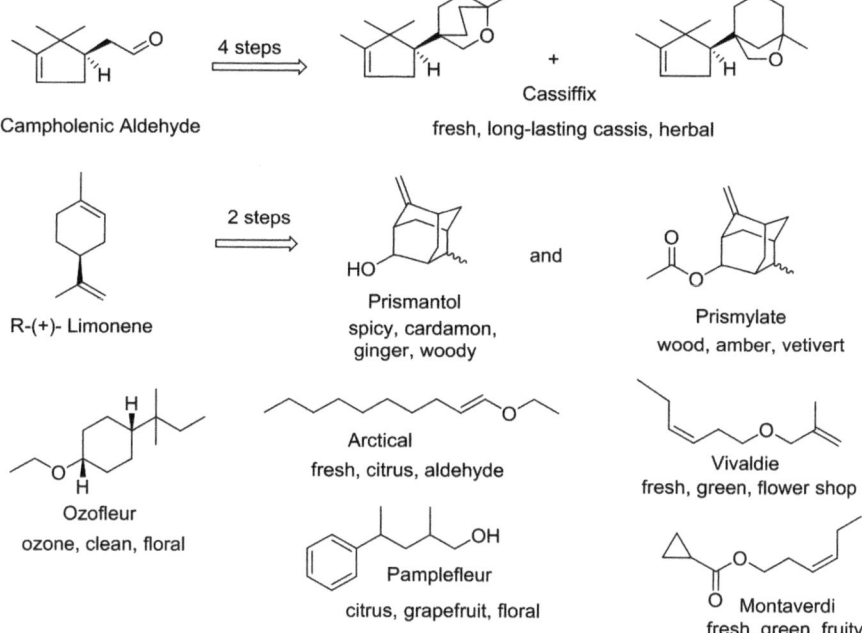

Figure 6. Fragrance ingredients for consumer fragrances.

## Search for New Amber Notes

Amber notes are essential to the performance and attraction of a fragrance. Amber notes are widely used in perfume creation and there are very few fragrances that do not contain an amber odorant. In addition, amber odorants display an enormous functional diversity, and many amber molecules contain a functional moiety such as ketone, ether, epoxide, ketal, and primary or secondary or tertiary hydroxyl groups. Some molecules contain hydroxyl ether functionality.

Of the amber molecules used in fragrances, (−)-l-Ambrox is considered one of the most precious aroma ingredients and is known for its beautiful, enticing, woody, amber, soft, and velvety smell and for its long-lasting power both on skin and cloth. Because of this, several years ago, intensive effort was mounted to find a new amber molecule that would compete and complement the performance and utility of Ambrox. Figure 7 delineates the structure of key amber molecules such as l-Ambrox, Grisalva, and Galaxolide that contain an ether ring. It may be pointed out that an ether ring or functionality plays a key role in the performance of a few other desirable benchmark fragrance ingredients like Cedramber or Cassifix.

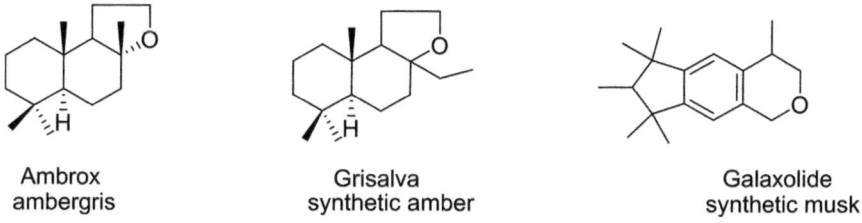

Figure 7. Structure of key fragrance ingredients with ether moiety.

# Genesis of an Exploratory Idea in Quest of New Amber Molecules

Since the presence of a tetrahydrofuran ether ring in Ambrox is a key structural necessity for the performance of amber odor, we envisioned a new technology to prepare THF-ether-like molecules derived from the cyclization of γ, δ-unsaturated ketones. γ, δ-unsaturated ketones are readily accessible via the Claisen rearrangement of ketones.

To test this hypothesis, we took Herbac, a perfumery ingredient and converted it into allyl Herbac first and then to the desired THF-ether structure, A, using a two-step sequence shown in Scheme 1. Lithium aluminum hydride was used to reduce allyl herbac ketone to its alcohol derivative, which on acid cyclization produced structure A. Similarly, Galbaniff was converted to Structure B using the same technology.

Surprisingly, such a transformation led to a change in the odor of allyl Herbac and Galbaniff from green and galbanum to woody and amber. Inspired by this observation, this technology was used to prepare new THF-ether-like molecules on diverse structural backbones. For this discussion, please refer to our previously published (15) work.

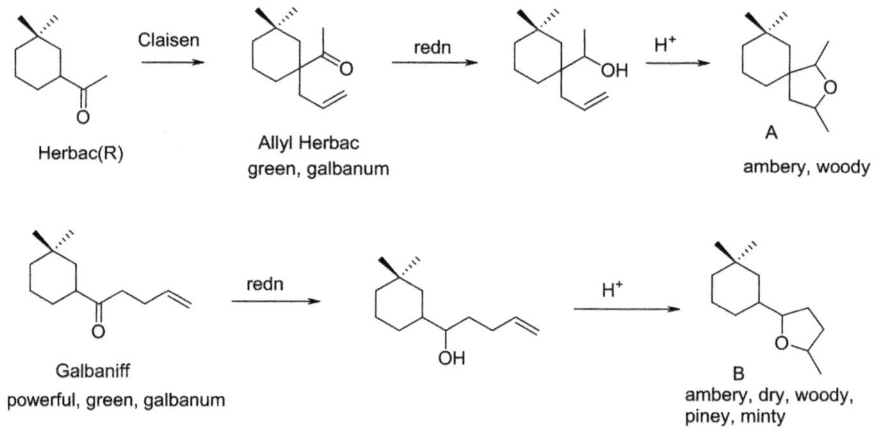

Scheme 1. Cyclization of allyl Herbac and Galbaniff to THF-ether derivatives.

# Discovery of Amber Xtreme and Trisamber—Two New Amber Molecules

The goal of this exploration was to discover a new amber molecule. Therefore, we prepared new THF-ether-like molecules using this technology on diverse structural backbones and building blocks. One backbone that led to the discovery of Amber Xtreme and Trisamber was derived from Pentamethylindane, a key building block of Galaxolkide, an IFF musk molecule. Pentamethylindane was converted into Dihydrocashmeran in three steps using the hydrogenation of a Cashmeran intermediate as depicted in Scheme 2.

Dihydrocashmeran was converted into Amber Xtreme and Trisamber using a multi-step synthetic technology (16, 17) employing a Claisen rearrangement to produce methallyl and allyl Dihydrocashmeran intermediates. These two intermediates were then converted to Amber Xtreme and Trisamber using a two-step sequence involving reduction followed by acid cyclization as delineated in Scheme 2.

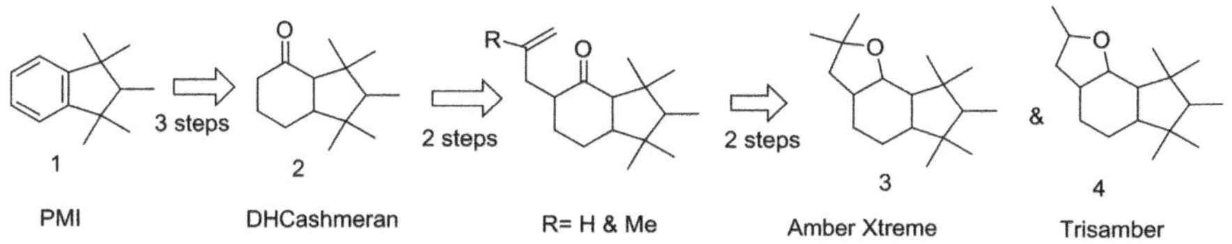

Scheme 2. Route to Amber Xtreme and Trisamber.

It is worth mentioning that Amber Xtreme is primarily a mixture of two isomers: the *cis* isomer (Structure A) which is a much more powerful woody, amber odor profile, and the *trans* isomer (Structure B), which is described in Figure 8. From the structure-odor (*18*) point of view, note the stereo chemical comparison between the Amber Xtreme structure and the structure of (–)-l-Ambrox. It is no wonder that Amber Xtreme is ambery and complements (–)-l- Ambrox in its odor and performance profile.

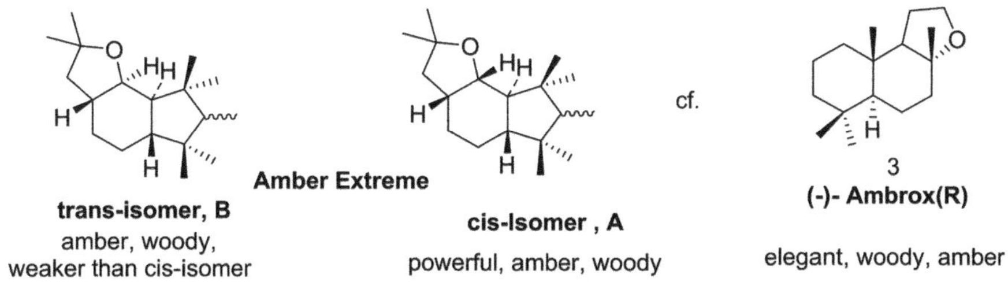

Figure 8. Odor differences between Amber Xtreme isomers.

## Chirality in Fragrance Ingredients

Let me briefly share a few examples of how chirality (*18*) influences odor preferences or attraction for chiral isomers of certain fragrance ingredients. It is shocking and surprising to find that certain chiral isomers of well-known aroma chemicals can have a completely different odor. For example, enantiomer ingredients like (R)-Limonene has an orange odor versus (S)-Limonene, which has a lemony smell, while (–)-Carvone smells like spearmint versus (+)-Carvone, which smells like caraway. Among other diastereomeric chiral ingredients such as menthol, which has 3 chiral centers and 8 possible stereoisomers, only l-menthol has a cool, minty smell. For this reason, it is widely used in both perfumery and flavors.

Another dramatic observation of a chiral odor difference is noted with methyl jasmonate, a key component of jasmine essential oil. Methyl jasmonate has two chiral carbons and thus can exist in 4 diastereoisomers as shown in Figure 9. It was later determined that only the 1R, 2S-(+)-methyl jasmonate epimer has a true jasmine floral smell which is 70 times stronger than the other three diastereomers. In addition, strangely enough, the 1S, 2S-diastreomer was found to be odorless (*19*).

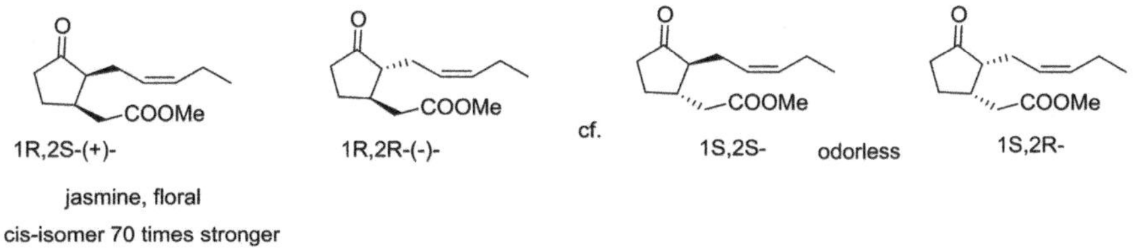

Figure 9. Odor differences among methyl jasmonate chiral diastereomers.

**11**

Among all the musk fragrance ingredients, Galaxolide is one of the largest produced musk molecules due to its outstanding performance and long-lasting odor. It was discovered by Dr. Beets, an IFF scientist, in 1957. Later, it was found that only the 4S, 7R- diastereomer (*20*) had the most powerful musk smell with a low threshold (0.63ng/L), which is depicted in Figure 10.

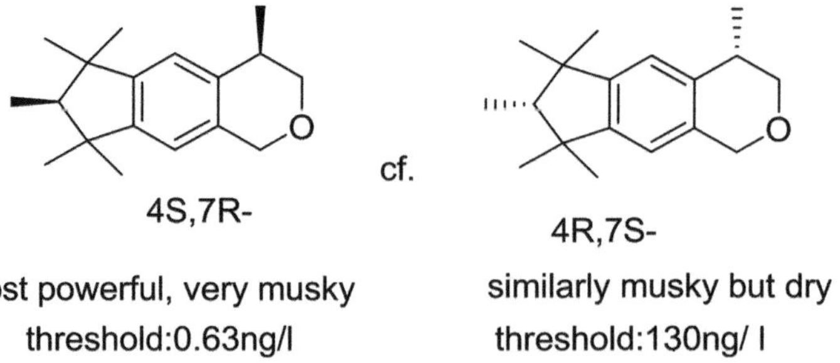

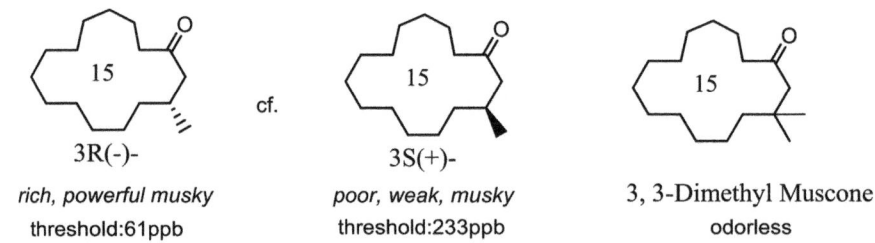

4S,7R-

most powerful, very musky
threshold:0.63ng/l

cf.

4R,7S-

similarly musky but dry
threshold:130ng/ l

*Figure 10. Odor differences among Galaxolide key chiral diasteromers.*

## 3R(–)-Muscone

Muscone is another fragrance ingredient for which striking odor differences are observed among its enantiomers. Muscone is loved by consumers and perfumers for its rich, powerful, powdery musk smell. (–)-muscone has a desirable musky smell with a low threshold value of 61ppb versus 3S-(+)-muscone which has a very weak musk odor as shown in Figure 11. If one incorporates another methyl group at the 3-position of muscone structure then 3,3-dimethyl muscone becomes odorless. Such is the unpredictable power of chirality and structure-odor relationship.

3R(-)-

*rich, powerful musky*
threshold:61ppb

cf.

3S(+)-

*poor, weak, musky*
threshold:233ppb

3, 3-Dimethyl Muscone

odorless

*Figure 11. Odor differences of Muscone chiral enantiomers.*

## Cassiffix vs. Ent-Cassiffix

Another example of dramatic odor differences among chiral fragrance ingredients comes from our work which was carried out during the discovery of Cassiffix. The Cassiffix that was prepared from S-(–)-campholenic aldehyde, which is derived from 1S, 5S-(-)-α-pinene, and was found to be strongly cassis in odor. In contrast, ent-Cassiffix was prepared from R-(+)-campholenic aldehyde, which is made from 1R, 5R-(+)-α-pinene as shown in Figure 12, and was found to have a very weak cassis odor.

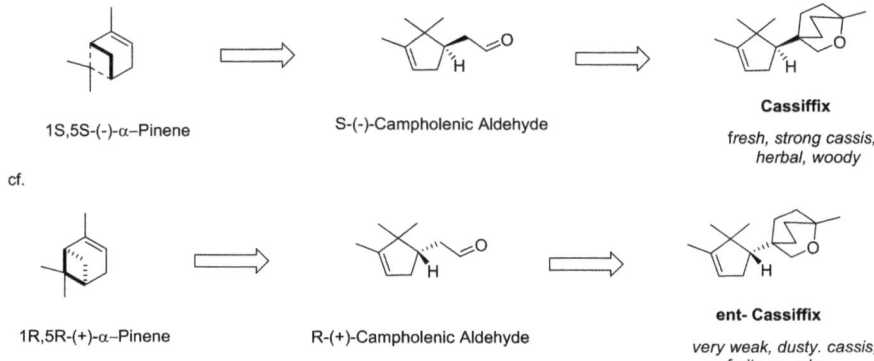

1S,5S-(-)-α–Pinene     S-(-)-Campholenic Aldehyde

**Cassiffix**

*fresh, strong cassis, herbal, woody*

cf.

1R,5R-(+)-α–Pinene     R-(+)-Campholenic Aldehyde

**ent- Cassiffix**

*very weak, dusty. cassis, fruity, woody*

*Figure 12. Odor differences among Cassiffix and enantiomer ent-Cassiffix.*

In conclusion, it is fair to state that in spite of tremendous advances in biological and olfactory research and artificial intelligence in computer modeling, even today, no one can predict with certainty the odor of a new structure. As a result, structure-odor relationship prediction remains empirical in nature.

## Living Flower Technology: A Tool for Capturing True to Nature Aroma of a Flower's Fragrance

Living Flower technology was pioneered by Dr. B. D. Mookherjee of the IFF in the 1980s. Use of such Living scents lead to the creation of many attractive fragrances. The idea behind this technology was to create a true natural scent of a blooming flower as shown in Figure 13.

This transformative technology was based on the fact that when a flower is plucked from the branch, the nutrition that is being provided by the root system of the flowering plant is severed, thus directly impacting the enzymatic chemistry of the flower leading to an observed change in the amount of chemical constituents of the essence. Using this technique led to the recreation of a natural flower scent. IFF has a database of hundreds of living accords of diverse flowers, fruits, and spices. In fact, the IFF botanical garden as shown in Figure 14 has a collection of 1500 unique plants from all over the globe. This has allowed IFF perfumers to create attractive and legendary fragrances.

*Figure 13. IFF headspace technology.*

*Figure 14. IFF botanical garden.*

## Future of Fragrances: Natural, Organic, Healthy, and Sustainable

Before concluding this article, let me express briefly a few thoughts about the future of fragrance design. Going forward, more and more consumers will use fragrances not just for sensory and esthetic benefit but they will also prefer to purchase scents that are healthy, holy, and sustainable. This trend of wellness is already gaining momentum. It is no wonder that there is an increasing interest in the use of natural essential oils in perfumes and aromatherapy. Concurrently, the use of niche fragrances is growing exponentially in the industry.

With regard to holy and healthy fragrances, Asian cultures have relied on the use of holy basil, sandal, natural spices and attars of select flowering plants for healing purposes for many years. There are numerous fragrance healing gardens in Kashmir, India dating back to the 1700's. Figure 15 depicts the Butchart Garden of Victoria, Canada, which is well known for its healing properties.

Many people regularly visit fragrant gardens like Butchart and surround themselves with the natural pleasant flora and fauna to rejuvinate and relieve stress. Many argue that visiting these gardens is a healthy alternative to using anti-anxiety medication.

*Figure 15. Butchart Garden of Victoria, Canada.*

## Fragrance Technologies of the Future

In moving forward toward the goal of assuring sustainable fragrances, synthetic biotechnology will be a major source of key natural flavor and fragrance ingredients. In addition to using natural ingredients, synthetic biotechnology will also be used for the production of sustainable raw materials and essential oils needed for creating attractive fragrances. This is primarily due to dwindling resources, climate changes, and a lack of abundant, cost-effective key fragrance ingredients without which modern fragrances cannot be created.

## Acknowledgments

The author would like to offer his sincere thanks to all of his associates and collaborators at IFF (whose names are cited in the references) for their passion and dedication to advancing the Art of Perfumery by discovering new fragrances molecules. I am also deeply appreciative of the IFF legal department for granting permission to speak at and publish information from the "Chemistry of Sex" symposium that was organized by the Agricultural and Food Chemistry Division of the ACS meeting that took place in the New Orleans, from 18–24 March 2018.

## References

1.  Ataman, D. 2018 Flavor and Fragrance Leaderboard. *Perfum. Flavor.* **2018**, *43*, 14.
2.  *Jicky, Women's Fragrances.* https://www.guerlain.com/us/en-us/fragrance/womens-fragrances/jicky (accessed March 8, 2019).
3.  Arctander, S. *Perfume and Flavor Chemicals*; Steffen Arctander: Elizabeth, NJ, 1969; Vol. 1–2.
4.  Guenther, E. *The Essential Oil*; Van Nostrand Reinhold: New York, 1985; Vol. 1–6.
5.  Arctander, S. *Perfume and Flavor Materials of Natural Origin*; Steffen Arctander: Elizabeth, N.J, 1960.
6.  Acree, T. E.; Barnard, J.; Cunningham, D. G. A Procedure for the Analysis of Gas Chromatographic Effluents. *Food Chem.* **1984**, *14*, 273.
7.  Ohloff, G.; Pickenhagen, W.; Kraft, P. *Scent and Chemistry. The Molecular World of Odors*; Ohloff, G., Pickenhagen, W., Kraft, P., Eds.; Wiley-VCH: Zurich, 2012.
8.  Overman, L. E. *Organic Reactions*; Wiley-VCH: Weinheim, 2009; Vol. 1–74.
9.  Trost, B. M.; Fleming, I. *Comprehensive Organic Synthesis*; Elsevier: Pergamon, 1991; Vol. 1–8.
10. Narula, A. P. S. The Search for New Functional Ingredients for Functional Perfumery. *Chem. Biodiversity.* **2004**, *1*, 1992.
11. Narula, A. P. S.; De Virgilio, J. D.; Benaim, C.; van Ouwerkerk, A.; Gillotin, O. *Substituted Cyclopentenyl Oxabicyclooctanes Processes for Preparing Same and Organoleptic Uses Thereof.* U.S. Patent 5,087,707, Feb 11, 1992.
12. Gillaspey, W.; Hagedorn, M. L.; Hanna, M. R.; Boardwick, K. L.; Beck, C. E. J.; Futoshi, F.; Bronco, A. G.; Narula, A.; Boden, R. M. *Adamantane Derivatives and Organoleptic and Deodorancy Uses of Said Adamantane Derivatives and Said Compositions.* U.S. Patent 4,956,481, Sep 11, 1990.
13. Narula, A. P. S.; Arruda, E. M.; Koestler, J. J.; Molner, E. A.. *Allyl Ethers.* U. S. Patent 6,340,666, Jan 22, 2002.

14. Narula, A. P. S.; Mookherjee, B. D.; Patel, S. M.; Arruda, E. M.; Merritt, P. M. *Acyclic Enol Ethers Isomers Thereof, Organoleptic Uses Thereof and for Processes for Preparing the Same.* U.S. Patent 7,175,871, Feb 13, 2007.

15. Narula, A. P. S. The Search for New Amber Ingredients. *Chem. Biodiversity.* **2014**, *11*, 1629.

16. Narula, A. P. S.; Arruda, E. M. *Polyalkylbicyclic Derivatives.* U.S. Patent 7,312,187, Dec 25, 2007.

17. Narula, A. P. S.; Arruda, E. M.; Levorse Jr., A. T.; Beck, C. E. J. *Polyalkylbicyclic Derivatives.* U.S. Patent 7,160,853, Jan 9, 2007.

18. Ohloff, G.; Pickenhagen, W.; Kraft, P. *Structure-Odor Relationship in Scent and Chemistry: The Molecular World of Odors*; Wiley-VCH: Weinheim, 2012, pp 61.

19. Acree, T. E.; Nishida, R.; Fukami, H. Odor Thresholds of the Stereoisomers of Methyl Jasmonate. *J. Agric. Food Chem.* **1985**, *33*, 425.

20. Frater, G.; Muller, U.; Kraft, P. Preparation and Olfactory Characterization of the Enantiomerically Pure Isomers of the Perfumery Synthetic Galaxolide. *Helv. Chim. Acta.* **1999**, *82*, 1656.

# Chapter 2

# "Candy Is Dandy": The Mind of Sexuality as Suggested by a Mind Genomics Experiment

Attila Gere,[1] Ryan Zemel,[2] Petraq Papajorgji,[3] and Howard Moskowitz*,[4]

[1]Szent István University, Villányi út 29-43, H-1118 Budapest, Hungary
[2]Independent Researcher, Downers Grove, Illinois 60515, United States
[3]University of New York Tirana, Rruga Kodra e Diellit, Selitë 1046, Tirana, Albania
[4]Mind Genomics Advisors, 11 Sherman Ave., White Plains, New York 10605, United States
*E-mail: mjihrm@gmail.com.

This chapter presents the mind genomics approach to study a very important social issue: interpersonal relationships. This topic has been studied for a long time and by a number of scientists. Different studies have used different approaches with different levels of success. We have used Mind Genomics as a new approach to analyze the problem. Our approach is based on experimental design theory and statistical models. We have used data from three different countries: India, the United States, and Albania. Finally, the chapter introduces the Viewpoint Identifier and shows how it can be used to evaluate individuals who have participated in the study. We find that the Viewpoint Identifier is an excellent tool to be used to create databases of personalized clients.

## Introduction

Interpersonal relationships are and have always been very important in any society. Studies have been undertaken, and books, movies, and poems have been written about this central human behavior in order to explain, present, and make known to large audiences the remarkable effects of interpersonal relationships in our everyday lives. Although a large number of studies have already been undertaken, this important topic is still under investigation from different researchers and perspectives (1). The popularity of the topic is reflected in the number of studies to which one can refer. The state of the art in understanding the topic is reflected in the fact that virtually all of the studies use self-reported questionnaires about attitudes and behaviors. This chapter presents an alternative approach, mind genomics, presented in full detail below, as the basis for the data reported here.

## Research on Love and Attraction: A Wealth of Data from Questionnaires

A holistic study about love and attraction, relationships and families, love and the difficult task of choosing a partner, family and stress, and the reasons for divorce is provided by Lamanna and Riedmann (2). These authors use a rigorous scientific approach to study the multifaceted problem of "making choices in a diverse society." They offer interesting and insightful perspectives on how the diversity of our modern society affects the institution of marriage.

Love and attraction are central to human activities, and have been under the strong influences of the social, political, and economic changes in society. As such, the concept of intimacy has seen transformations, and this phenomenon has been the subject of many studies, such as those by Giddens and Muniruzzaman (3, 4). One of the most relevant publications in this topic is Anthony Giddens's famous book, *The Transformation of Intimacy: Sexuality, Love, and Eroticism in Modern Societies* (3). Giddens discusses how the concept of intimacy has gradually evolved during different periods of our history, focusing on the four types of expression: physical, emotional, cognitive, and experiential,.

A similar perspective on the concept of intimacy but specifically for developing countries is provided by Muniruzzaman (4). It is important to stress that an approach to intimacy is a product of many factors, such as economic development of the country, the level of societal emancipation of the country, the religious practices of the society, and so on. There is more, of course. There must be some type of attraction, on either a physical or a personality level, in order to create a relationship among people. Several studies demonstrate (not unexpectedly) that initial attraction to a potential future partner is highly associated with physical attractiveness (1, 5–9).

There is a widely accepted theory that men are more inclined to appreciate physical attractiveness than women. The results of meta-analyses conducted from five research paradigms examining the hypothesis is presented by Feingold (6), who found the anticipated sex difference emerged in all five meta-analyses.

A similar study on the role of external looks was conducted by Snyder et al. (8) The authors conclude that low self-monitoring individuals paid a greater amount of attention to and placed greater weight on information about interior personal attributes than did high self-monitoring individuals. In contrast, high self-monitoring individuals paid more attention to and put greater weight on exterior physical appearance than did low self-monitoring individuals.

It is believed that the socioeconomic status of a potential future partner is relevant for a number of people. In order to evaluate the role of the effects of potential partners' physical attractiveness and socioeconomic status on sexuality and partner selection, Townsend and Levy undertook such a study in detail (9). In the study, a number of students were shown photos of three opposite-sex individuals. After collecting input from the subjects, information about the socioeconomic standing of the same individuals was added. According to Townsend and Level, compared to men, women are more likely to prefer or insist that sexual intercourse occur in relationships that involve affection and marital potential, and women place more emphasis than men do on partners' socioeconomic status in such relationships.

## The Conventional Method: Asking Questions

In the literature of consumer and social research, the notion of experiments is not a popular one. A preferred method is to ask the respondents questions about a topic and then tabulate the results. We see an example of this in Table 1, which was part of the self-profiling classification of this study. The respondents had just completed an evaluation of vignettes that presented different pieces of information about a situation, and were asked about the likelihood of an erotic encounter within 12 h (see Figure 1).

Following this experiment, we asked a number of questions of the respondents (in addition to those pertaining to demographics, such as age, gender, country, etc.), to find out what each felt was the most important aspect of the four presented that would increase intimacy. This question was formatted consistently with the sociological approach of asking survey questions. We see the results in Table 1, which shows the answers to survey question 11 in the classification part of the study. Each respondent was asked to select which of four aspects was most important to increase intimacy. There was no information explaining the aspects, other than a few words, such as "setting location," "fragrance of the romantic interest," and so on. The respondent could choose "none of these" by selecting "not applicable."

Table 1 gives us a sense of what is important when the respondent has to focus on the topic, and select one answer. We get a sense of people-to-people differences, but nothing further. There is no sense of content in the question, content that mirrors the richness of experience. The results in Table 1 are shown by total, and by the three countries, the United States, Hungary, and Albania, respectively.

**Table 1. Selection by Test Respondents Regarding Which of Four Factors or Aspects Is the Most Important to Increase Intimacy**

| Q11: What is most important to increase intimacy? | Total | United States N = 100 | Hungary N = 100 | Albania N = 53 |
|---|---|---|---|---|
| Setting location | 45% | 49% | 43% | 40% |
| Fragrance of the romantic interest | 35% | 25% | 48% | 28% |
| Not applicable | 9% | 7% | 5% | 21% |
| Persons in surrounding atmosphere | 7% | 15% | 2% | 0% |
| Meat/drink choice of romantic interest | 5% | 4% | 2% | 11% |

## Mind Genomics

In contrast to today's focus on questionnaires, inspired by sociology and by increasingly available computing power, the approach taken in this chapter is one of a simple experiment. Table 1 showed us what we could get from a simple questionnaire. There is no depth in the information, because the test stimuli are general questions.

Mind Genomics differs from conventional surveys because it is an experiment in which we present the respondent with test stimuli of a cognitively rich nature. By cognitively rich we mean that the stimulus has meaning in everyday life. The stimulus is of the type one would encounter in ordinary experiences, and might even be the stuff of description in a novel or a short story. The stimuli are phrases, ideas, or in our case, phrases representing sentences as answers to questions.

We present combinations of ideas (elements and messages), combinations created by experimental design, or a statistical recipe specifying which elements should appear together in order to create a set of stimuli that can be analyzed by regression modeling. The combinations comprise messages of different types, such as who the people are, what they eat, and so forth. The combinations create vignettes. The vignettes are thus the equivalent of a very short story, or at least a mis-en-scène. The respondent, reading this combination created by a computer, is instructed to rate her or his guess about what will transpire in 12 h. For this specific experiment, the respondent was asked to guess about the likelihood of an erotic encounter.

The science behind this experimentation can be traced back more than 50 years, to seminal work on the measurement of values proposed by a mathematical psychologist (R. Duncan Luce) and a statistician (John Tukey). Their objective was to create a measurement system of percepts through the presentation of combinations of messages or ideas, the measurement of responses to such combinations, and then the estimation of the underlying values of the elements (the individual messages) from measures of the responses. They called this approach conjoint measurement (*10*).

The adaptation of conjoint measurement over the ensuing years moved to the presentation of combinations of messages or elements, and either paired comparison (choice measurement) or rating of the combination. When the respondent rates the combination, for example, of the vignettes that we present here, it becomes very straightforward using ordinary least-squares regression to estimate the contribution of each element or message.

## The Mind Genomics Approach

We have taken the method of conjoint measurement and created a knowledge-development system to understand what people value and what they do not. Rather than focusing on the topic and letting the topic drive the research approach, we have created what might be called a "knowledge creation machine." The objective of this machine is to uncover how the mind responds to different aspects of a topic, and how there might be different groups in the population who respond in different but understandable ways. These groups could be called mind alleles, to paraphrase a term from genomics.

Mind Genomics begins with a structured set of questions that tell a story. Reactions to different versions of the story will enable us to understand the mind of the individual. For this chapter, the focus is on the application of Mind Genomics to the nature of intimate relationships, specifically, and how they emerge out of situations. That is our topic. Table 2 presents an example of four questions dealing with the situation before the relationship may occur. The reader may notice that these questions are not the typical questions that one would ask in a questionnaire about relationships, but rather questions that allow us to write different test stories. These stories have more meaning than do single questions, although they are simply combinations of answers put together, one atop of the other, rather than the result of an author crafting a coherent story in (hopefully) felicitous prose.

We limit the number of questions and answers to four each, with the questions telling a story and the answers to each question filling in the particulars of the story. It will be the answers to which the respondent will react, or more correctly the combinations of answers. These combinations, vignettes, or short concepts, present a meaningful albeit very short story, to which a respondent can react.

Mind Genomics then proceeds to combine the answers into small, easy-to-read combinations, as shown in Figure 1. The combinations are created according to experimental design, which puts together elements that might never appear together at any other time. The experimental design for Mind Genomics is skeptical of irrational or seemingly impossible combinations, but previous work did have a modified version with constraints (*11*). That version, not presented here, enabled the researcher to prevent incompatible elements from appearing together. Such constraints can be implemented in this BimiLeap version by putting these incompatible elements into answers for a single question. The experimental design never allows two elements or answers from the same question to appear in a single test vignette.

**Table 2. The Four Questions and the Four Answers to Each Question[a]**

|  |  |
|---|---|
| | *Question 1: Who is in the scene?* |
| A1 | Private … just you two in scene |
| A2 | Group … out with friends |
| A3 | Double date … with a couple |
| A4 | Around strangers … no one familiar in scene |
| | *Question 2: Where is the scene taking place?* |
| B1 | Romantic setting … walk in the park, picnic, fancy restaurant, etc. |
| B2 | Casual setting … movie, casual restaurant, etc. |
| B3 | Social setting … bar, club, etc. |
| B4 | Home setting |
| | *Question 3: What fragrance is he/she wearing?* |
| C1 | No fragrance |
| C2 | Pleasant smelling cologne/perfume … floral, oriental, etc. |
| C3 | Deodorant only… common name brands |
| C4 | Shampoo/conditioner smell … nice smell, not overwhelming |
| | *Question 4: What are they eating/drinking?* |
| D1 | Appetizer with multiple alcoholic beverages |
| D2 | Entree with glass or glasses of wine/alcoholic beverage |
| D3 | Only drinks |
| D4 | Entree with nonalcoholic beverages |

[a] Combinations of the answers will become vignettes, to be rated by the respondents.

Figure 1 shows that the elements are simply stacked, one atop of another, without any effort to create connectives. The underlying principle is that the respondent is asked to look at the combination and give a single answer.

Each respondent evaluates 24 unique combinations. The same element repeats several times through the combinations. The fact that the combinations tested by one respondent differ from the combinations tested by another respondent means that the analysis focuses on the individual, specifically on the contribution of the individual element to the rating. The vignettes (the test concepts) are simply vehicles by which the elements are presented, and the reality that the vignette comprises many different elements defeats the respondents' attempt to be rational and consistent. It is important to get the real feeling of each respondent without the interference of one's mental editor. The answers are shown (Figure 1) stacked upon one another and centered, but the underlying questions are not shown.

Most respondents feel nervous at first when faced with the task of looking at a combination that seems to have been put together in a way that cannot be easily understood. That is, in most writing, the writer attempts to put together a combination of sentences that reinforce a single idea.

Mind Genomics does something quite different. Mind Genomics poses different elements in a stark manner, forcing the respondent to make a decision based upon compound and possibly complex information. There is no hint about the right answer. The respondent simply guesses, at first trying to be consistent, but eventually simply in a manner that is virtually automatic. When respondents are asked "How did you make your decision?" many respondents answer that they could not discern a pattern to help them, and therefore guessed. It is the guessing that we want, the abandonment of the attempt to be rational and consistent.

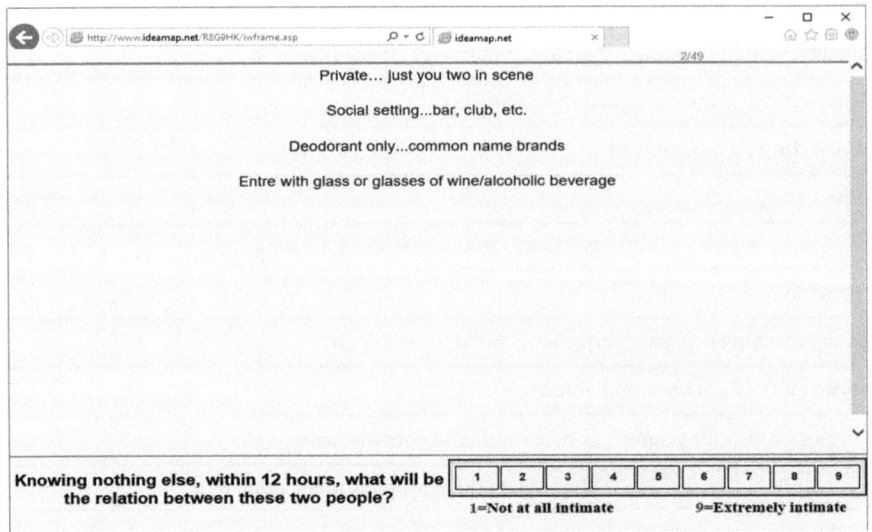

*Figure 1. Example of a test vignette showing four answers to the questions.*

### Making Sense of the Results: How Do Different Countries and the Two Genders Rate These Vignettes?

At the simplest level, we can look at the distribution of the ratings by country, or by any other convenient arrangement of the data. The distribution tells us the degree to which respondents in the different countries are likely to feel that these types of situations lead to erotic encounters. The distribution of the ratings shows us the responses from the outside of the person's mind, simply providing us with a sense of the predisposition to perceive a likely encounter, but it does not provide any additional information about why. Recall that each vignette comprised 2–4 elements, combined by an underlying experimental design, and that each respondent rated 24 different combinations or test vignettes.

We can divide the 9-point scale into three regions to denote likelihood of intimacy after 12 h. Table 3 presents the raw data in terms of the ratings for the different vignettes.

- Ratings 1–3 correspond to low likelihood.
- Ratings 4–6 correspond to moderate likelihood.
- Ratings 7–9 correspond to high likelihood.

The raw data (the distribution of ratings by country and by gender) hint that there may be some differences across countries, but the reality of the data is that the percentages tell us very little, other than that the respondents in the United States are more likely to believe that the vignettes describe a situation that will lead to an erotic encounter within 12 h. We know little else from this very superficial analysis (really just a count of the results) as one might for standard surveys. We understand the surface behavior (externalities, suggested by the vignette) but we have no idea what is driving the decisions. We know only the outside, not the inside, of the mind.

**Table 3. The Raw Data, Showing the Percent of Respondents from Country and Gender Rating the Vignettes Low, Medium, or High on the Likelihood That an Erotic Encounter Will Occur Within 12 h**

|               | R1–3 Low | R4–6 Moderate | R7–9 High | Total |
|---------------|----------|---------------|-----------|-------|
| United States | 24%      | 39%           | 37%       | 100%  |
| Hungary       | 21%      | 47%           | 32%       | 100%  |
| Albania       | 35%      | 36%           | 29%       | 100%  |
| Total         | 27%      | 41%           | 33%       | 100%  |
|               | R1–3 Low | R4–6 Moderate | R7–9 High | Total |
| Male          | 25%      | 43%           | 32%       | 100%  |
| Female        | 31%      | 36%           | 33%       | 100%  |
| Total         | 27%      | 41%           | 33%       | 100%  |

## Deeper Understanding by Deconstructing the Ratings into the Contributions of the Elements

Recall that each respondent evaluated 24 unique vignettes. Although the 16 elements were the same, the program created a set of 500 unique experimental designs/recipe booklets. Each experimental design is structurally identical to every other design, except that the specific combinations of elements intentionally varies from one design to the other. These experimental designs are so-called isomorphic permutations of each other. Each design, one to a respondent, is set up to allow the ordinary least-squares regression model to relate the presence or absence of the elements to the rating, or more commonly, to a transformation of the rating that creates a binary scale.

The first step of the analysis transforms the ratings of the 9-point scale into a binary scale, with ratings of 1–6 transformed to 0 and ratings of 7–9 transformed to 100. The transformation is done because 35 years of experience with the scientific and business communities, as well as with non–science savvy academics, suggests that no one really understands the meaning of the 9-point, or for that matter, any other scale. People understand the mechanics of the scale, but to give a simple example, it is actually quite hard to explain what it means when a vignette is assigned the rating of 6 on the 9-point scale, for exmple. The question generally posed by most people who are given the value 6 is "What does that mean in a practical sense, a meaning that I can understand?"

A small random number ($<10^{-5}$) is added to each transformed number. The addition of the small random number to every one of the transformed ratings ensures that the analysis by ordinary least-squares regression will not crash. The underlying experimental design used here works at the level of the individual respondent, ensuring as it does that the 16 answers to the question (our elements, the independent variables) are all statistically independent of each other. There is never a problem with the independent variable. In contrast, a respondent might confine all ratings to the range 1–6 (transformed to 0), or all ratings to the range 7–9 (transformed to 100). In that not uncommon case, there would be no variation in the dependent variable. Adding the random number ensures some small variation in the dependent variable.

As every respondent evaluated an appropriate array of test stimuli created by experimental design, it is straightforward to run the regression models in one of two ways. The typical approach creates a single model for each respondent, comprising the 16 messages or answers as the independent variable and the transformed rating (including the addition of the small random number) as the dependent variable. The corresponding parameters of the individual-level models (the additive constant, the 16 corresponding coefficients) are then separately averaged across all the respondents in the relevant subgroup. The subgroup may be defined by age, gender, country, and so on. Another option incorporates all of the relevant data into one data set, and runs a grand model with all of these relevant data, independent of the other defining variables. In general, these two methods generate similar patterns of numbers, although actual values of the numbers change somewhat. The decisions are generally the same, whether one uses the average of the individual-level models or the grand model, respectively.

Table 4 shows the coefficients from the average model by relevant subgroup, and suggests that for the most part the pattern that we see from the total panel repeats in the individual subgroups. That is, there may be some group-to-group differences, but to a great degree, we are not dealing with different ways of thinking about the outcomes.

Table 5 presents the pairwise correlation coefficients, and reveals a remarkably high correlation between the pattern of coefficients for all subgroups. The lowest correlation exceeds 0.90. Therefore, when we think of group to group differences, we are probably thinking more of the additive constant, the base estimate or likelihood of an erotic encounter, and not the individual elements. Yet, except for age, the additive constants are quite similar.

## Mind Types That Transcend Subgroups

A hallmark of Mind Genomics research is the postulate that in any population, one can generally find different sets of ideas that move together. In simpler terms, this fundamental notion means that when we look at individual patterns of responses for a specific study, we will often find two, sometimes three, and perhaps even four or more, patterns of coefficients. We call these patterns mind types or mind genomes to highlight the fact that we believe that there are basic patterns of ideas in any topic area. We discover the nature of these mind genomes by doing the experiment, such as the study here on perception of future erotic encounters.

There is no reason to assume that there are fixed mind genomes. The mind genomes emerge as a byproduct of the study, and are operationally developed for the elements of the study. That is, they are emergent groupings of ideas.

How then do we find these mind genomes? If they exist, they certainly do not distribute by gender, age, or country. The pattern of the coefficients for our study on expected erotic encounters is virtually identical when we look at the pairwise Pearson correlations in Table 5. We must use a different approach, based in the data, and accept the hypothesis that this mind genome is an emergent division of the population, operationally defined for a specific topic and a specific set of elements tested by mind genomics.

**Table 4. Summary Results of the Likelihood of an Erotic Encounter After 12 h for Specific Subgroups[a]**

| | | Total | Albania | Hungary | United States | Male | Female | Age <35 | Age 35–55 | Age 55+ |
|---|---|---|---|---|---|---|---|---|---|---|
| | CONSTANT | 26 | 26 | 22 | 32 | 26 | 28 | 33 | 25 | 12 |
| A1 | Private … just you two in scene | 17 | 13 | 24 | 14 | 16 | 20 | 14 | 21 | 19 |
| B1 | Romantic setting … walk in the park, picnic, fancy restaurant, etc. | 11 | 7 | 16 | 9 | 10 | 12 | 8 | 12 | 16 |
| D1 | Appetizer with multiple alcoholic beverages | 9 | 8 | 7 | 11 | 8 | 11 | 8 | 9 | 11 |
| B4 | Home setting | 8 | 4 | 15 | 5 | 7 | 12 | 5 | 13 | 10 |
| D2 | Entree with glass or glasses of wine/alcoholic beverage | 7 | 6 | 11 | 5 | 7 | 8 | 6 | 8 | 10 |
| C2 | Pleasant smelling cologne/perfume … floral, oriental, etc. | 6 | 5 | 6 | 7 | 7 | 3 | 6 | 7 | 6 |
| D3 | Only drinks | 3 | 2 | 6 | 0 | 3 | 1 | 1 | 4 | 5 |
| C4 | Shampoo/conditioner smell … nice smell, not overwhelming | 1 | 1 | −1 | 3 | 2 | −2 | 0 | 2 | 1 |
| D4 | Entree with nonalcoholic beverages | 0 | −3 | 3 | −1 | −1 | 2 | −2 | 2 | 1 |
| B2 | Casual setting … movie, casual restaurant, etc. | −1 | −1 | −2 | 1 | −1 | 0 | −4 | 0 | 3 |
| B3 | Social setting … bar, club, etc. | −2 | −2 | −3 | −2 | −2 | −3 | −4 | −1 | 1 |
| C3 | Deodorant only … common name brands | −2 | −4 | −1 | −1 | 0 | −5 | −2 | −4 | 1 |
| C1 | No fragrance | −5 | −3 | −9 | −2 | −3 | −9 | −4 | −6 | −4 |
| A3 | Double date … with a couple | −6 | −6 | −8 | −5 | −5 | −9 | −9 | −3 | −5 |
| A4 | Around strangers … no one familiar in scene | −8 | −5 | −5 | −14 | −8 | −8 | −5 | −13 | −7 |
| A2 | Group … out with friends | −9 | −7 | −11 | −7 | −9 | −8 | −11 | −8 | −3 |

[a] The numbers are average parameters from the individual-level models, with the averages computed for the respondents who belong to the subgroup. Strong performing elements (>7.51) are shaded.

**Table 5. Pearson Correlation Coefficients Between the Pairs of Subgroups Defined at the Start of the Research (Country, Gender, and Age)[a]**

| | Total | Albania | Hungary | United States | Male | Female | Age <35 | Age 35–55 | Age >55 |
|---|---|---|---|---|---|---|---|---|---|
| Total | 1.00 | | | | | | | | |
| Albania | 0.98 | 1.00 | | | | | | | |
| Hungary | 0.96 | 0.91 | 1.00 | | | | | | |
| United States | 0.94 | 0.92 | 0.82 | 1.00 | | | | | |
| Male | 0.99 | 0.97 | 0.94 | 0.95 | 1.00 | | | | |
| Female | 0.97 | 0.94 | 0.97 | 0.88 | 0.94 | 1.00 | | | |
| Age <35 | 0.97 | 0.98 | 0.94 | 0.90 | 0.98 | 0.93 | 1.00 | | |
| Age 35–55 | 0.97 | 0.93 | 0.94 | 0.93 | 0.96 | 0.96 | 0.91 | 1.00 | |
| Age >55 | 0.97 | 0.94 | 0.95 | 0.92 | 0.96 | 0.97 | 0.93 | 0.95 | 1.00 |

[a] The correlation coefficient is based on the pattern of coefficients for the 16 elements in the models, and computed from the data in Table 4.

We use conventional statistics to identify these mind genomes, which will operationally turn out to be reflected by clusters of respondents. These clusters are based upon individuals who show a similar pattern of coefficients. We cluster the respondents based upon the values of the 16 coefficients, ignoring the additive constant. As a measure of distance between pairs of respondents, we choose any of a variety of distance measures appropriate for clustering. There is no *right* distance measure, because clustering by itself is simply a data-exploration tool that attempts to divide a set of items (in this case, people) in a way that makes sense numerically. The measure of distance between two people is defined for mind genomics as $(1 - \text{Pearson } R)$, where the Pearson $R$ (linear correlation coefficient between two respondents) is calculated using the 16 corresponding coefficients. The correlation goes from a minimal distance of 0 ($R = 1$) when two patterns of coefficients are perfectly correlated, to 1 ($R = 0$) when two patterns of coefficients are not related at all, to a maximal distance of 2 ($R = -1$) when two patterns of coefficients are perfectly inversely correlated.

Beyond the statistical definition, we impose two additional criteria:

1. Interpretability: The clusters must make intuitive sense, and tell a reasonably clear story. The meaning of the clusters comes from the commonality of the elements that show the highest coefficient.
2. Parsimony: Fewer clusters are always better than more clusters. It may be better to work with fewer clusters, even at the loss of some interpretability.

Following the statistical approach of $k$-means clustering and additional selection criteria, we end up with four clusters or mindsets, shown in Table 6. Each mindset reflects one of the four postulated mind genomes.

The coefficients in Table 6 are shown in sorted order. We begin with the segment or mindset showing the greatest number of strong performing coefficients (value >7.51). This happens to be Segment 4B. We then sort until we reach a point where the next segment or mindset looks like it is getting strong. The order of segments presented in the table is not important. What is important is the fact that we have four segments, which seem to allow for some simplistic labels, based on the coefficients that perform most strongly in that segment.

Table 6 shows the results from the segmentation. The segments all show approximately the same additive constant—from a low of 24 (alcohol) to a high of 33 (opportunist.) Without any additional information being provided, the respondents in the Opportunist Mindset (4A) believe more strongly that an erotic encounter will occur within 12 h.

- Mindset 4A: Opportunist. They respond to nothing, other than the two people alone.
- Mindset 4B: Sensory. They respond strongly to the elements that talk about fragrance and about alcohol.
- Mindset 4C: Alcohol. They respond strongly when there is a mention of drinks.
- Mindset 4D: Setting. They respond strongly to the setting described in the vignette.

**Table 6. Performance of the 16 Elements for Total Panel and for the Four MindSet Segments[a]**

| | | Total | Mindset 4B (Sensory) | Mindset 4D (Setting) | Mindset 4C (Alcohol) | Mindset 4A (Opportunist) |
|---|---|---|---|---|---|---|
| | **Additive constant** | 26 | 26 | 25 | 24 | 33 |
| | **Strongest – Mindset 4B (Sensory seekers)** | | | | | |
| C2 | Pleasant smelling cologne/perfume … floral, oriental, etc. | 6 | 19 | 2 | –4 | 5 |
| C4 | Shampoo/conditioner smell … nice smell, not overwhelming | 1 | 16 | –12 | –10 | 3 |
| C3 | Deodorant only … common name brands | –2 | 14 | –10 | –11 | –4 |
| D1 | Appetizer with multiple alcoholic beverages | 9 | 11 | 8 | 21 | –8 |
| D2 | Entree with glass or glasses of wine/alcoholic beverage | 7 | 10 | 8 | 17 | –7 |
| | **Strongest – Mindset 4D (Setting seekers)** | | | | | |
| B1 | Romantic setting … walk in the park, picnic, fancy restaurant, etc. | 11 | 9 | 28 | 4 | 6 |
| B4 | Home setting | 8 | 8 | 27 | –4 | 6 |
| B3 | Social setting … bar, club, etc. | –2 | –5 | 17 | –7 | –7 |
| | **Strongest – Mindset 4C (Alcohol seekers)** | | | | | |
| A1 | Private … just you two in scene | 17 | 7 | 14 | 16 | 31 |
| D3 | Only drinks | 3 | 5 | 1 | 15 | –13 |
| | **Strongest – Mindset 4A (Opportunist)** | | | | | |
| B2 | Casual setting … movie, casual restaurant, etc. | –1 | –2 | 12 | –9 | –2 |
| A3 | Double date …with a couple | –6 | –18 | –14 | 3 | 5 |
| A4 | Around strangers … no one familiar in scene | –8 | –18 | –16 | –1 | 2 |
| A2 | Group … out with friends | -9 | –21 | –12 | –2 | 1 |
| C1 | No fragrance | –5 | 5 | –16 | –14 | 1 |
| D4 | Entree with nonalcoholic beverages | 0 | 6 | 1 | 6 | –16 |

[a] The labels for the mindset segments were chose based upon the elements that score highly for each segment.

The mindset segmentation draws apart the respondents in a way that the conventional subgroups could not. That is, we know that there are different ways of thinking about the possibility of the erotic encounter after 12 h, but we find no differences by age, gender, or country. Dividing people by how they think produces four dramatically different groups, and those coefficients are far less correlated with each other (Table 7). The pairwise correlations are quite low, suggesting four different, unrelated mindsets in the population.

**Table 7. Pairwise Correlations between the Total and the Four MindSets**

|  | Total | Mindset 4B | Mindset 4C | Mindset 4D | Mindset 4A |
|---|---|---|---|---|---|
| Total | 1.00 |  |  |  |  |
| Mindset 4B | 0.67 | 1.00 |  |  |  |
| Mindset 4C | 0.73 | 0.34 | 1.00 |  |  |
| Mindset 4D | 0.57 | 0.08 | 0.30 | 1.00 |  |
| Mindset 4A | 0.40 | –0.01 | 0.15 | –0.02 | 1.00 |

Finally, the distribution of the segments is almost the same across country and across gender (Table 8). The fact that we cannot really see covariation of clear mindsets with conventional ways of dividing the population means that we have a phenomenon that can be clearly established, but whose distribution in the population must be ascertained in a different way.

**Table 8. Percent of Respondents from Each Segment Falling into Each Subgroup**

|  | Mindset 4A | Mindset 4B | Mindset 4C | Mindset 4D |
|---|---|---|---|---|
| Albania | 33 | 33 | 35 | 31 |
| Hungary | 30 | 35 | 26 | 45 |
| United States | 37 | 32 | 39 | 24 |
| Total | 100% | 100% | 100% | 100% |
|  | Mindset 4A | Mindset 4B | Mindset 4C | Mindset 4D |
| Male | 64 | 73 | 66 | 78 |
| Female | 36 | 27 | 34 | 22 |
| Total | 100% | 100% | 100% | 100% |

### The Personal Viewpoint Identifier: The Metaphor of Color Primaries and the Colorimeter

Whether we select two, three, four, or even more mindsets, it has no impact on the observation that the distribution of the mindsets does not simply covary with the gender or the country of the respondents. In most other studies of this type, we discover again and again that the mindsets distribute somewhat evenly across the conventional subgroups that researchers use to classify people. These conventional subgroups are geodemographic, but often in the subgroups emerging from preestablished psychographic groups, so-called lifestyle and belief segmentation fail to covary with the mindset segments in any dramatic fashion.

The failure to find simple covariation of these segments with conventional ways to divide people means that we must find another way to assign a person to a mindset segment. Furthermore, the method of assigning people must be easy to do and affordable; there may be many of these groups of mindsets with each mindset grouping emerging from a study or set of related studies. With one study, we need one assignment tool. With hundreds or thousands of these studies, we need hundreds or thousands of assignment tools.

We have created a mindset tool to uncover the mindset of a new person. The tool uses the coefficients for each of the mindsets, as shown in Table 6.

It is important to make the assignment tool easy and fun. Figure 2 shows the assignment tool at work. The assignment tool is web based. The researcher receives the tool automatically from the Mind Genomics app, after the study has been completed, and the segments identified. The tool is created separately for the solutions comprising two, three, and four segments, respectively, meaning that in actuality there are three such assignment tools (one for the two-segment or mindset solution, one for the three-segment solution, and one for the four-segment solution).

The assignment tool looks for the elements showing the largest differences between the segments and creates a set of a few elements that express the largest between-segment differences within the data set. In order to do so, the elements are combined multiple times according to a Monte Carlo simulation method. However, thousands of element sets are created, and the researcher sees the final result (e.g., the most discriminating set of elements) immediately on the screen due to the cloud-based computing used by the app. The assignment tool is then ready to use for different tasks.

The assignment tool only works with the average of the segment scores. It is the job of the researcher to give the tool the text for the user:

1. *Orient each respondent:* Introduce the assignment tool and present the respondent with a set of questions, to be answered on a no/yes (i.e., binary) scale. Figure 2 shows the basic tool, which for the purpose of our study comprises six questions generated from the most discriminating elements. The questions are answered on a binary scale: "How do these statements predict slightly later intimacy between two people, say in about 12 h or less?" The two endpoints of the binary scale are *Little* and *Significantly*. We used the binary response to make the task less focused on judgment and more focused on immediate or gut reaction.

2. *Require the respondent to give his or her email:* Other information may also be collected at the discretion of the researcher. It is important to assure the respondent that the data obtained will be treated confidentially. The nature of the end-use of the assignment data will dictate the nature of the information requested of the respondent.

3. *Rephrase the questions into easy thoughts:* The questions are simply the relevant elements that best differentiate among the segments, and produce the best assignment. The elements or answers from the study can be rephrased into simpler questions.

4. *Give the mindset segments names:* Mind Genomics simply creates the mindset segments but does not name these segments. We generally choose to name the segments using the elements to which each segment most strongly responds in a positive manner. See Figure 3.

5. *Give feedback text, regarding what to say to the mindset segments, and what not to say:* Again, this is left to the researcher, who may want to use the assignment tool to help people know themselves, in which case the person being typed would want to know about himself or herself, and so feedback is welcome. Alternately, the feedback may be about how to treat the person doing the typing (i.e., a patient), in terms of what to say and what not to say.

6. *Describe the segment membership (Segment 4A) in a reader-friendly format:* Here the respondent is assigned to Mindset 4A (Setting). The language is respondent driven, as follows: The important aspect of this segment is *where* the encounter takes place. Members of this mindset segment place a great deal of emphasis on the location and setting leading up to the intimate encounter.

Welcome to Shiny Typing Tool of study:

*The Mind of Sexual Attraction*

Your answers will help us to understand the human mind of sexual attraction

**How do these statements predict slightly later intimacy between two people, say in about 12 hours or less?**

| | | |
|---|---|---|
| Appetizer with multiple alcoholic beverages | Little | Significantly |
| Social setting...bar, club, etc. | Little | Significantly |
| Shampoo/conditioner smell...nice smell, not overwhelming | Little | Significantly |
| Deodorant only...common name brands | Little | Significantly |
| Home setting | Little | Significantly |
| Double date...with a couple | Little | Significantly |

**Please provide your email address.**

We will not share your e-mail address with third-parties and it will not be used for marketing purposes.

Submit

*Figure 2. The six questions of the assignment tool for the current study of expected intimacy within 12 h. The answers to the assignment questions are shown to the right.*

Setting

The most important question is 'WHERE', members of this segment place a great emphasis on the location and setting...

| positive | negative |
|---|---|
| Romantic setting; walk in the park, picnic, fancy restaurant, etc. | Shampoo/conditioner smell; nice smell, not overwhelming |
| Home setting | Deodorant only; common name brands |
| Social setting; bar, club, etc. | Double date; with a couple |
| Private; just you two in scene | Around strangers; no one familiar in scene |
| Casual setting; movie, casual restaurant, etc. | Group; out with friends |
| | No fragrance |

*Figure 3. The expected segment membership for one respondent.*

## Perspectives on the Future

We have introduced a new approach to looking at the mind. What then are the prospects for this new approach, perhaps even new science? What relation might this approach have for the world of chemistry and the physical and biological sciences?

One answer to the above question is that we have structured a way to look at behavior and attitudes. The structure that we develop is purely psychological. The interesting extensions are whether there are physical and biological correlates to these uncovered mind genomes (our four segments) as well as whether one can induce a mindset by external physical stimuli such as chemicals.

The foregoing answer is purely structural, moving beyond the nature of these mindsets to the nature of covariation with other physical variables. The chemist is not necessarily interested in the mind as much as in the interrelation between chemical structure (or physical structure) and mental state. We believe that Mind Genomics provides a structure by which these covariations might be discovered, should they actually exist.

## Acknowledgments

Attila Gere is thankful for the support of the Premium Postdoctoral Research Program of the Hungarian Academy of Sciences.

## References

1. Braxton-Davis, P. The Social Psychology of Love. *McNair Sch. J.* **2010**, *14*Article 2.
2. Lamanna, M. A.; Riedmann, A. *Marriages, Families, and Relationships: Making Choices in a Diverse Society*, 10th ed.; Thomson Wadsworth: Belmont, CA, 2011.
3. Giddens, A. *The Transformation of Intimacy: Sexuality, Love, and Eroticism in Modern Societies*; Stanford University Press: Stanford, 1992.
4. Muniruzzaman, M. D. Transformation of Intimacy and Its Impact in Developing Countries. *Life Sci. Soc. Policy* **2017**, *13*, 10.
5. Simon, V. A.; Aikins, J. W.; Prinstein, M. J. Romantic Partner Selection and Socialization During Early Adolescence. *Child Dev.* **2008**, *79*, 1676–1692.
6. Feingold, A. Gender Differences in Effects of Physical Attractiveness on Romantic Attraction: A Comparison Across Five Research Paradigms. *J. Pers. Soc. Psychol.* **1990**, *59*, 981–993.
7. Sakalli-Ugurlu, N. How Do Romantic Relationship Satisfaction, Gender Stereotypes, and Gender Relate to Future Time Orientation in Romantic Relationships? *J. Psychol.* **2003**, *137*, 294–303.
8. Snyder, M.; Berscheid, E.; Glick, P. Focusing on the Exterior and the Interior: Two Investigations of the Initiation of Personal Relationships. *J. Pers. Soc. Psychol.* **1985**, *48*, 1427–1439.
9. Townsend, J. M.; Levy, G. D. Effects of Potential Partners' Costume and Physical Attractiveness on Sexuality and Partner Selection. *J. Psychol.* **1990**, *124*, 371–389.
10. Luce, R. D.; Tukey, J. W. Simultaneous Conjoint Measurement: A New Type of Fundamental Measurement. *J. Math. Psychol.* **1964**, *1*, 1–27.
11. Moskowitz, H.; Krieger, B. Consumer Requirements for a Mid-Priced Business Hotel: Insights from Analysis of Current Messaging by Hotels. *Tour. Hosp. Res.* **2003**, *4*, 268–288.

# Chapter 3

# The Chemistry of Chocolate and Pleasure

Michael H. Tunick[*] and Jennifer A. Nasser

College of Nursing and Health Professions, Drexel University, Philadelphia, Pennsylvania 19104, United States

[*]E-mail: mht39@drexel.edu.

Chocolate is celebrated as a highly craved food with sugar, fat, caffeine, and other compounds all contributing to its desirability. The sensory properties of chocolate, including aroma, sweetness, and texture, originate in the many chemical changes that occur during fermentation, roasting, and other manufacturing steps. Hundreds of odor-active compounds, including pyrazines, esters, acids, alcohols, aldehydes, and ketones, are present in chocolate, but only a few of these substances contribute to the way it feels in the mouth. Most people enjoy chocolate, many crave it, and some believe it is an aphrodisiac. These behaviors have been attributed to some of the previously mentioned compounds, but the evidence shows that chocolate consumption is purely a sensory experience.

Chocolate has been described as the most craved substance in the United States (*1, 2*). This desirability has been attributed to its texture, sweetness, fat, and aroma. Some have speculated that certain compounds in chocolate are responsible for increases in craving, pleasure, and even libido. This chapter will describe the production of chocolate, some of the compounds present, and how these compounds may relate to emotions generated upon consuming the product.

## Manufacture

Chocolate may be defined as a semi-solid suspension of fine solid particles from sugar and cocoa (around 70% total) in a continuous fat phase (*3*). The cocoa solids are derived from beans obtained from the fruit of *Theobroma cacao,* which has several cultivars. The primary cultivar is Forastero (Spanish for "stranger"), which represents roughly 80% of the world output, is native to the Amazon basin, and produces a relatively bland product. Other cultivars include Criollo ("Creole", from Mesoamerica), Trinitario ("Trinidad", a natural hybrid of Forastero and Criollo), and Nacional ("national", from Ecuador) (*4, 5*). Criollo is particularly susceptible to disease and is considered the most prized and most expensive of the common cultivars. Many varieties of these cultivars are grown in tropical regions across the world (*5*).

The first people to cultivate *T. cacao* were Mesoamericans, who brewed bitter drinks from the beans and often flavored them with chili. Spanish conquerors began to export the beans to Europe in the 16th century, and the derived chocolate eventually became popular as a sweet food (6). Through trial and error, a series of steps used for producing chocolate was established (Figure 1). The tree, flower, pod, seed, and postharvest handling are all different from the processing of any other food ingredient, and manufacturers can work with chocolate in ways that no other foods permit.

Cocoa pods are first harvested from trees by machete or knife. Each pod, which reaches maturity in four to six months, contains 30–50 cocoa beans in the form of white pulp-covered seeds. The beans contain two nibs (kernels) that yield cocoa mass for chocolate manufacture. Depending on local custom the beans are fermented in baskets, wooden boxes, or heaps stored away from light. Turning of the beans ensures an even fermentation process lasting five to seven days, depending on the variety of cultivar. Anaerobic exothermic reactions take place during the first two days as yeasts ferment the glucose, fructose, and sucrose in the pulp. The temperature increases to 40 °C as ethanol is produced. Aeration from turning the beans leads to a further elevation of the temperature to 50 °C as bacteria produce lactic and acetic acids. The germ within the bean dies due to the heat, alcohol, and acetic acid during this stage; enzymes within the bean, which are important for flavor, are released (7). In the final three days, browning reactions involving polyphenols, proteins, and peptides produce colors characteristic of cocoa.

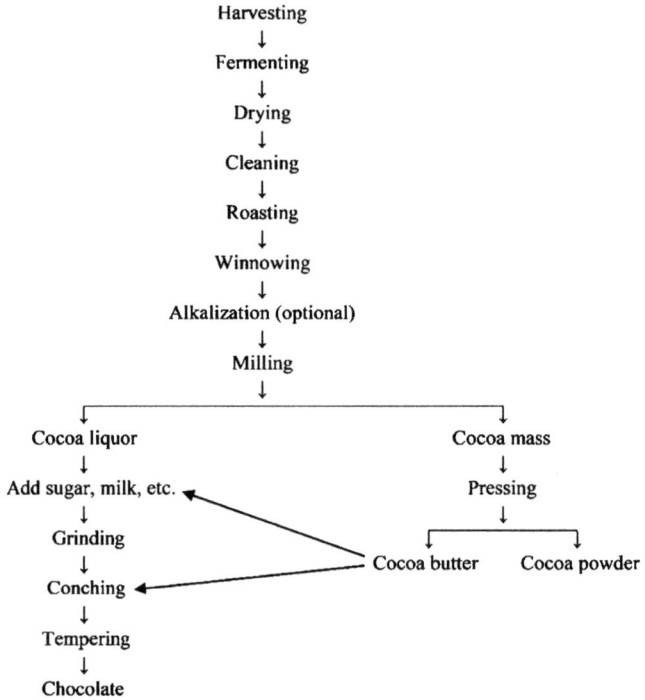

Figure 1. Flow chart of chocolate manufacture.

Fermentation removes some of the natural tannins and acids that are present; tannins, which are 5%–15% of the bean's weight, bring astringent and bitter flavors to the final product. The beans are then sun-dried for one to four weeks or are artificially dried below 60 °C for over 48 h to limit mold growth with the moisture content dropping from 60% to 7.5%. The dried beans are shipped to the processor in bags and are inspected for foreign matter. The beans are then roasted for 5–120 min at 120–150 °C, during which volatile acids are removed and Maillard reactions and the Strecker degradation begin to produce aldehydes and other products with chocolate flavor notes. The shells are cracked and winnowed (air-separated) from the nibs. The roasted beans are crushed under heat into cocoa liquor, which is then cooled to produce cocoa mass. At this stage, or before roasting, the cocoa may be treated with alkali (a process known as Dutching) to improve the color and flavor, to increase the dispersability of cocoa powder in beverages, and to decrease astringency and bitterness. Cocoa made from superior beans does not have the high acid and bitterness of Forastero beans and, as such, does not need to be Dutched (5).

Some of the fat (cocoa butter) is pressed out of the cocoa mass to produce cocoa cake, which may be ground into a powder for some applications. The cocoa mass can be mixed with milk products (for milk chocolate), emulsifiers (lecithin, sometimes in combination with polyglycerol polyricinoleate), and sweeteners. Milk products are not used in the production of dark chocolate. The cocoa mass is beaten in a mixer (conched) at 40°C–80 °C for a few days, coating the particles with fat and developing the final flavor of the chocolate (5, 7).

Cocoa butter is the most critical raw material for chocolate. The triglycerides in cocoa butter are primarily palmitic acid (27%), stearic acid (34%), and oleic acid (34%) (4). The structure leads to unique solidifying and liquefying properties with six different polymorphic forms. The chocolate must undergo careful heating and cooling cycles (tempering) to produce the crystal form of cocoa butter necessary for the proper texture and melting characteristics. Tempering has to be altered when producing milk chocolate since the milk fat creates a eutectic mixture with the cocoa butter (3).

In the U.S., milk chocolate must contain at least 10% chocolate liquor and 12% whole milk (usually in dried form). Bars of fine milk chocolate generally contain 30%–45% cacao while cheaper chocolate can have as little as 5% cacao. Dark chocolate contains 15%–35% chocolate liquor with cocoa butter, vanilla, sweetener, and usually lecithin as an emulsifier. White chocolate is composed of at least 20% cocoa butter along with sugar, milk solids, and optional flavorings such as vanilla. White chocolate contains no cocoa mass or cocoa liquor (8). A comparison of the composition of dark, milk, and white chocolate is shown in Table 1.

### Table 1. Typical Composition (g/100 g) of Chocolate Types (9)

|                     | Dark | Milk | White |
|---------------------|------|------|-------|
| Protein             | 7    | 7    | 7     |
| Saturated fat       | 24   | 20   | 19    |
| Unsaturated fat     | 19   | 11   | 12    |
| Sugar               | 29   | 56   | 60    |
| Other carbohydrates | 19   | 4    | 0     |

Types of chocolate used in baking but not consumed directly include baking chocolate (also known as bitter chocolate and unsweetened chocolate), which is made from pure chocolate liquor (100% cacao with no sugar added); bittersweet chocolate, which is sweetened dark chocolate with added sugar and cocoa butter and at least 35% chocolate liquor (70%–100% cacao); and semisweet chocolate, which is dark, sweetened chocolate made with at least 15% chocolate liquor (10).

### Sensory Attributes

Much of the pleasure derived from eating chocolate comes from the ways in which chocolate affects the five senses. Careful processing and selection of ingredients are necessary to produce desirable properties. Key compounds relating to sensory properties are shown in Table 2.

The appearance should be smooth and shiny with a color ranging from mahogany to black. A piece should make a clear and crisp snap sound when broken off a bar. A good, clean snap indicates a high cacao content and a well-tempered chocolate. Chocolate is considered to be too dry if it splinters and too waxy if it resists breaking. Milk chocolate, which has lower levels of cocoa solids, and white chocolate, with no cocoa solids, do not have the same snap. A crumbly break is undesirable (4). This tactile sense is vital, as a piece should quickly start to melt in the hand with no graininess in the mouth. Mouthfeel and textural properties are determined by the unique properties of cocoa butter (11). The viscosity can be controlled by adding cocoa butter and surface-active ingredients, especially lecithin. Optimization of texture requires consideration of palate sensitivity: a product is perceived as gritty or coarse in the mouth if the maximum particle size is is greater than 30 µm. Chocolate milled to a maximum particle size of 20 µm will have a creamier taste and texture than chocolate with 30 µm (3).

### Table 2. Compounds Responsible for Sensory Attributes in Chocolate

| Class | Specific Compounds | Attributes |
|---|---|---|
| Various | Various[a] | Brown appearance |
| Fatty acids | Palmitic, stearic, oleic acids | Creamy mouthfeel |
| Pyrazines | 2,3,5-Trimethylpyrazine | Cocoa/nutty aroma |
| Pyrazines | 2,3,5,6-Tetramethylpyrazine | Cocoa/chocolate aroma |
| Esters | Various[b] | Fruity aroma/flavor |
| Alcohols | 2-Phenylethanol | Honey/floral aroma/flavor |
| Aldehydes | 2-Phenyl acetaldehyde | Honey/floral aroma/flavor |
| Aldehydes | 2-Methylpropanal | Chocolate aroma/flavor |
| Aldehydes | 3-Methylbutanal | Chocolate aroma/flavor |
| Acids | 3-Methylbutyric acid | Sweaty aroma |
| Furanones | Furaneol | Caramel flavor |
| Pyrones | Maltol | Caramel flavor |
| Pyrroles | 2-Acetyl-1-pyrrole | Caramel/chocolate flavor |

[a] Products of enzymatic and nonenzymatic browning reactions.    [b] Mostly ethyl esters.

The aroma of chocolate may have hints of fruits, nuts, spices, flowers, or sugar. The most important odor active compounds in chocolate are pyrazines; around 80 of these compounds contribute to overall flavor. Pyrazines generally originate from α-aminoketones through the Strecker degradation and Maillard reactions that occur during roasting. Esters, which arise from amino acid degradation and fermentation, are the second most important components contributing to characteristic aromas. Alcohols, aldehydes, and ketones are also generated from fermentation and amino acids. Acids include acetic acid (the most odor-active) and various fatty acids. Short chain acids are mostly removed during processing since these lead to undesirable odors. Other compounds include furanones and pyrones, which are produced by degradation of monosaccharides during drying and roasting, confer pleasant caramel notes, and enhance flavor impression (10).

The basic taste of chocolate should include sweet, bitter, and slight notes of acid, sour, and salt. The taste is determined by the processing variables and inherent characteristics of the cocoa bean. More than 600 volatile compounds, some of which contribute to flavor, have been detected in chocolate. Flavor precursors develop during fermentation and primarily interact at roasting temperatures. Complex browning reactions also occur during roasting. The numerous heterocyclic flavor compounds produced then contribute to the characteristic and complex chocolate flavor (5).

## Specific Compounds

About 80% of the cells in cacao beans contain protein and cocoa butter for the nourishment of the plant, and the remainder contain defensive compounds meant to ward off animals and microbes. These deterrents include bitter alkaloids such as methylxanthines, astringent phenolic compounds, and anthocyanin pigments (12). Methylxanthines, particularly theobromine and caffeine, have been implicated in causing cravings. The word theobromine is derived from the Greek words θεός and βρῶμα meaning "god" and "food". One hundred grams of milk chocolate and dark chocolate can contain around 150 mg and up to 440 mg of theobromine, respectively (13). However, significant mood and behavior changes are not seen in people who consume less than 560 μg of theobromine in most people (14).

Caffeine (Figure 2), which comes from the French word café meaning "coffee", is noted for increasing alertness, mental energy, and cognitive and psychomotor performance and is also responsible for a small increase in feelings of well-being. One hundred grams of milk chocolate and dark chocolate can contain around 20 mg and 43 mg of caffeine, respectively; white chocolate contains no caffeine (4). By way of comparison, the amount of caffeine per 100 g in coffee, tea, and Coca-Cola is 40 mg, 11 mg, and 8 mg, respectively (4). Caffeine's mechanism of action is related to adenosine, a central nervous system neuromodulator with specific receptors. Neural activity slows when adenosine binds to receptors, which creates sleepiness. Blood vessels are also dilated to ensure good oxygenation during sleep. Caffeine acts as an adenosine-receptor antagonist; it binds to the same receptors but doesn't reduce neural activity, meaning that fewer receptors are available for the natural braking action of adenosine, which results in an increase in neural activity (15). Moreover, caffeine causes the pituitary gland to secrete hormones that result in an increase in adrenalin production, which increases a person's energy level and alertness while increasing production of dopamine in the pleasure circuits of the brain (15).

Phenethylamine (Figure 2; also called phenylethylamine) is a phenolic compound found in cacao. It is derived from phenylalanine by microbes and acts as a neuromodulator and neurotransmitter in humans. Typically, 50–100 mg of phenylethylamine are found in a 100 g chocolate bar. However, this compound is quickly metabolized by monoamine oxidase B, preventing significant concentrations from reaching the brain (14).

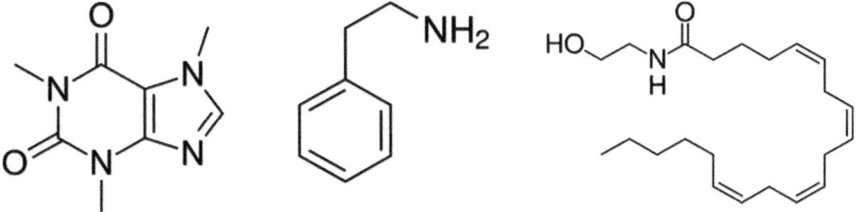

Figure 2. (Left to right) structures of caffeine, phenethylamine, and anandamide.

Anandamide (Figure 2), which comes from the Sanskrit word ananda meaning bliss, was not described until 1992. It is a metabolite of arachidonic acid (a 20-carbon fatty acid containing four double bonds) and appears to serve as a fatty acid neurotransmitter. Around 50 μg of anandamide are found in 100 g of cocoa beans (16) and less than 100 μg has been measured per g of chocolate or cocoa powder (*17*). Neural receptors for anandamide are the same receptors to which THC (the main psychoactive compound in cannabis) binds. It has been hypothesized that the endogenous cannabinoid system plays a role in regulating appetite and food intake and is involved in reward processes that mediate the incentive or hedonic value of food. Experiments conducted on rats injected with anandamide and fed sugar and quinine solutions showed that anandamide specifically amplified the hedonic impact of sweetness, a prototypical sensory pleasure. Rewarding and euphoric effects of exogenous cannabinoid drugs (such as THC) were mediated by the same endocannabinoid hedonic hotspot that amplifies taste "liking" (*18*). The authors concluded that anandamide might contribute to a feeling of well-being, but is rapidly broken down by fatty acid amide hydrolase. They estimate that more than 30 kg of chocolate must be consumed to experience effects comparable to one dose of cannabis (*18*). A recent study showed that circulating levels of ghrelin (the "hunger hormone" that helps regulate appetite), anandamide, and 2-arachidonoyl-glycerol (another endogenous cannabinoid) in the blood become elevated upon eating food for pleasure, but the presence of the latter two compounds in chocolate produce no effects since they are immediately broken down in the gastrointestinal tract (*19*).

One hundred grams of milk chocolate and dark chocolate contain around 63 and 146 mg of magnesium (Mg), respectively. Mg deficiency has been indicated as contributing to major depression (*20*). Craving chocolate has been seen as a response to a Mg deficit. Foods high in Mg are not craved, however, and do not satisfy a craving for chocolate (*21*).

## Neurotransmitters

Simultaneous activation of dopamine and opioid systems are seen in the ingestion of high fat and high sugar foods and are similar to effects by dopamine agonists (such as amphetamines and cocaine) and opioid agonists (such as heroin and morphine). Chocolate may interact with some neurotransmitter systems such as dopamine (chocolate contains tyrosine, a dopamine precursor), serotonin, and endorphins (contained in cocoa and chocolate). These systems contribute to appetite, reward, and mood regulation. However, the contribution of the dopaminergic system to chocolate craving and eating is likely a result of the general effects of the system and are not specific to chocolate (*21, 22*).

The intake of sweet food is increased by opiate agonists and decreased by opiate antagonists. Opioids can stimulate the immediate release of β-endorphin in the hypothalamus, which produces an analgesic effect. The opioid system plays a role in palatability of preferred foods: Endorphins released during eating could enhance the pleasure of eating, and chocolate stimulates endorphin release (*21, 22*).

After ingesting carbohydrates, a person's brain serotonin concentrations rise only when the protein component of a meal is less than 2%. Per its calorie contents, a standard serving of chocolate contains 5% protein, which negates any serotonin effect. Extreme dietary manipulations of tryptophan (a serotonin precursor) result in physiological changes, but these are too slow to account for the mood effects described during or soon after eating chocolate (*22*).

## Human Studies

Nasser et al. (*23*) tested 280 passersby by giving each 12.5 g of chocolate containing various levels of cocoa, sugar, and fat. They found that tasting had measurable psychoactive effects associated with a desire to consume more. The desire is proportional to the sugar and fat contents and the percentage of cocoa. Samples containing sorbitol instead of sucrose produced the same results as samples with 0% and 38% sucrose (all had 56% sugar). It appeared that binding to sweetness receptors, rather than the taste of sugar, plays a role in triggering psychoactive effects. The authors also found that men had significantly lower chocolate craving and liking scores, but when asked how much more chocolate they would like to consume, men asked for four pieces and women asked for three. There may be numerous reasons for this: The women may have been more concerned with weight, or the men may have been heavier and hungrier.

In another study, when regular chocolate consumers were deprived of chocolate for one week but otherwise ate normally, they wanted, liked, and ate chocolate more than other high-energy savory foods after their abstinence period was finished. This hedonic deprivation may have implications for people who are trying to lose weight (24). Other studies indicate that women tend to crave chocolate more than men and that a depressed mood may increase this craving (25). Research has shown that merely imagining the consumption of chocolate activates areas of the brain that are associated with hedonic effects of food and feeling full (26).

Composite sensory properties are more likely to play a prominent role in a person liking or craving chocolate. If a caloric deficit motivates chocolate craving, both milk chocolate and white chocolate should, but do not appeal equally to all subjects. If psychoactive substances or a deficit in Mg motivate chocolate craving, both milk chocolate and unsweetened cocoa powder also should, but do not appeal equally. When subjects who said they experienced chocolate cravings once a week were given milk chocolate, white chocolate, cocoa powder, and a non-cocoa control, the milk chocolate was associated with the greatest reduction in craving, significantly more than white chocolate and white chocolate plus cocoa (1). The appeal of eating chocolate appears to be solely for sensory gratification, and craving it is likely triggered by its sight and smell (21).

## Chocolate as an Aphrodisiac

An aphrodisiac is a substance that increases libido. The word is derived from Aphrodite, the Greek goddess of love. Aztecs were the first to correlate cocoa beans and sexual desire, and the idea was brought to Europe by the Spanish in the 16th century. From then until the early 20th century, Europeans claimed chocolate had a variety of medicinal uses, including improving poor sexual desire (27). However, the amounts of compounds like phenylethylamine or anandamide in chocolate are too small to have any measurable effect on libido, as these compounds are broken down too quickly. Reports on this topic in the scientific literature are scarce, but none have found a direct link between chocolate consumption and heightened sexual arousal. The concentration of epicatechin, a flavonoid that acts as an antioxidant, increases after eating chocolate but is associated only with a decrease in the activity of low-density lipoprotein (28). Aphrodisiac qualities of chocolate are probably psychological and not physiological (29).

## Conclusions

Compounds in chocolate do not contribute to craving (caffeine, Mg) and are not present in high enough concentrations (theobromine) and are broken down too quickly (phenylethylamine, anandamide) or too slowly (carbohydrates, serotonin) to produce an aphrodisiac effect. Emotions associated with eating chocolate likely stem from flavor, texture, aroma, and psychology more than specific compounds.

# References

1. Michener, W.; Rozin, P. Pharmacological Versus Sensory Factors in the Satiation of Chocolate Craving. *Physiol. Behav.* **1994**, *56*, 419–422.

2. Osman, J. L.; Sobal, J. Chocolate Cravings in American and Spanish Individuals: Biological and Cultural Influences. *Appetite* **2006**, *47*, 290–301.

3. Afoakwa, E. O.; Paterson, A.; Fowler, M. Factors Influencing Rheological and Textural Qualities in Chocolate – A Review. *Trends Food Sci. Technol.* **2007**, *18*, 290–298.

4. Beckett, S. T. *The Science of Chocolate*; RSC Publishing: Cambridge, UK, 2008.

5. Aprotosoaie, A. C.; Luca, S. V.; Miron, A. Flavor Chemistry of Cocoa and Cocoa Products – An Overview. *Compr. Rev. Food Sci. Food Saf.* **2015**, *15*, 73–91.

6. Atkinson, A.; Banks, M.; France, C.; McFadden, C. *The Chocolate and Coffee Bible*; Hermes House: London, UK, 2009.

7. Schwan, R.; Wheals, A. The Microbiology of Cocoa Fermentation and its Role in Chocolate Quality. *Crit. Rev. Food Sci. Nutr.* **2004**, *44*, 205–221.

8. U.S. Food and Drug Administration, *Code of Federal Regulations, 21CFR163*. https://www.gpo.gov/fdsys/pkg/CFR-2011-title21-vol2/pdf/CFR-2011-title21-vol2-part163.pdf (accessed Oct 9, 2018).

9. U.S. Department of Agriculture, Agricultural Research Service, Nutrient Data Laboratory, *USDA National Nutrient Database for Standard Reference*. https://ndb.nal.usda.gov/ndb/ (Oct 9, 2018).

10. McGee, H. *On Food and Cooking*; Scribner: New York, 2004; pp 694–712.

11. Hoskin, J. C. Sensory Properties of Chocolate and Their Development. *Am. J. Clin. Nutr.* **1994**, *60*, 1068S–1070S.

12. Owusu, M.; Petersen, M. A.; Heimdal, H. Effect of Fermentation Method, Roasting and Conching Conditions on the Aroma Volatiles of Dark Chocolate. *J. Food Proc. Preserv.* **2012**, *36*, 446–456.

13. Nehlig, A. The Neuroprotective Effects of Cocoa Flavanol and its Influence on Cognitive Performance. *Br. J. Clin. Pharmacol.* **2013**, *75*, 716–727.

14. Smit, H. J.; Gaffan, E. A.; Rogers, P. J. Methylxanthines are the Psycho-Pharmacologically Active Constituents of Chocolate. *Psychopharmacology* **2004**, *176*, 412–419.

15. Nehlig, A.; Daval, J. L.; Debry, G. Caffeine and the Central Nervous System: Mechanisms of Action, Biochemical, Metabolic and Psychostimulant Effects. *Brain Res. Rev.* **1992**, *17*, 1139–1170.

16. Wishart, D. S.; Feunang, Y. D.; Marcu, A.; Guo, A. C.; Liang, K. HMDB 4.0 – The Human Metabolome Database for 2018. *Nucleic Acids Res.* **2018**, *46* (D1), D608–617.

17. di Tomaso, E.; Beltramo, M.; Piomelli, D. Brain Cannabinoids in Chocolate. *Nature* **1996**, *382*, 677–678.

18. Mahler, S. V.; Smith, K. S.; Berridge, K. C. Endocannabinoid Hedonic Hotspot for Sensory Pleasure: Anandamide in Nucleus Accumbens Shell Enhances "Liking" of a Sweet Reward. *Neuropsychopharmacology* **2007**, *32*, 2267–2278.

19. Rigamonti, A. E.; Piscitelli, F.; Aveta, T.; Agosti, F.; De Col, A.; Bini, S.; Cella, S. G.; Di Marzo, V.; Sartorio, A. Anticipatory and Consummatory Effects of (Hedonic) Chocolate

Intake are Associated with Increased Circulating Levels of the Orexigenic Peptide Ghrelin and Endocannabinoids in Obese Adults. *Food Nutr. Res.* **2015**, *59*, 29678.

20. Eby, G. A.; Eby, K. L. Rapid Recovery from Major Depression Using Magnesium Treatment. *Med. Hypoth.* **2006**, *67*, 362–370.

21. Parker, G.; Parker, I.; Brotchie, H. Mood State Effects of Chocolate. *J. Affect. Disord.* **2006**, *92*, 149–159.

22. Bruinsma, K.; Taren, D. L. Chocolate: Food or Drug? *J. Am. Diet. Assoc.* **1999**, *99*, 1249–1256.

23. Nasser, J. A.; Bradley, L. E.; Leitzsch, J. B.; Chohan, O.; Fasulo, K.; Haller, J.; Jaeger, K.; Szulanczyk, B.; Del Parigi, A. Psychoactive Effects of Tasting Chocolate and Desire for More Chocolate. *Physiol. Behav.* **2011**, *104*, 117–121.

24. Blechert, J.; Naumann, E.; Schmitz, J.; Herbert, B. M.; Tuschen-Caffier, B. Startling Sweet Temptations: Hedonic Chocolate Deprivation Modulates Experience, Eating Behavior, and Eyeblink Startle. *PloS One* **2014**, *9* (1), e85679.

25. Katz, D. L.; Doughty, K.; Ali, A. Cocoa and Chocolate in Human Health and Disease. *Antioxid. Redox Sign.* **2011**, *15*, 2779–2811.

26. Kiortsis, D. N.; Spyridonos, P.; Margariti, P. N.; Xydis, V.; Alexiou, G.; Astrakas, L. G.; Argyropoulou, M. I. Brain Activation During Repeated Imagining of Chocolate Consumption: A Functional Magnetic Resonance Imaging Study. *Hormones* **2018**, *17*, 367–371.

27. Dillinger, T. L.; Barriga, P.; Escárcega, S.; Jimenez, M.; Salazar Lowe, D.; Grivetti, L. E. Food of the Gods: Cure for Humanity? A Cultural History of the Medicinal and Ritual Use of Chocolate. *J. Nutr.* **2000**, *130*, 2057S–2072S.

28. Wilson, P. K. Centuries of Seeking Chocolate's Medicinal Benefits. *Lancet* **2010**, *376*, 158–159.

29. Afoakwa, E. O. Cocoa and Chocolate Consumption – Are There Aphrodisiac and Other Benefits for Human Health? *S. Afr. J. Clin. Nutr.* **2008**, *21*, 107–113.

# Chapter 4

# Smoky, Vanilla, or Clove-Like?

## Structure-Odor Activity Relationships of Guaiacol Derivatives

R. Ghadiriasli,[1,2] K. Lorber,[1] M. Wagenstaller,[2] and A. Buettner*,[1,2]

[1]Department of Chemistry and Pharmacy, Emil Fischer Center,
Friedrich-Alexander-Universität Erlangen-Nürnberg,
Erlangen 91054, Germany
[2]Department Sensory Analytics,
Fraunhofer Institute for Process Engineering and Packaging IVV,
Freising 85354, Germany
*E-mail: andrea.buettner@ivv.fraunhofer.de.

Smoky smells, primarily comprising guaiacols, phenols, and cresols, are enjoyed in diverse foods and beverages, such as whiskey, smoked ham, cheese, or baked goods, such as pizza. They contribute to the flavor and smell experience when smoking tobacco products or when sitting at a fireplace. In other cases, smoky smells indicate potential locations of pyrolytic and combustion processes and may act as alarm signals. In the animal kingdom, interestingly, some compounds related to common constituents of smoke may serve as chemo-communication tools or even pheromones. However, previous studies on the odor properties of alkylated and halogenated guaiacols, phenols, and cresols demonstrate that smoky smell is not the same for everybody. When resolving the underlying chemical structures and carrying out systematic structure-odor activity relationship investigations, one realizes that there are obviously more complicated considerations for each of us in the smell of smoke. This chapter gives an overview of the sensory data of guaiacol and selected structurally related odorants, such as alkylated, alkenylated, methoxylated, and halogenated guaiacol derivatives.

## Introduction

Phenols are one of the largest groups of compounds present in the environment. Guaiacol (2-methoxyphenol or 1-hydroxy-2-methoxybenzene) and syringol (2,6-dimethoxyphenol) belong to a group of compounds called methoxyphenols, which commonly exist in the environment (1). Guaiacol was extracted for the first time via distillation from Guaiacum resin or wood creosote, hence its common name (2). Guaiacol and its derivatives have widespread presence in flora and fauna and are used as aroma components in the food and perfume industry and as local anesthetics and antiseptics in medicine (3). Guaiacol is responsible for the flavor of smoke in smoked and roasted foods such as bacon, fish, and coffee. Although guaiacol is the main chemical compound responsible for the smoky odor, wood smoke comprises a wide range of components and can be condensed to form liquid smoke, which is added to many foods as a flavorant (4).

Alcoholic beverages, such as wine and spirits, are usually matured or aged by storing them in wooden barrels or by their contact with wood chips for varying durations. Barrel aging is an important step for the quality of alcoholic beverages. More specifically, oxygen permeation through the wood promotes redox processes and several wood compounds transfer from wood to the products. This process is finalized to improve limpidity, stabilize the color, and enrich the sensory and chemical characteristics of the product. Among the phenolic aldehydes derived from lignin through hydrolysis, oxidation reactions, or pyrolysis, vanillin plays an essential role owing to its vanilla-like odor. Other volatile phenols, like guaiacol and its derivatives, eugenol and isoeugenol, contribute to smoky, clove-like, and spicy notes and have an important sensory impact on aged alcoholic beverages. These compounds have been identified in wine, spirits, and some other alcoholic beverages (5–8). Oakwood is the most commonly used material in barrel manufacture for aging, although other species such as chestnut, false acacia, cherry, and more rarely, ash and mulberry, have been considered as possible wood sources for the production and maturation of wine and derived products like vinegar, cider, and whisky (9, 10).

Guaiacol may serve as the starting component in the synthesis of vanillin (11). 4-Vinyl-, 4-ethyl-, 4-allyl-, and 5-propylguaiacol are derivatives responsible for the aroma of smoked food and have been identified inter alia in smoked ham (12). Moreover, guaiacol and 4-vinylguaiacol have been found in wheat beers as aroma compounds, and 4-ethyl-, 4-allyl-, and 4-vinylguaiacol have been found in brandy (13, 14). In addition, many plants produce these compounds. For example, the most common derivative is eugenol (4-allylguaiacol), which is known as a main compound of clove oil and has a clove-like smell. Another component of clove oil is isoeugenol [4-(1-propanyl)-guaiacol], which is also a clove-like, smoky smelling odorant. These compounds also influence the aromatic composition and quality of wines aged in oak barrels (15, 16). In a study on natural, untreated oakwood, eugenol and acetyleugenol were identified as spicy, clove-like smelling compounds (17). These components also have been excessively used in medicine (18, 19). 5-Allylguaiacol, also known as chavibetol, has an exotic and unusual odor and is produced by different plants (20). Other derivatives, such as 4-methyl-, 4-ethyl-, 6-ethyl-, 4-propyl-, and 4-allylguaiacol, have been identified in some Japanese plants (21). 5-Ethylguaiacol, also known as locustol, is a pheromone that was shown to be responsible for the aggregation behavior of locusts (22). This pheromone is produced by intestinal tract bacteria within locusts, which convert lignin from vegetal material to locustol (23).

Halogenated derivatives of guaiacol are considered toxic above certain concentrations. For instance, chlorinated derivatives can easily penetrate skin and epithelium, which can lead to damage and necrosis of tissue, heart disease, asthma, and lung cancer in humans. 4-Chloro-, 5-chloro-, 6-chloro-, or 4,5-dichloroguaiacol derivatives, were identified in bleaching filters and pulp mill waste water or in sediments of rivers around pulp mills of pinewood, eucalyptus, wheat straw, rice straw, or bamboo processing units (24, 25). Furthermore, 4-chloro- and 4,5-dichloroguaiacol were also found in tissue of aquatic animals living downstream of the bleaching pulp mills, which might endanger animal health and pose a risk to consumer health (26).

# Guaiacol and Its Biosynthesis

Besides primary plant compounds, which are essential for the plant's survival, secondary plant compounds, known as so-called phytochemicals, are metabolized. Those compounds are not essential, but remain very important (e.g., for defense or coloring). Guaiacol is one example of such a phytochemical. Its ground-structure comes from phenol and anisol and occurs in different woods (tars and resins), like in the resin of Guajak. Unsurprisingly, a main formation process of guaiacol occurs during microbial or thermal degradation of lignin, cross-linked phenolic polymers in plants, and is present especially in wood and bark (27–30).

Wood is composed of an interconnected structure of three large biopolymers, namely cellulose, hemicellulose, and lignin. Phenolic compounds stem from the degradation of the biopolymer lignin in the cell wall of wood. Mono- and dimethoxylated phenols are mostly formed by thermodegradation, resulting in high levels of these compounds in toasted wood. One application of this process is its use in barrels that are employed in the aging of wine and other alcoholic beverages (31, 32).

Guaiacol can also be biosynthesized via enzymatic processes, as is done in the cyanogenic gland of a polydesmid millipede, *Oxidus gracilis* (33), or in the gut of desert locusts, *Schistocerca gregaria*, where guaiacol is one of the main components of the pheromones responsible for swarming (23). Furthermore, some fungi and microbes can also degrade complex structures of lignin to low molecular weight compounds including methoxyphenols (34).

## Relationship between Guaiacol-Derived Structure Compounds and Smoky Odor

Guaiacol and its diverse derivatives are naturally occurring compounds that are well-known for their smoky, clove-like, and vanilla-like notes. In order to investigate the structure-odor activity relationship of the alkylated, alkenylated, and methoxylated derivatives of guaiacol, Schranz et al. (35) studied the different guaiacol derivatives that were substituted in position 3 (meta), 4 (para), 5 (meta), and 6 (ortho). The odor thresholds (OTs) and odor qualities of these derivatives were determined by means of gas chromatography-olfactometry (GC-O) and gas chromatography-mass spectrometry analysis. Guaiacol itself was reported as a smoky, vanilla-like compound. The alkylated and alkenylated groups of guaiacol in positions 3, 4, 5, and 6, such as 4-propylguaiacol, 5- and 6-methylguaiacol, 6-ethylguaiacol, 3-, 5-, and 6-allylguaiacol, and 6-methoxyguaiacol were found to exhibit a smoky odor. Others, like 4-methyl-, *trans*-5-propyl-, 5-ethyl-, and 5-methoxyguaiacol were characterized by a vanilla-like, sweet odor. The vanilla-like, sweet note, therefore, was attributed mostly to the derivatives that are substituted in the meta position from the hydroxyl function. A clove-like smell was predominantly associated with the unsaturated derivatives, such as 4-vinyl-, *cis*-4-, *trans*-4-, and *cis*-6-propenylguaiacol, 4- and 5-allyl, and 4-methoxyguaiacol. Furthermore, some compounds exhibit two or three odor qualities, namely smoky and sweet for 5-ethylguaiacol, or clove- and vanilla-like for 5-propylguaiacol. Interestingly, 6-allylguaiacol exhibited plastic-like, clove-like, and smoky odor qualities. The median OTs in air, comparative OTs, and the odor quality for guaiacol and its related substances are listed in Table 1 (35, 36).

**Table 1. Median OTs and Odor Qualities of Guaiacol and Its Alkylated, Alkenylated, and Methoxylated Derivatives[a]**

| Odorant | OT [ng/L$_{air}$][b] | Comparative OT [ng/L$_{air}$][c,d,e,f] | Odor quality[g] |
|---|---|---|---|
| Guaiacol | 0.084 | 1.5[c] 0.28[d] | smoky, vanilla-like, ham-like |
| 4-Ethylguaiacol | 0.15 | 0.017[e] | clove-like, smoky |
| 4-Vinylguaiacol | 0.16 | 2.8[c] 0.57[e] | clove-like, smoky |
| 6-Vinylguaiacol | 0.2 | - | smoky, ham-like |
| 4-Propylguaiacol | 0.22 | - | smoky, clove-like, sweet |
| 5-Vinylguaiacol | 0.23 | - | smoky, ham-like, clove-like, sweet, vanilla-like |
| 4-Methoxyguaiacol | 1.0 | - | clove-like, sweet, smoky, vanilla-like, ham-like |
| 5-Allylguaiacol | 1.1 | - | smoky, ham-like, clove-like, sweet |
| 4-Methylguaiacol | 1.4 | - | vanilla-like, sweet, ham-like, smoky |
| 5-Ethylguaiacol | 1.9 | - | smoky, sweet, ham-like |
| 5-Methylguaiacol | 2.2 | - | vanilla-like, sweet, smoky |
| 5-Propylguaiacol | 2.6 | - | clove-like, vanilla-like |
| cis-4-Propenylguaiacol | 2.7 | - | clove-like |
| trans-4-Propenylguaiacol | 2.8 | - | clove-like |
| 4-Allylguaiacol | 3.2 | 0.24[f] | clove-like |
| 6-Methylguaiacol | 6.2 | - | smoky, plastic-like, sweet, bacon-like |
| 6-Allylguaiacol | 21 | - | plastic-like, clove-like, vanilla-like |
| 6-Ethylguaiacol | 26 | - | smoky |
| 6-Propylguaiacol | 33 | - | plastic-like, sweet |
| 3-Vinylguaiacol | 33 | - | smoky, clove-like |
| cis-6-Propenylguaiacol | 44 | - | smoky, ethereal, clove-like |
| trans-6-Propenylguaiacol | 111 | - | ham-like, smoky |

[a] Investigated by Schranz et al. (35) and listed in order of ascending OT. [b] OTs in air were determined as described by Ullrich and Grosch (36). [c] OTs by Yang (37). [d] OT reported by Guth and Grosch (38). [e] OTs reported by Blank and Grosch (39). [f] OT reported by Blank (40). [g] Odor qualities as perceived at the GC-O sniffing port.

Maga investigated the contribution of phenolic compounds to smoke aroma in foods and named sweet, smoky odor descriptors for guaiacol, 3-methyl-, and 4-ethylguaiacol. In the same study, 4-allylguaiacol was reported as the major contributor to the woody odor (41). Further, Prez-Silva et al. (42) studied the aroma compounds in cured vanilla beans and described guaiacol as eliciting chemical, sweet, and spicy odor impressions. 4-Methylguaiacol and 4-vinylguaiacol were reported to be sweet, woody, and chemical phenolic, respectively. In a study by Blank and Grosch on the potent odorants of the roasted powder and brew of Arabica and Robusta coffees, the odor quality of the potent odorants guaiacol, 4-ethylguaiacol, and 4-vinylguaiacol were reported as phenolic, spicy (39, 43). Czerny and Grosch (44) demonstrated that 4-ethylguaiacol and 4-vinylguaiacol compounds exhibited sweet and clove-like odors, respectively. According to Lorjaroenphon et al. (45), 4-allylguaiacol was described as spicy, clove-like, and sweet in the study dealing with character-impact odorants in a cola-flavored carbonated beverage. Finally, 4-propylguaiacol was found to smell clove-like, smoky by Cadwallader (46).

Juhlke et al. (47) investigated the influence of the chemical structure on the odor qualities of halogenated guaiacol derivatives substituted in position 3 (meta), 5 (meta), and 6 (ortho). Also disubstituted derivatives in position 4 and 5 or 5 and 6, in relation to the hydroxyl function, were analyzed in this study. The halogenated derivatives of guaiacol were reported to exhibit smoky, vanilla-like, sweet, plaster-like, and medicinal odor qualities. The mono- and dichloro-substituted derivatives such as 3-chloro-, 5-chloro-, 6-chloro-, 4,5-dichloro-, 5,6-dichloro-, and 5-bromoguaiacol mainly showed smoky odors, often followed by a sweet and vanilla-like note. 4-Chloro-, 4-bromo-, 4-iodo, and 5-iodoguaiacol were reported to smell sweet and vanilla-like. In contrast, 3-bromo- and 3-iodoguaiacol derivatives were perceived to elicit a musty odor. A medicinal note was shown by bromo- and iodo-substitutions at position 6 of guaiacol. Predominantly the odorants were found to elicit two to four odor qualities. For some derivatives like 4,5-dichloroguaiacol the odor quality of the respective compound was found to be a mix of several attributes, such as smoky, vanilla-like, and sweet. While medicinal, smoky, and plaster- and plastic-like attributes were elicited at equal intensities for 6-bromoguaiacol, the 6-iodoguaiacol derivative was found to elicit a medicinal odor only. The OT of halogenated guaiacol-derived odorants and their odor qualities are reported in Table 2 (47).

**Table 2. Median Odor OTs and Odor Qualities of Halogenated Guaiacol Derivatives[a]**

| Odorant | OT [ng/L$_{air}$][b] | Odor quality[c] |
|---|---|---|
| 5-Chloroguaiacol | 0.00072 | smoky, ham-like |
| 5-Bromoguaiacol | 0.0023 | smoky, sweet |
| 4,5-Dichloroguaiacol | 0.0025 | smoky, sweet, vanilla-like |
| 6-Chloroguaiacol | 0.00251 | smoky, sweet |
| 6-Bromoguaiacol | 0.0046 | medicinal, smoky, plaster-like, plastic-like |
| 5-Iodoguaiacol | 0.0048 | sweet, smoky |
| 5,6-Dichloroguaiacol | 0.0068 | smoky, medicinal, plaster-like |
| 4-Bromoguaiacol | 0.029 | vanilla-like, sweet, smoky |
| 6-Iodoguaiacol | 0.036 | medicinal |
| 4-Chloroguaiacol | 0.35 | sweet, vanilla-like |

**Table 2. (Continued). Median Odor OTs and Odor Qualities of Halogenated Guaiacol Derivatives[a]**

| Odorant | OT [ng/$L_{air}$][b] | Odor quality[c] |
|---|---|---|
| 4-Iodoguaiacol | 4.1 | vanilla-like, smoky, sweet |
| 3-Chloroguaiacol | 5.7 | smoky, medicinal |
| 3-Bromoguaiacol | 11 | musty |
| 3-Iodoguaiacol | 23 | musty, moldy |

[a] Investigated by Juhlke et al. (47) and listed in ascending order of OT.    [b] OTs in air were determined as described by Ullrich and Grosch (36).    [c] Odor qualities as perceived at the GC-O sniffing port.

The variation and overlap of odor qualities of selected guaiacol derivatives are shown in Figure 1.

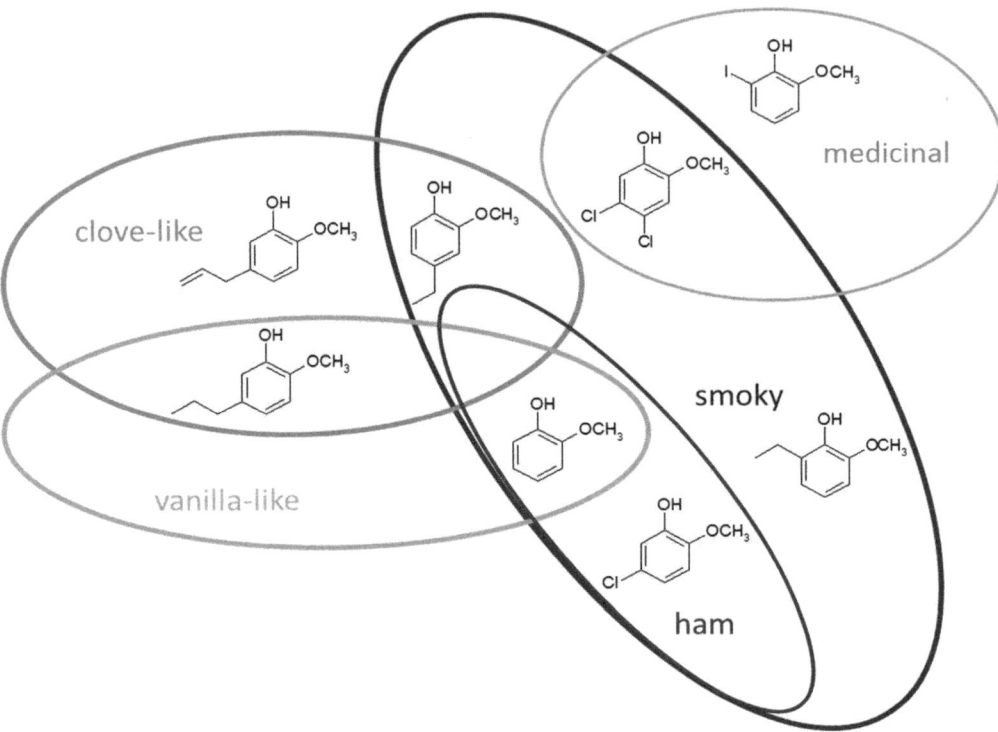

*Figure 1. The odor quality of guaiacol-derived odorants.*

## OTs of Guaiacol-Derived Odorants

OTs in air are most commonly determined by means of GC-O with (*E*)-2-decenal as an internal standard (36). To account for large interindividual variations, the median OTs are sometimes reported (48). For the compounds studied by Schranz et al. (37), guaiacol derivatives, substituted in the para and meta position to hydroxyl functional group, showed lower OTs than the ortho-substituted compounds. For the ethyl- and propylguaiacols substituted in the para position, the reported values were lower than for meta-substituted derivatives reported in the same study. In the case of *trans*- and *cis*-propenyl and allyl- and methoxyguaiacols, the lowest median values were shown for derivatives with the additional side-chain in position 5. Therefore, the derivatives substituted in para- and meta-position were found to have lower OTs or higher odor potencies than derivatives substituted in the ortho-position. In unsaturated compounds, the presence of a double-bond, in the case of vinylguaiacol derivatives, resulted in lower OTs in comparison to the respective saturated compound ethylguaiacol. In contrast, the double-bond in the propenyl derivatives seemed to cause the opposite effect. The median OT values of the propenylguaiacols were reported to be higher than those of the respective propylguaiacol derivatives. The OTs of allylguaiacol derivatives were shown to be between the propyl- and propenyl-derivatives. Therefore, it may be concluded that the double-bond on the end of a molecule in the hydrocarbon side-chain can cause a decrease in the OT while a double-bond with closed position to the aromatic ring leads to an increase in OT (36). However, these conclusions should be corroborated by investigations with further compounds. Moreover, for propenylguaiacols it was reported that the values of OTs in *cis* derivatives were slightly lower than the values of the corresponding *trans* isomer (Table 1; (35)).

In the study by Yang et al. (41), guaiacol and 4-vinylguaiacol showed OT values of 1.5 and 2.8 ng/$L_{air}$, respectively. By comparison, the OT of guaiacol was determined by Guth and Grosch (38) to be 0.28 ng/$L_{air}$ in their study on the odorant in virgin olive oils. 4-Ethylguaiacol and 4-vinylguaiacol were studied by Blank and Grosch (39), who reported relatively low OTs of 0.017 ng/$L_{air}$ and 0.57 ng/$L_{air}$, respectively, for these compounds. Additionally, Blank et al. (40) reported an OT of 0.24 ng/$L_{air}$ for 4-allyguaiacol.

A comparison of the OT data from our group with the limited values reported in the literature reveals a lower median OT of guaiacol (0.084 ng/$L_{air}$) compared to the literature values of 0.28–1.5 ng/$L_{air}$ (38, 41), which is a difference of between a factor of 3 and 17. In the case of 4-ethylguaiacol (0.15 ng/$L_{air}$), the value reported in our study differs from the literature value by a factor of ~10. Additionally, 4-vinylguaicol (0.16 ng/$L_{air}$) varied by a factor of 4–18 in relation to the values determined Blank et al. (40), furthermore, 4-allyguaiacol (3.2 ng/$L_{air}$) exhibited a comparatively lower value in the literature (0.24 ng/$L_{air}$). Despite the use of the same methodology for OT determination in air, the differences between the OTs determined in our studies compared with the literature data are possibly a result of the differing number of panelists between the studies, whereby the number of panelists was small in some studies.

Overall, the OT investigation of guaiacol and its alkylated, alkenylated, and methoxylated derivatives showed that no specific substituent caused a universal and significant OT effect. However, the positioning of the substituents to the hydroxyl function in the molecule seems to have highest impact. For example, the vinyl group was shown to exhibit the lowest median OTs for the compounds substituted in positions 4, 6, and 5 with the values 0.16, 0.20, and 0.23 ng/$L_{air}$, respectively. Surprisingly, the same group led to a much higher OT in position 3, with 3-vinylguaiacol having a threshold of 33 ng/$L_{air}$. However, with regard to the saturated hydrocarbon substituent groups methyl, ethyl, and propyl, it was found that OTs increased from para- over meta- to ortho-substituted derivatives.

Among the halogenated guaiacol derivatives, substituted derivatives in position 3 (meta) to the hydroxyl functional group were found to show the highest median OTs, whereas the lowest OT was found in derivatives substituted in position 5. The OTs for para- and ortho-substituted derivatives were reported to be between those substances substituted in the meta-position and the 5-substituted analogs. Regarding the three substituents chlorine, bromine, and iodine, it was found that the chloroguaiacols showed the lowest median OT compared to bromo and iodo derivatives. In the case of dichlorinated guaiacol derivatives, Juhlke et al. (47) reported very low OTs (Table 2).

Comparing halogenated to alkylated, alkenylated, and methoxylated guaiacol derivatives, some of the halogenated compounds showed OTs that were several orders of magnitude lower than all the other guaiacol compounds analyzed by Juhlke et al. (47) and Schranz et al. (35). Overall, it can be concluded that the chlorine substituent in position 5 (5-chloroguaiacol) exhibits the lowest OT, with a value of 0.00072 $ng/L_{air}$, followed by mono-substituted bromo- and iodoguaiacol derivatives in position 5, with values 0.0023 and 0.0048 $ng/L_{air}$ for 5-bromo- and 5-iodoguaiacol, respectively. However, dichlorinated guaiacol compounds exhibit similarly low OTs that are slightly higher than the aforementioned substituents.

## Individual Odor Qualities of Guaiacol-Derived Odorants

In the odor quality investigation of guaiacol derivatives by a sensory panel, the odor qualities of individual compounds were generally described as the same by the panelists, while the OTs varied considerably. However, changes in concentration have been found to affect the odor quality. It was reported by Schranz et al. (35), for example, that the odor of 5-methylguaiacol changed from vanilla-like and sweet to clove-like when the concentration decreased, while *cis*-5-propenylguaiacol altered its odor from vanilla-like and smoky to clove-like.

Large differences in OTs were additionally reported for individual halogenated derivatives, whereas the odor qualities were generally described to be the same, with the exception of some derivatives (47, 49). Therefore, there is seemingly no direct correlation between odor impressions and specific substitution patterns.

## Conclusion

Overall, guaiacol derivatives exhibit very low OTs. Alkylated, alkenylated, and methoxylated substances elicit odor impressions of smoky, vanilla-like, clove-like, and sweet. The majority of guaiacol derivatives has been reported in food, wood, plants, nature, or even in industrial products. The main odor descriptors for the halogenated derivatives of guaiacol are smoky, sweet, vanilla-like, and medicinal. With regards to the structure-odor relationship, compounds substituted in position 3 show smoky and medicinal smells for chlorinated derivatives, whereas brominated and iodinated components elicit musty and moldy odor notes.

In general, a series of factors might influence the perception of odorants between individuals. For example, physiological or psychological state can modify the way an odorant is perceived. OTs have been shown to increase after food intake but to decrease during hunger. Other factors such as health, gender, emotional state, pregnancy, and menstrual cycle can also influence odor perception (50). Furthermore, genetic preposition in individual receptors and various expressed metabolizing enzyme systems may cause interindividual differences in odor perception (51). Before docking to the receptor protein in the cilia, odor-active compounds must pass through the aqueous milieu of the nasal cavity. Enzymes (e.g., members of the cytochrome p450 system) in this aqueous viscous colloidal mucus are able to metabolize odorants before reaching the receptor sites. This process, which is called "peri-receptor events", may impact the OT and perceived odor quality of the odorants (52, 53).

While many studies have been conducted on structure-odor relationships, we are still far from understanding the underlying principles, as is evident for the guaiacol derivatives reported in this chapter. More interdisciplinary studies are needed to unveil these unknown mechanisms with the final aim to predict odor quality and OTs of derivatives.

# References

1. Michalowicz, J.; Duda, W. Occurrence of Chlorophenols, Chlorocatechols, Chlorinated Methoxyphenols and Monoterpenes in Soils of the River Brda in the North West Part of the Tucholski Landscape Park. *Pol. J. Soil Sci.* **2004**, *37*, 121–130.

2. Unverdorben, O. Ueber das Guajakharz. *Annalen der Physik* **1829**, *92*, 369–376.

3. Kumbar, S. M.; Shanbhag, G. V.; Halligudi, S. B. Synthesis of Monoallyl Guaiacol via Allylation Using HY Zeolite. *J. Mol. Catal. A: Chem.* **2006**, *244*, 278–282.

4. Simon, R.; de la Calle, B.; Palme, S.; Meier, D.; Anklam, E. Composition and Analysis of Liquid Smoke Flavouring Primary Products. *J. Sep Sci.* **2005**, *28*, 871–882.

5. Fernández de Simón, B.; Cadahía, E.; Sanz, M.; Poveda, P.; Perez-Magariño, S.; Ortega-Heras, M.; González-Huerta, C. Volatile Compounds and Sensorial Characterization of Wines from Four Spanish Denominations of Origin, Aged in Spanish Rebollo (*Quercus pyrenaica* Willd.) Oak Wood Barrels. *J. Agric. Food Chem.* **2008**, *56*, 9046–9055.

6. Li, S.; Crump, A. M.; Grbin, P. R.; Cozzolino, D.; Warren, P.; Hayasaka, Y.; Wilkinson, K. L. Aroma Potential of Oak Battens Prepared from Decommissioned Oak Barrels. *J. Agric. Food. Chem.* **2015**, *63*, 3419–3425.

7. Setze, W. N. Volatile Components of Oak and Cherry Wood Chips used in Aging of Beer, Wine, and Sprits. *American Journal of Essential Oils and Natural Products* **2016**, *4*, 37–40.

8. Alañón, M. E.; Ramos, L.; Díaz-Maroto, M. C.; Pérez-Coello, M. S.; Sanz, J. Extraction of Volatile and Semi-Volatile Components from Oak Wood Used for Aging Wine by Miniaturised Pressurised Liquid Technique. *Int. J. Food Sci. Technol.* **2009**, *44*, 1825–1835.

9. Culleré, L.; Fernández de Simón, B.; Cadahía, E.; Ferreira, V.; Hernández-Orte, P.; Cacho, J. Characterization by Gas Chromatography–Olfactometry of the Most Odor-Active Compounds in Extracts Prepared from Acacia, Chestnut, Cherry, Ash and Oak Woods. *LWT-Food Science and Technology* **2013**, *53*, 240–248.

10. De Rosso, M.; Cancian, D.; Panighel, A.; Dalla Vedova, A.; Flamini, R. Chemical Compounds Released from Five Different Woods Used to Make Barrels for Aging Wines and Spirits: Volatile Compounds and Polyphenols. *Wood Sci. Technol.* **2009**, *43*, 375–385.

11. Esposito, L. J.; Formanek, K.; Kientz, G.; Mauger, F.; Maureaux, V.; Robert, G.; Truchet, F. Vanillin. In *Kirk-Othmer Encyclopedia of Chemical Technology*, 4th ed.; Kirk, R. E.; Othmer, D. F., Eds.; John Wiley & Sons: New York, 1997; Vol. 24, pp 812–825.

12. Baloga, D. W.; Reineccius, G. A.; Miller, J. W. Characterization of Ham Flavor Using an Atomic Emission Detector. *J. Agric. Food. Chem.* **1990**, *38*, 2021–2026.

13. Coghe, S.; Benoot, K.; Delvaux, F.; Vanderhaegen, B.; Delvaux, F. R. Ferulic Acid Release and 4-Vinylguaiacol Formation During Brewing and Fermentation: Indications for Feruloyl Esterase Activity in *Saccharomyces cerevisiae*. *J. Agric. Food. Chem.* **2004**, *52*, 602–608.

14. Willner, B.; Granvogl, M.; Schieberle, P. Characterization of the Key Aroma Compounds in Bartlett Pear Brandies by Means of the Sensomics Concept. *J. Agric. Food Chem.* **2013**, *61*, 9583–9593.

15. Diaz-Plaza, E. M.; Reyero, J. R.; Pardo, F.; Alonso, G. L.; Salinas, M. R. Influence of Oak Wood on the Aromatic Composition and Quality of Wines with Different Tannin Contents. *J. Agric. Food Chem.* **2002**, *50*, 2622–2626.

16. Fernández de Simón, B.; Cadahía, E.; Jalocha, J. Volatile Compounds in a Spanish Red Wine Aged in Barrels Made of Spanish, French, and American Oak Wood. *J. Agric. Food Chem.* **2003**, *51*, 7671–7678.

17. Ghadiriasli, R.; Wagenstaller, M.; Buettner, A. Identification of Odorous Compounds in Oak Wood Using Odor Extract Dilution Analysis and Two-Dimensional Gas Chromatography-Mass Spectrometry/Olfactometry. *Anal. Bioanal. Chem.* **2018**, *410*, 6595–6607.

18. Jadhav, B. K.; Khandelwal, K. R.; Ketkar, A. R.; Pisal, S. S. Formulation and Evaluation of Mucoadhesive Tablets Containing Eugenol for the Treatment of Periodontal Diseases. *Drug Dev. Ind. Pharm.* **2004**, *30*, 195–203.

19. Koeduka, T.; Fridman, E.; Gang, D. R.; Vassao, D. G.; Jackson, B. L.; Kish, C. M.; Orlova, I.; Spassova, S. M.; Lewis, N. G.; Neol, J. P.; Baiga, T. J.; Dudarera, N.; Pichersky, E. Eugenol and Isoeugenol, Characteristic Aromatic Constituents of Spices, are Biosynthesized via Reduction of a Coniferyl Alcohol Ester. *Proc. Natl. Acad. Sci. U. S. A.* **2006**, *103*, 10128–10133.

20. Rathee, J. S.; Patro, B. S.; Mula, S.; Gamre, S.; Chattopadhyay, S. Antioxidant Activity of *Piper betel* Leaf Extract and its Constituents. *J. Agric. Food. Chem.* **2006**, *54*, 9046–9054.

21. Yusuhara, A.; Sugiura, G. Volatile Compounds in Pyroligneous Liquids from Karamatsu and Chishima-sasa. *Agr. Bioi. Chem.* **1987**, *51*, 3049–3060.

22. Nolte, D. J.; Eggers, S. H.; May, I. R. A Locust Pheromone: Locustol. *J. Insect. Physiol.* **1973**, *19*, 1547–1554.

23. Dillon, R. J.; Vennard, C. T.; Charnley, A. K. Exploitation of Gut Bacteria in the Locust. *Nature* **2000**, *403*, 851–852.

24. Sequeira, A. J.; Taylor, L. T. Identification and Quantification of Some Chlorinated Phenolics in Wood Pulp Extracts by Gas Chromatography-Time Varied Selected Multiple Ions Mass Spectra. *J. Chromatogr. Sci.* **1991**, *29*, 351–356.

25. Furtado, J. L. F.; Peralba, M.; Do, C.; Zini, C. A.; Caramao, E. B. Chlorinated Phenolic Compounds in Bleaching Filtrates from a Mixed Eucalyptus and Acacia Pulp Using Different Sequences. *Holzforschung* **2000**, *54*, 159–164.

26. Owens, J. W.; Swanson, S. M.; Birkholz, D. A. Environmental Monitoring of Bleached Kraft Pulp Mill Chlorophenolic Compounds in a Northern Canadian River System. *Chemosphere* **1994**, *29*, 89–109.

27. Wagenführ, A.; Scholz, F. *Taschenbuch der Holztechnik*. Carl Hanser Verlag GmbH & Company KG: Munich, Germany, 2012.

28. Hartmann-Schreier, J. *Guaiacol. Römpp Catchword Entry Goeorg Thieme Verlag.* https://roempp.thieme.de/roempp4.0/do/data/RD-07-02087 (accessed Sept 7, 2003).

29. Burdock, G. A. *Fenaroli's Handbook of Flavor Ingredients*, 6th ed.; CRC Press: Boca Raton, FL, 2010; p 2159.

30. Kawamoto, H. Lignin Pyrolysis Reactions. *J. Wood Sci.* **2017**, *63*, 117–132.

31. Cutzach, I.; Chatonnet, P.; Henry, R.; Dubourdieu, D. Identification of Volatile Compounds with a "Toasty" Aroma in Heated Oak Used in Barrelmaking. *J. Agric. Food Chem.* **1997**, *45*, 2217–2224.

32. Flamini, R.; Dalla Vedova, A.; Cancian, D.; Panighel, A.; De Rosso, M. GC/MS-Positive Ion Chemical Ionization and MS/MS Study of Volatile Benzene Compounds in Five Different Woods Used in Barrel Making. *J. Mass Spectrom.* **2007**, *42*, 641–646.

33. Duffey, S. S.; Blum, M. S. Phenol and Guaiacol: Biosynthesis, Detoxication, and Function in a Polydesmid Millipede, *Oxidus gracilis*. *Insect Biochem.* **1977**, *7*, 57–65.

34. Crawford, D. L.; Crawford, R. L. Microbial Degradation of Lignin. *Enzyme Microb. Technol.* **1980**, *2*, 11–22.

35. Schranz, M.; Lorber, K.; Klos, K.; Kerschbaumer, J.; Buettner, A. Influence of the Chemical Structure on the Odor Qualities and Odor Thresholds of Guaiacol-Derived Odorants, Part 1: Alkylated, Alkenylated and Methoxylated Derivatives. *Food Chem.* **2017**, *232*, 808–819.

36. Ullrich, F.; Grosch, W. Identification of the Most Intense Volatile Flavour Compounds Formed During Autoxidation of Linoleic Acid. *Zeitschrift für Lebensmittel-Untersuchung und Forschung* **1987**, *184*, 277–282.

37. Yang, D. S.; Shewfelt, R. L.; Lee, K. S.; Kays, S. J. Comparison of Odor-Active Compounds from Six Distinctly Different Rice Flavor Types. *J. Agric. Food Chem.* **2008**, *56*, 2780–2787.

38. Guth, H.; Grosch, W. A. Comparative Study of the Potent Odorants of Different Virgin Olive Oils. *Lipid/Fett* **1991**, *93*, 335–339.

39. Blank, I.; Sen, A.; Grosch, W. Potent Odorants of the Roasted Powder and Brew of Arabica Coffee. *Zeitschrift für Lebensmittel-Untersuchung und Forschung* **1992**, *195*, 239–245.

40. Blank, I.; Fischer, K. H.; Grosch, W. Intensive Neutral Odourants of Linden Honey Differences from Honeys of Other Botanical Origin. *Zeitschrift für Lebensmittel-Untersuchung und Forschung* **1989**, *189*, 426–433.

41. Maga, J. A. Contribution of Phenolic Compounds to Smoke Flavor. In *Phenolic Compounds in Food and Their Effects on Health*; Huang, M.-T., Lee, C. Y., Ho, C.-T., Eds.; ACS Symposium Series 506: American Chemical Society: Washington, DC, 1992; Vol. 1, pp 170–179.

42. Pérez-Silva, A.; Odoux, E.; Brat, P.; Ribeyre, F.; Rodriguez-Jimenes, G.; Robles-Olvera, V. GC–MS and GC–Olfactometry Analysis of Aroma Compounds in a Representative Organic Aroma Extract from Cured Vanilla (*Vanilla planifolia* G. Jackson) Beans. *Food Chem.* **2006**, *99*, 728–735.

43. Semmelroch, P.; Laskawy, G.; Blank, I.; Grosch, W. Determination of Potent Odourants in Roasted Coffee by Stable Isotope Dilution Assays. *Flavour Fragrance J.* **1995**, *10*, 1–7.

44. Czerny, M.; Grosch, W. Potent Odorants of Raw Arabica Coffee. Their Changes During Roasting. *J. Agric. Food Chem.* **2000**, *48*, 868–872.

45. Lorjaroenphon, Y.; Cadwallader, K. R. Identification of Character-Impact Odorants in a Cola-Flavored Carbonated Beverage by Quantitative Analysis and Omission Studies of Aroma Reconstitution Models. *J. Agric. Food Chem.* **2015**, *63*, 776–786.

46. Cadwallader, K. R. *Potent Odorants in Hickory and Mesquite Smokes and Liquid Smoke Extracts.* Proceedings of the Annual Meeting of the Institute of Food Technologies, New Orleans, LA, USA. June 22–26, 1996; pp 34–36.

47. Juhlke, F.; Lorber, K.; Wagenstaller, M.; Buettner, A. Influence of the Chemical Structure on Odor Qualities and Odor Thresholds of Halogenated Guaiacol-Derived Odorants. *Front. Chem.* **2017**, *5*, 120.

48. Lorber, K.; Buettner, A. Structure–Odor Relationships of (E)-3-Alkenoic Acids, (E)-3-Alken-1-ols, and (E)-3-Alkenals. *J. Agric. Food Chem.* **2015**, *63*, 6681–6688.

49. Strube, A.; Buettner, A. The Influence of Chemical Structure on Odour Qualities and Odour Potencies in Chloro-Organic Substances. In *Expressions of Multidisciplinary Flavour Science*.

Proceedings of the 12th Weurman Aroma Symposium, 2010; Blank, I.; Wüst, M.; Yeretzian, C., Eds.; ZHAW Zürcher Hochschule für Angewandte Wissenschaften: Switzerland, 2010; pp 486–489.

50. Schmidt, R. F.; Lang, F.; Heckmann, M. *Physiologie des Menschen*; Springer: Berlin, Heidelberg, Germany, 2011.

51. Keller, A.; Zhuang, H.; Chi, Q.; Vosshall, L. B.; Matsunami, H. Genetic Variation in a Human Odorant Receptor Alters Odour Perception. *Nature* **2007**, *449*, 468.

52. Nagashima, A.; Touhara, K. Enzymatic Conversion of Odorants in Nasal Mucus Affects Olfactory Glomerular Activation Patterns and Odor Perception. *J. Neurosci.* **2010**, *30*, 16391–16398.

53. Chougnet, A.; Woggon, W. D.; Locher, E.; Schilling, B. Synthesis and In Vitro Activity of Heterocyclic Inhibitors of CYP2A6 and CYP2A13, Two Cytochrome P450 Enzymes Present in the Respiratory Tract. *ChemBioChem* **2009**, *10*, 1562–1567.

# Chapter 5

# Thermal Decomposition of Wood-Derived Organic Matter under Specific Industrial Process Conditions

Johannes Kiefl,[*,1] Sandra Boerding,[1] Birgit Kohlenberg,[1] Michael Backes,[1] Petra Slabizki,[1] Smita Raithore,[2] and Gerhard Krammer[1]

[1]Symrise AG, Flavor Division Research & Technology, P.O. Box 1253, 37601 Holzminden, Germany

[2]Symrise Inc., Flavor Division Research Analysis, 300 North Street, Teterboro, New Jersey 07608, United States

[*]E-mail: johannes.kiefl@symrise.com.

Pyrolysis of wood and other ligneous biomass is a complex degradation process that produces charcoal and tar and is accompanied by the release of volatiles. Various parameters like wood composition, structure, heating rate, and residence time affect the overall yield and composition of these volatiles. Degradation of wood typically starts at 200 °C, but higher temperatures of 450–600 °C are typically used in industrial pyrolysis applications. During this process, complex reactions lead to the formation of characteristic volatiles that are of specific interest for the flavor industry. Within this study, we investigated the formation of aroma components produced by heating beech wood to 300 °C; this is in contrast to European Union flavor legislation where only temperatures below 240 °C are permissible. Therefore, a fingerprint analysis using comprehensive gas chromatography-mass spectrometry (GC×GC-MS) to qualitatively describe and conventional GC-MS to quantitatively describe the release of aroma components by thermal degradation in an oxygen-free atmosphere at 180 °C, 230 °C, and 300 °C was performed. The results were compared with other ligneous biomass of hickory, oak, bamboo, corn, and straw.

## Introduction

Since ancient times, charcoal has been produced as metallurgical fuel and for other processes where intense heat is required. Kilns of log wood were heated for days or weeks to produce charcoal while any byproducts got lost as pyroligneous acid (also known as wood vinegar). Nowadays, various industries use retorts to process large volumes of preferably beech wood (1). The smoke generated during this process is captured, condensed, purified, concentrated, and made available as liquid smoke (2). Traditional processes of smoking food include direct smoking, where smoke is generated in a smoke chamber, or indirect smoking, where smoke is generated in a separate smoke generator and then introduced into the smoke chamber. The technologies used to generate common smoke can also be used to produce liquid smoke if the condensate is separated and processed (for example, fractionated, purified and concentrated smoke). This processing is also used to reduce undesired and unhealthy byproducts. (1, 2).

The original purpose of smoking food was to preserve the food and avoid spoilage; however, this is only of minor importance in modern industrialized countries. Smoking imparts a pleasant aroma and color to food due to the complex mixture of carboxylic acids, carbonyls, and phenolic compounds generated during the smoking process. In this context, phenols have been extensively studied and are used as flavoring substances in many flavor creations like tea, cocoa, red fruit, coffee, or meat flavors (3).

According to the European Union (EU) legislation, smoke flavorings (4, 5) and natural flavoring materials (6) are regulated by different directives. Flavoring materials with smoky notes carrying the adjective "natural" can be made, for example, by heating plant materials to 240 °C. Wood and other ligneous biomass such as hickory, oak, bamboo, corn, and straw are permissible starting materials, and heating to 240 °C is regarded as a traditional method to prepare food and is therefore permitted for generating natural aroma compounds (6). However, there are only few reports confirming the formation of aroma components under these "natural" conditions. In contrast, liquid smoke flavors do not fall within the flavor regulation and cannot be labeled as such.

Many scientific studies have discussed the pyrolysis, or heating of wood at temperatures higher than 240 °C, and its main components hemicellulose, cellulose, and lignin (3, 7–9). Candelier et al. treated beech sawdust at 210 °C, 230 °C, and 280 °C with helium in a thermodesorption (TD)-gas chromatography-mass spectrometry (GC-MS) and studied the reaction kinetics of acetic acid, furfural, and vanillin. The study additionally identified methylfurfural, hydroxymethylfurfural, guaiacol, syringol, acetovanillone, syringaldehyde and acetosyringone during 60 min of heating time (10). Wittkowski et al. studied the course of ferulic acid degradation under oxygen atmosphere and found maxima at 242 °C and 380 °C. Methylguiacol, ethylguiacol, vinylguaiacol, vanillin, acetovanillone, isopropylguaiacol, eugenol, and isoeugenol were identified between 230 °C and 260 °C (3). However, to the best of our knowledge, the formation of volatiles from beech wood below pyrolysis temperatures has yet to be systematically studied. Therefore, we investigated the formation of phenolic compounds as well as other aroma components generated by degrading beech wood at 180 °C, 230 °C, and 300 °C and compared that data with the thermal decomposition of hickory, oak, bamboo, corn and straw at 230 °C.

## Materials and Methods

### Materials

The purity of the internal standard 2,3-dimethoxytoluene (Sigma-Aldrich, Steinheim, Germany) was 99%, and methyl *tert*-butyl ether (MTBE) was Chromasolv quality with purity higher than 99.8% (Honeywell, Seelze, Germany).

### Thermal Decomposition of Biomass

Commercially available ground beech wood (ca. 1×1 mm pieces, *Fagus sylvatica L.*, Thomsen Räucherspäne Räucherholz GmbH & Co. KG, Handewitt, Germany) with a residual moisture of 12% was extracted in a nitrogen gas flow with 10 mL/min using the ThermoExtractor equipment

(Gerstel, Mülheim a. d. Ruhr, Germany). A glass tube with a 13 mm i.d. and 17.6 mm length was filled with 0.1 g of the beech wood fixed with one round metal grid on each side and heated to 180 °C, 230 °C, and 300 °C for 1 h. To calibrate the temperature controller, a temperature sensor was inserted at the position of the sample via the gas outlet before the measurements. A temperature difference of 9 °C between the sensor and controller was recorded and corrected in ongoing experiments by adjusting the controller settings. The temperature was held constant for 90 min at a target temperature of 230 °C with an uncertainty of 229.9 °C $\pm$ 0.03%. The low boiling fraction of the thermally decomposed wood was recovered by directing the nitrogen gas through a 17.8-cm-long glass tube filled with 0.3 g of deactivated glass wool followed by two Tenax-filled adsorption tubes (4 mm i.d., 60 mm length, 80 mg adsorbens, Gerstel). The glass tube was used to condense the high boiling fraction (tar) before trapping the low boiling fraction (wood vinegar) on Tenax at room temperature. The Tenax tubes were analyzed using the comprehensive gas chromatography/mass spectrometry (GC×GC-MS) system. The chromatographic run of the first tube was used for data processing whereas the chromatographic run of the second tube was used to check a possible breakthrough of volatile components. The high boiling fraction was condensed in a second experiment at room temperature by directing the nitrogen gas through an empty u-shape glass tube (preparative fraction collector PFC, Gerstel) at -10 °C. A viscous brown condensate was formed, weighed, and dissolved in a 4 g solution of 2.4 mg/kg of 2,3-dimethoxytoluene in MTBE. The liquid sample was analyzed using GC-MS. Hickory (*Carya sp.*, Hahne Raeucherzwerg, Brunsbuettel), oak (*Quercus alba L.*, Hahne Raeucherzwerg, Brunsbuettel), bamboo (*Fargesia murielae* [Gamble] T.P.Yi), corn (*Zea mays L.*, dried stem and fruit of young plant) and straw (dry straw of *Hordeum vulgare L.*) were separately extracted in the same way as beech wood at 230 °C and only the high boiling fractions were recovered and analyzed using GC-MS.

## TD-GC×GC-MS Analysis of Low Boiling Fraction

The low boiling fractions obtained by extracting beech wood at 180 °C, 230 °C, and 300 °C were analyzed by comprehensive two-dimensional GC in triplicate.

A Shimadzu GC system (GC 2010 Plus, Duisburg, Germany) equipped with a CIS 4, TDU 2 (both Gerstel, Mühlheim a. d. Ruhr, Germany), and a loop-type modulator (Zoex Corporation, Houston, TX, U.S.A.) was coupled to a triple quadrupole MS (GCMS-TQ8040, Shimadzu). The low boiling fractions trapped on Tenax were thermodesorbed at 280 °C for 10 min with a desorption flow of 100 mL/min and a total split ratio of 1:50. Wood samples heated in the themoextractor at 180 °C were desorbed with 40 mL/min flow to achieve a lower total split ratio of 1:20. During thermodesorption, the CIS was set to -20 °C and the desorbed components were trapped on a packed liner with Tenax. After thermodesorption, the CIS temperature was raised at 12 °C/s to 280 °C (10 min isothermally) with a closed split valve for 4 min. The transfer line between TDU and CIS was set to 300 °C.

The analytical column system consisted of a 30 m × 0.25 mm i.d. fused silica capillary column coated with a 0.25-μm polyethylene glycol (ZB-Wax, Phenomenex, Aschaffenburg, Germany) connected via press-fit connector (Restek, Bad Homburg, Germany) to the two-dimensonal analytical column of 2 m × 0.15 mm i.d. coated with 0.15 μm of a 5 % phenylmethylpolysiloxane phase (Rxi-5SilMS, Restek). The carrier gas was helium at a constant flow of 1.8 mL/min. The initial oven temperature of 40 °C was held for 2 min, then raised at 4 °C/min to the final temperature of 240 °C (15 min isothermally). The modulation time was 6 s. The MS transfer line and the ion source temperature were set to 240 °C and 230 °C, respectively. Mass spectrometric detection was performed in positive electron ionization (EI) mode at 70 eV in scan mode (m/z 33–500).

The thermodesorption was automated using an MPS robotic autosampler controlled by Maestro Software (version 1.5.2.9, Gerstel). Data acquisition was done with a GC-MS solution (version 4.41, Shimadzu), and GC Image-software version 2.7 (Lincoln, NE, USA) was used to integrate the chromatographic runs and align the data sets. Compound identification was done in the same way as the GC-MS analysis.

## GC-MS Analysis of High Boiling Fraction

An autosampler (MPS 2XL Autosampler, Gerstel) was used to inject 5 µL of the solubilized high boiling condensate into a cooled injection system inlet (CIS 4, Gerstel), which was heated at 12 °C/s from 40 °C to 180 °C (5 min isothermally) to transfer the sample to a DB-Wax ultra-inert column (60 m x 0.32 mm i.d., df = 0.25 µm, Agilent, Waldbronn, Germany) placed in an Agilent 7890A gas chromatograph using helium as the carrier gas.

The effluent was transferred into deactivated fused silica capillaries directed to a flame ionization detector (FID) set at 300 °C and MS (MSD 5977B, Agilent) for simultaneous detection using a Graphpack-3D/2 sulfonated crosspiece (Gerstel, Muelheim, Germany). A constant flow of 2 mL/min was applied. The GC oven temperature was programmed starting at 40 °C, held for 2 min, then raised by 4 °C/min to 240 °C, and held for 75 min. Retention indices (RIs) were calculated using chromatography of a homologous series of $n$-alkanes from $C_6$–$C_{30}$. The effluent was transferred to the MS set in scan mode (EI ionization 70 eV, $m/z$ 25–350) via a 2.2-m deactivated capillary column held at 280 °C. Data were acquired by Agilent Mass Hunter-Software (Agilent B07.05.2479) and elaborated by AMDIS (V 3.2.13.03.08).

The volatile analytes were identified by matching the RI (constraint of $\pm5$ set for positive match) and mass spectrum (constraint of >800 or 1000 match factor set for positive match) with an in-house database that was built by analyzing authentic references on standard instruments with defined protocols. Estimation of the analyte concentration was performed using 2,3-dimethoxytoluene as an internal standard, the GC-FID response, and a one-point calibration.

## Results and Discussion

The thermoextraction was optimized by adopting the amount of beech wood, temperature, and nitrogen flow. As a result, 0.1 g of beech wood was used to cover only a small cross section of the glass tube to ensure a homogenous temperature profile. The temperature controller was calibrated at each temperature point and showed, for example, 229.9 °C $\pm$ 0.03% to produce the set temperature of 230 °C. The flow of nitrogen was optimized to avoid a breakthrough and finally set at 10 mL/min to efficiently trap volatiles on Tenax tubes.

A fingerprint comparison of the low boiling fractions at 180 °C, 230 °C, and 300 °C using GC×GC-MS indicates the formation of acetic acid, furfural, 5-methylfurfural, 5-methylenfuran-2-one, 2-hydroxymethylfuran, cyclopent-2-en-1-one, and 2-oxopropanal at 180 °C. Phenols such as guaiacol, 4-methylguaiacol, 4-ethylguaiacol, and $o$-cresol were identified at 230 °C, in addition to the furan components detected at 180 °C. These results indicate that a degradation of the hemicellulose fraction of beech wood–forming furans starts below a temperature of 200 °C and is followed by a degradation of the lignin monomers, $p$-coumaryl, coniferyl, and sinapyl alcohol, forming phenolic compounds. Compounds such as 2-methyl-5-prop-1-enyl-furan, 2-ethyl-5-vinyl-furan, ethylbenzene, 2-hydroxy-3-methyl-benzaldehyde, 2,3-dimethylbenzofuran, and 3-methyl-2H-furan-5-one show a high increase in volume percentage from 230 °C to 300 °C. When comparing the chromatograms in Figure 1, we must consider that the split ratio has been reduced from 1:50 to

1:20 to increase the detectability of analytes obtained by extracting and trapping at 180 °C (Figure 1).

The number of peaks and the peak intensity increase with temperature (Figure 1) highlighting that the degradation of wood components is slower at lower temperatures and increases with increasing temperatures. It is reported that the relative concentrations of acetic acid and vanillin released from beech wood increase as the extraction temperature rose from 210 °C to 230 °C by a factor 10 and 2.5, respectively, and from 230 °C to 280 °C by a factor 3.6 and 1.6, respectively, for 1 h of heating time. Thereby, the formation of both aroma components was found to be faster in the early stage of heating beech hardwood (*10*).

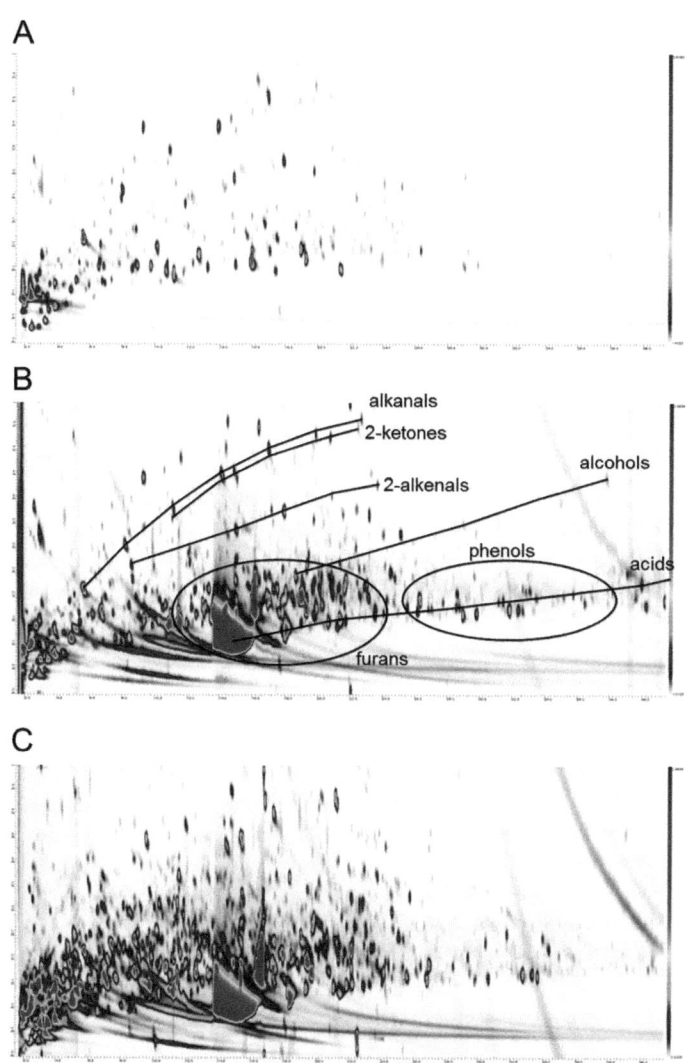

*Figure 1. Two-dimensional GC chromatograms (x-axis one-dimensional retention time in min, y-axis two-dimensional retention time in s) of the low boiling fractions after heating beech wood (A) at 180 °C, (B) at 230 °C elution of substance classes is marked, and (C) at 300 °C for 1 h.*

Apart from the qualitative view on the formation of volatiles, the quantitation of phenolic aroma components is of particular interest due to the way components such as 2-methoxyphenol, 2-methoxy-4-methylphenol, 2-methoxy-4-vinylphenol, 4-vinylguaiacol and vanillin impart characteristic aroma notes to food. The high boiling fraction rich in phenolic components was condensed on a glass tube, solubilized with MTBE, mixed with a 2,3-dimethoxytoluene standard solution, and quantitated with GC-FID-MS (Table 1).

**Table 1. Constituents of the Beech Condensate Obtained by Heating 0.1 g Wood Pieces with Nitrogen Gas at 230 °C for 1 h**

| No. | Name | Conc.[a] [mg/kg] | RSD[b] [%] |
|---|---|---|---|
| 1 | Acetic acid | 553.2 | 18.6 |
| 2 | (*E*)-3-(4-hydroxy-3,5-dimethoxyphenyl)prop-2-enal (sinapyl aldehyde) | 270.0 | 1.3 |
| 3 | (*E*)-3-(4-hydroxy-3-methoxyphenyl)prop-2-enal (coniferyl aldehyde) | 206.5 | 1,7 |
| 4 | Formic acid | 124.9 | 18.5 |
| 5 | 2-Methyl-3(2H)-furanone | 83.4 | 11.0 |
| 6 | *Trans*-isoeugenol | 66.7 | 0.1 |
| 7 | 2,6-Dimethoxy-4-(*E*)-propenylphenol | 61.5 | 6.9 |
| 8 | 5-Hydroxymethylfurfural | 59.4 | 5.1 |
| 9 | Furan-2-carbaldehyd (furfural) | 47.1 | 31.8 |
| 10 | 4-Hydroxy-3,5-dimethoxybenzaldehyd (syringyl aldehyde) | 43.5 | 5.2 |
| 11 | Furfurylalkohol | 40.2 | 11.0 |
| 12 | 4-Hydroxy-3-methoxybenzaldehyd (vanillin) | 36.6 | 6.4 |
| 13 | 2-Methoxy-4-vinylphenol (4-vinylguaiacol) | 33.5 | 8.5 |
| 14 | 2-Hydroxy-2-cyclopentenone | 32.6 | 3.8 |
| 15 | 2,6-Dimethoxyphenol | 31.6 | 8.3 |
| 16 | 2(5H)-furanone | 18.3 | 5.3 |
| 17 | 3-Hydroxy-2-pyrone | 17.5 | 5.1 |
| 18 | 1-(4-Hydroxy-3,5-dimethoxyphenyl)-2-propanone | 17.0 | 1.9 |
| 19 | 3-Hydroxy-tetrahydrofuran-2-one | 16.1 | 9.6 |
| 20 | 4-Allyl-2,6-dimethoxyphenole | 14.6 | 1.3 |
| 21 | 3,5-Dimethoxy-4-hydroxy-acetophenone | 14.1 | 5.4 |
| 22 | 3-Hydroxy-2-methyl-4-pyrone (maltol) | 14.0 | 3.6 |
| 23 | 1-(4-Hydroxy-3-methoxy-phenyl)propan-2-one | 12.9 | 4.6 |
| 24 | 4-Allyl-2-methoxyphenol (eugenol) | 12.6 | 6.7 |
| 25 | 2-Methoxyphenol (guaiacol) | 9.8 | 6.7 |

**Table 1. (Continued). Constituents of the Beech Condensate Obtained by Heating 0.1 g Wood Pieces with Nitrogen Gas at 230 °C for 1 h**

| No. | Name | Conc.[a] [mg/kg] | RSD[b] [%] |
|---|---|---|---|
| 26 | 4-(E)-3-hydroxypropenyl-2-methoxy-phenol (coniferyl alcohol) | 9.0 | 9.4 |
| 27 | 4-Hydroxy-5-methyl-furan-3-one (norfuraneol) | 8.8 | 12.8 |
| 28 | 1-(4-Hydroxy-3-methoxyphenyl)ethanone | 7.8 | 4.1 |
| 29 | 5-Methylfuran-2-carbaldehyde | 7.3 | 15.0 |
| 30 | 1-(4-Hydroxyphenyl)propan-1-one | 5.9 | 58.2 |
| 31 | 2,6-Dimethoxy-4-methylphenol | 5.5 | 19.5 |
| 32 | 2,6-Dimethoxy-4-(Z)-propenylphenol | 5.5 | 7.2 |
| 33 | 2-Methoxy-4-(Z)-propenylphenol | 5.3 | 1.7 |
| 34 | 3,5-Dihydroxy-6-methyl-2,3-dihydropyran-4-one | 5.0 | 6.0 |
| 35 | 2-Hydroxy-3-methyl-cyclopent-2-en-1-one (methyl cyclopentenolone) | 5.0 | 5.4 |
| 36 | 1-(2-Furyl)-2-hydroxyethanone | 4.8 | 2.8 |
| 37 | 4-Ethyl-2-methoxyphenol (4-ethylguaiacol) | 4.8 | 12.7 |
| 38 | 4-(E)-propenylphenol | 4.3 | 1.7 |
| 39 | O-cresol | 4.2 | 3.3 |
| 40 | 2-Hydroxy-3-methyl-cyclohex-2-en-1-one (methyl cyclohexenolone) | 4.2 | 7.1 |
| 41 | 3,4,5-Trimethoxyphenol | 4.0 | 57.3 |
| 42 | 2-Methoxy-4-methylphenol (4-methyl guaiacol) | 4.0 | 6.1 |

[a] Concentration was estimated by using 2,3 dimethoxytoluene as an internal standard.   [b] RSD = relative standard deviation.

On average 2.5 mg of condensate was obtained by heating 0.1 g of ground beech wood at 230 °C, which consisted of furan- and pyran-type volatiles (**8, 5, 9, 11, 16, 17, 19, 22, 27, 29, 34,** and **36**) and phenolic volatiles (**2, 3, 6, 7, 10, 12, 13, 15, 18, 20, 21, 23-26, 28, 30-33, 37-39, 41,** and **42**). The detected volatiles account for almost 0.2% of the condensate's mass. Low boiling organic acids **1** and **4** are the main components of the condensate's volatiles, although they have not been quantitatively recovered in the high boiling fraction. Lignin monomers **2, 3, 10,** and **26** account for more than 500 mg/kg and exceed the concentration of the most abundant monomer degradation components *trans*-isoeugenol (**6**) and 2,6-dimethoxy-4-(E)-propenylphenol (**7**) (Table 1). At elevated temperatures of about 300 °C or more, secondary pyrolysis products can be formed by radical cleavages and recombination; for example, guaiacols are decomposed to pyrocatechols, alkylphenols, and dialkylphenols (3). Only traces of pyrocatechol and alkylated phenols were detected in condensates obtained at 230 °C. The lack of major secondary pyrolysis products indicates that aroma components such as guaiacol are not degraded to a significant degree at 230 °C. This indicates that degradation of the lignin monomers is not complete. Hence, the concentration optimum given by an equal degree of forming target aroma components and their subsequent degradation is neither reached nor overcome at this temperature. This observation is supported by findings from Candelier et al. who found a more rapid increase of vanillin concentration after increasing the heating temperature range of beech sawdust from 210–230 °C to 230–280 °C, and by findings from Wittkowski et al. who identified a maximum of ferulic acid degradation in a model system at 242 °C (3, 10).

The phenolic volatiles were formed by degrading the lignin monomers p-coumaryl (abbreviation h for 4-hydroxyphenyl-), coniferyl alcohol (abbreviation g for guaiacyl-), and sinapyl alcohol (abbreviation s for syringyl-) (Figure 2). The mass ratio of these degradation components is known as the h:g:s ratio (derived from the chemical names of the benzoyl rings of the corresponding lignin monomers) and is characterized by different wood types and thermal extraction conditions (11, 12). The thermal extraction of beech wood at 230 °C delivered a h:g:s ratio of 6:48:46 wt% by summing the quantities of all detectable phenols with respective h, g, and s substitution. Ratios of 5:49:46 wt% and 3:45:52 mol% are reported for the pyrolysis of beech wood (11, 12). Our study has demonstrated that similar ratios are obtained at temperatures far below that which is required for pyrolysis.

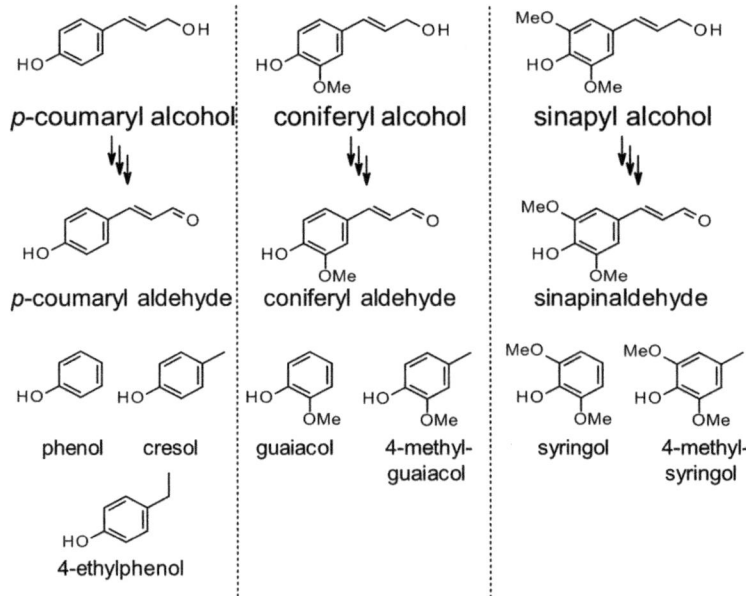

Figure 2. Formation of phenolic aroma substances via degradation of lignin monomers. The ratio of degradation components (h:g:s ratio) is specific to different woods and their lignin monomer compositions.

The total concentration of phenolic components in the condensate gained at 230 °C was nearly 900 mg/kg and distinctively shaped the sensory profile of the material, which was described as phenolic and very smoky by expert panelists. Aroma components reported in this study that are preferred in the flavor industry due to their low odor threshold and sensory profile are eugenol (24), guaiacol (25), isoeugenol (6), 4-methylguaiacol (42), 4-ethylguaiacol (37), maltol (22), methylcyclopentenolone (35), methylcyclo-hexenolone (40), norfuraneol (27), vanillin (12), and 4-vinylguaiacol (13) (Table 1). Thus, at temperatures lower than 240 °C, important aroma components are formed in relevant amounts from beech. Hickory and oak are other hardwoods frequently used for smoke flavoring production (5). Thermoextraction of both materials and comparing the GC-MS chromatograms (Figure 3) show a similar pattern of released volatiles in the high boiling fraction. 2,6-Dimethoxy-4-(E)-propenylphenol, syringyl aldehyde, coniferyl aldehyde, and sinapyl aldehyde are among the most abundant volatile phenols across all three hardwoods (Figure 3).

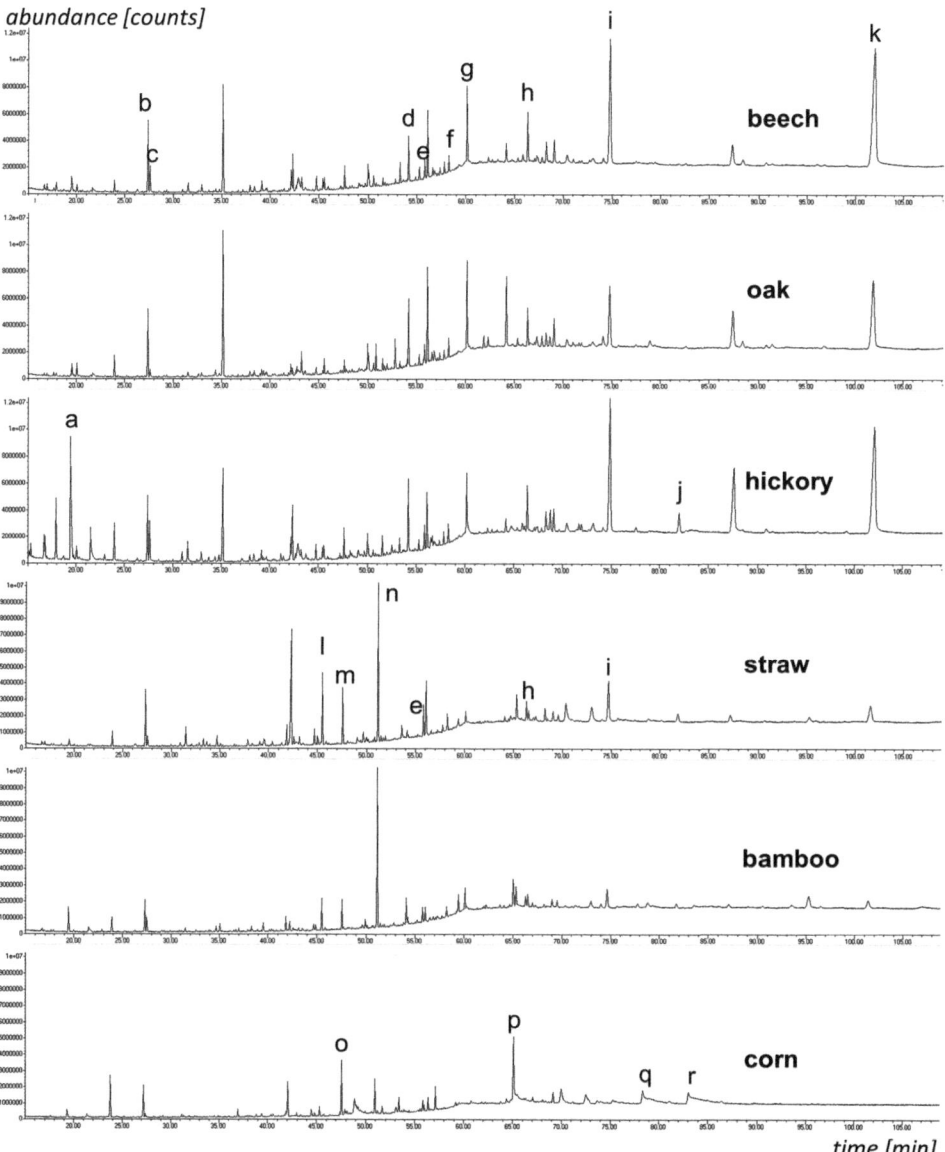

*Figure 3. Comparison of GC-MS chromatograms of the high boiling fractions of beech, oak, hickory, straw, bamboo, and corn obtained by thermoextraction at 230 °C for 1 h in a nitrogen atmosphere. Peaks are: (a) acetic acid, (b) 2,3-dimethoxytoluene (internal standard), (c) furfural, (d) 5-hydroxymethylfurfural, (e) vanillin, (f) 4-allylsyringol, (g) 2,6-dimethoxy-4-(E)-propenylphenol, (h) syringyl aldehyde, (i) coniferyl aldehyde, (j) coniferyl alcohol, (k) sinapyl aldehyde, (l) 4-vinylguaiacol, (m) 2,6-dimethoxyphenol, (n) 4-vinylphenol, (o) hydroxydihydromaltol, (p) palmitic acid, (q) linolic acid, and (r) linoleic acid.*

Lignin from wood as well as agricultural by-products like straw is a valuable resource for chemicals or hydrocarbon fuel. Their thermal conversion is the subject of many studies focused on fuel production from renewable sources rather than petroleum (1, 7). These sources are relevant starting materials for thermal decomposition to produce aroma compounds for straw, bamboo, and corn (Figure 3). Straw and bamboo are characterized by the release of 4-vinylguaiacol, 2,6-dimethoxyphenol, and 4-vinylphenol, the latter of which results from degradation of the lignin monomer p-coumaryl alcohol. The release of the aroma component hydroxydihydromaltol (3,5-dihydroxy-6-methyl-2,3-dihydropyran-4-one) and fatty acids (Figure 3) correlates with a low degree of lignification of the stem and fruit of corn used in this experiment. Overall, straw and bamboo can be considered as alternative starting materials to generate natural flavoring substances upon fractionation and purification.

In conclusion, the thermal decomposition of hardwood already yields phenolic components at temperatures below 240 °C. At elevated temperatures near 300 °C, secondary pyrolysis products from degradation of aroma components can occur. This confirms that the formation of phenolic aroma components under natural conditions, according to EU legislation on flavoring materials below 240 °C, is feasible.

## Acknowledgments

We would like to thank Margit Liebig and Stephanie Korte for extracting the samples and Melanie Behringer and Gerald Reinders for regulatory support.

## References

1.  Mohan, D.; Pittman, C. U.; Steele, P. H. Pyrolysis of Wood/Biomass for Bio-Oil: A Critical Review. *Energy Fuels* **2006**, *20*, 848–889.
2.  Bridgwater, A. V.; Meier, D.; Radlein, D. An Overview of Fast Pyrolysis of Biomass. *Org. Geochem.* **1999**, *30*, 1479–1493.
3.  Wittkowski, R.; Ruther, J.; Drinda, H.; Rafiei-Taghanaki, F. Formation of Smoke Flavor Compounds by Thermal Lignin Degradation. In *Flavor Precursors*; American Chemical Society: Washington, DC, 1992; Vol. 490, pp 232–243.
4.  *Official Journal of the European Union*, L309 EU 2065/2003, 2003.
5.  *Official Journal of the European Union*, L333 EU 1321/2013, 2013.
6.  *Official Journal of the European Union*, L354 EU 1334/2008, 2008.
7.  Brebu, M.; Vasile, C. Thermal Degradation of Lignin — A Review. *Cellul. Chem. Technol.* **2009**, *44*, 353–363.
8.  Gucho, E.; Shahzad, K.; Bramer, E.; Akhtar, N.; Brem, G. Experimental Study on Dry Torrefaction of Beech Wood and Miscanthus. *Energies* **2015**, *8*, 3903.
9.  Shen, D. K.; Gu, S.; Luo, K. H.; Wang, S. R.; Fang, M. X. The Pyrolytic Degradation of Wood-Derived Lignin from Pulping Process. *Bioresour. Technol.* **2010**, *101*, 6136–6146.
10. Candelier, K.; Dumarçay, S.; Pétrissans, A.; Pétrissans, M.; Kamdem, P.; Gérardin, P. Thermodesorption Coupled to GC–MS to Characterize Volatiles Formation Kinetic During Wood Thermodegradation. *J. Anal. Appl. Pyrolysis* **2013**, *101*, 96–102.

11. Faix, O.; Meier, D.; Grobe, I. Studies on Isolated Lignins and Lignins in Woody Materials by Pyrolysis-Gas Chromatography-Mass Spectrometry and Off-Line Pyrolysis-Gas Chromatography with Flame Ionization Detection. *J. Anal. Appl. Pyrolysis* **1987**, *11*, 403–416.

12. Freudenberg, K. Lignin: Its Constitution and Formation from p-Hydroxycinnamyl Alcohols. *Science* **1965**, *148*, 595–600.

# Chapter 6

# Changes in Aroma and Sensory Profile of Food Ingredients Smoked in the Presence of a Zeolite Filter

XinLing Chua,[1] Elizabeth Uwiduhaye,[1] Petroula Tsitlakidou,[1,2] Stella Lignou,[1] Huw D. Griffiths,[2] David A. Baines,[3] and Jane K. Parker[*,1]

[1]Department of Food and Nutritional Sciences, University of Reading, Whiteknights, Reading RG6 6AP, United Kingdom

[2]Besmoke Ltd, Unit B1, Ford Airfield Industrial Estate, Arundel BN18 0HY, United Kingdom

[3]Baines Food Consultancy Ltd., 22 Elisabeth Close, Thornbury, Bristol BS35 2YN, United Kingdom

[*]E-mail: j.k. parker@reading.ac.uk.

During smoking, formation of desirable smoky compounds and carcinogenic polycyclic aromatic hydrocarbons (PAH) are inextricably linked. We have previously developed a zeolite filter technology (PureSmoke Technology or PST) that reduces the PAH content of a smoke stream, particularly reducing the concentration of benzo[a]pyrene, a known carcinogen, by up to 93%. The aim of this work was to determine whether there were changes in the volatile and sensory profiles of ingredients smoked using PST compared to the traditional smoking process (Trad). Smoked tomato flakes (either PST or Trad) were added to either low-fat or full-fat cream cheese for sensory profiling and consumer preference tests, and volatile analysis was carried out using solid phase microextraction (SPME) followed by gas chromatography-mass spectrometry (GC-MS). The sensory analysis showed a significant decrease ($p < 0.01$) in bitterness when the PST was employed and a significant decrease in overall smoky aroma and flavor ($p < 0.001$), which resulted in an increase in the perception of cheesy aroma and flavor. This was consistent with a decrease in many of the smoky aroma compounds, particularly the guaiacols. However, consumer preference tests showed that there was no adverse effect on the flavor of the products, and there was even a tendency for the PST product to be preferred to the Trad product ($p = 0.096$). The smoke compounds were quantitated and compared in smoked tomato paste. Odor activity values (OAVs) calculated from the literature thresholds suggested that guaiacol and 4-alk(en)yl-substituted guaiacols are likely to be among the most highly odor-active compounds in these smoked ingredients.

# Introduction

The use of smoke for preservation has become secondary to its use in creating unique smoky aromas and flavors in foods. Smoked ingredients are used widely by the food industry to impart a characteristic smoky flavor to rubs, dips, marinades, soups, and snacks. The volatile components of aqueous smokes have been studied extensively (1) as have the smoky aroma compounds in various fish (2, 3) and cheeses (4–6). The highly desirable smoky flavor is generated by the burning of wood chips, of varying origin, at high temperatures (400–1000 °C). Phenolic compounds such as syringol and guaiacol are essential for the sensory characteristics of smoke, and the gas chromatography–olfactometry (GC–O) of smoked fish has shown many more important compounds that contribute to the smoky aroma (7, 8).

However, the smoking process also results in the formation of polycyclic aromatic hydrocarbons (PAHs). These PAHs are a series of fused benzene ring structures, and many of these are classified as Class 2 carcinogens. One of these PAHs, benzo[a]pyrene, is a known carcinogen, and epidemiological evidence has implicated smoked foods in an increased risk of cancer in humans (9, 10). In 2015, Griffiths, Baines, and Parker-Gray (11) developed a filtration technology based on zeolites (PureSmoke Technology or PST) whereby up to 93% of the PAHs could be removed from a smoke stream (12). Comparison of the headspace of oils smoked either traditionally (Trad) or through the filter (PST) showed that generally, the low molecular weight aroma compounds were not removed by the filter. However, many of the components of smoke most likely to contribute to the smoky flavor were partially removed by the filter, particularly the guaiacols and the eugenols. Preliminary sensory testing of smoked tomato ketchup suggested that the PST product had a sweeter aroma than the Trad product. Differences in the smoky, rubbery, and tar aroma and flavor were not observed, despite a decrease in smoky aroma compounds; these differences were possibly masked by the intensity of the neat ketchup. The changes in flavor warrant further investigation since it is important to establish that PST does not adversely affect the flavor of the product. In this study, we used tomato flakes that were Trad or PST smoked. The flakes were finely ground and added to cream cheese for sensory, consumer, and instrumental analysis.

# Materials and Methods

## Materials

Tomato flakes were purchased from Camstar Ingredients (Eye, U.K.), and tomato paste was purchased from Silbury Marketing (Banbury, U.K.). They were smoked with oak chips obtained from Ashwood Smoking Chips (Kettering, U.K.) using either PST or Trad smoking. Portions of the smoked tomato flakes were ground with a pestle and mortar and sieved. The fraction collected from sieve size 3 (355 μm–1 mm) was added to either low-fat or full-fat cream cheese for volatile analysis, sensory profiling, and consumer testing. The tomato paste was used to quantitatively compare the aroma compounds derived from either PST or Trad smoking.

Two types of cream cheese were used in order to provide four samples for sensory profiling, rather than just two. Tubs (180 g) of Philadelphia Original (21% fat) and Philadelphia Light (11% fat) (Mondelez, Uxbridge, UK) were purchased from one local supermarket, ensuring that the tubs for each product were from one batch. Ground smoked tomato flakes were added to the cheese (2.5% w/w), mixed thoroughly, and returned to the container to equilibrate overnight before tasting or analysis. Thus, four samples of cream cheese with smoked tomato flakes were prepared: low-fat

Trad, low-fat PST, full-fat Trad, and full-fat PST. All sensory references were purchased from a local supermarket.

2-Octanol, 2,3-butanedione (diacetyl), 3-hydroxybutanone (acetoin), acetic acid, benzeneacetaldehyde, 6-methyl-5-hepten-2-one, 2-acetylpyrrole, phenol, 5-butyl-4-methyloxolan-2-one (whiskey lactone, mix of two isomers), 4-methyl-2-methoxyphenol (4-methylguaiacol), 4-ethyl-2-methoxyphenol (4-ethylguaiacol), 2-methoxy-4-propylphenol (4-propylguaiacol), 4-ethenyl-2-methoxyphenol (4-vinylguaiacol), 2,6-dimethoxyphenol (syringol), 2-hydroxy-3-methyl-2-cyclopenten-1-one (cyclotene), and 2,6-dimethylphenol were purchased from Sigma-Aldrich (Poole, U.K.). 2-Furaldehyde (furfural), 1-(2-furyl)ethanone (2-acetylfuran), 5-methyl-2-furaldehyde (5-methylfurfural), 4-methylphenol (*p*-cresol), 2-methylphenol (*o*-cresol), and 2-methoxyphenol (guaiacol) were obtained from Fisher Scientific (Loughborough, U.K.). 2-Methoxy-4-[(1*E*)-1-propen-1-yl]phenol ([*E*]-isoeugenol) containing 1% of the *Z*-isomer and 4-allyl-2-methoxyphenol (eugenol) were purchased from Givaudan (Milton Keynes, U.K.). 2-Isopropyl-5-methylphenol (thymol) was purchased from Mane (London, U.K.).

## Volatile Analysis of Cream Cheese with Added Smoked Tomato Flakes

Samples of cream cheese (20 g) were incubated in a Duran bottle at 40 °C for 30 min. A triple-phase Stabilflex fiber (PDMS/Carboxen/DVB, 11 mm, from Supelco, Poole, U.K.) was exposed to the headspace for a further 30 min to extract the volatile compounds. Gas chromatography-mass spectrometry (GC-MS) analysis was conducted using a 5972 MS coupled to an Agilent Technologies 5890 GC (Agilent, Santa Clara, CA, U.S.A.). Each extraction was injected in splitless mode onto a J&W DB-WAX column (30 m × 250 μm × 1 μm film thickness) (Agilent, Santa Clara, CA), and the following temperature program was employed: 2 min at 40 °C, then raised to 250 °C at a rate of 5 °C/min. The flow rate of the helium carrier gas was 0.9 mL/min. Mass spectra were measured in electron ionization mode at 70 eV. The scan range was from $m/z$ 29–300. Samples (20 g) of unsmoked tomato flakes, unsmoked tomato paste, and unflavored full-fat and low-fat cheese were also analyzed for comparison purposes. Volatiles were identified by comparing each mass spectrum with the spectrum of the authentic compounds analyzed in our laboratory. To confirm the identification, the linear retention index (LRI) was calculated for each volatile compound using the retention times of a homologous series of $C_6$–$C_{25}$ *n*-alkanes and by comparing the LRI with those of authentic compounds analyzed under similar conditions. Samples were also analyzed on a non-polar DB5 column (30 m × 250 μm × 1 μm film thickness) (Agilent, Santa Clara, CA) using the same temperature program to further confirm their identity. Rather than adding an internal standard into a semi-solid cheese, an external standard of 2-octanol was injected every six samples. The deviation in the peak areas was no greater than 10% and there was no observed trend.

## Sensory Profiling of Cream Cheese with Added Smoked Tomato Flakes

A panel of nine trained assessors (90% women aged 35–60), each with a minimum of six months' experience, was used to develop a quantitative sensory profile for describing the sensory characteristics of the four different samples of cream cheese. Following an initial collection of terms, reference materials were provided to help assessors standardize the terms and reach a consensus vocabulary. The references included a range of smoked foods (smoked haddock, smoked mackerel, kippers, and smoked cheese) as well as smoky bacon snacks, burnt wood, burnt paper, and burnt matches. The final vocabulary consisted of 5 aroma terms, 18 taste/flavor terms, 2 mouthfeel terms, and 1 after-effect term. The quantitative sensory assessment took place in individual sensory booths (under red light) at 22 ± 0.5 °C. Assessors were provided with a glass of warm water and unsalted crackers (Carr's of Carlisle, Carlisle, U.K.) for palate cleansing between samples. Samples (~2–3 g) were presented to the assessors on a plastic teaspoon in a balanced order and randomly allocated. The assessors were asked to smell, taste, and swallow the samples and score them on appearance, odor, taste, flavor, and mouthfeel attributes. After a 45-s pause, they scored the samples for after-effects. The intensity of each attribute was recorded on a 150 mm unstructured line scale (scaled 0–100) and all data were collected using Compusense @ Hand (Compusense Inc., Guelph, Ontario, Canada). A duplicate assessment was carried out in a separate session.

### Consumer Preference of Cream Cheese with Added Smoked Tomato Flakes

Consumer testing was carried out as described by IFT-SED (13) in individual sensory booths. A total of 115 naïve consumers (70% women aged 19–63; mean age of 32) carried out a paired preference test. They were served two samples of full-fat cream cheese; approximately 2–3 g of cheese was placed on the tip of a plastic teaspoon. One cheese sample contained 2.5% Trad smoked tomato flakes and the other contained 2.5% PST smoked tomato flakes as described for sensory profiling purposes. The consumers tasted the two samples in a balanced and randomly allocated order and were asked to select their preferred sample.

### Quantitation of Volatiles in Smoked Tomato Paste

The aroma compounds generated during the smoking process were quantified in tomato paste using solid phase microextraction (SPME) followed by GC-MS analysis. External calibration curves were prepared with unsmoked tomato paste containing a cocktail of standards at appropriate concentrations. Single standards were prepared in methanol (or acetone for compounds in cocktail C) to form stock solutions (200 mg/L), from which four standard cocktails were prepared in high-performance liquid chromatography (HPLC) water containing analytes at 10 mg/L unless otherwise indicated. Cocktail A contained a mixture of guaiacol, 4-methylguaiacol, 4-ethylguaiacol, syringol (30 mg/L), and cyclotene (30 mg/L). Cocktail B contained a mixture of phenol, 2-methylphenol, 4-methylphenol, 2,6-dimethylphenol, eugenol, and (E)-isoeugenol. Cocktail C contained a mixture of furfural, 5-methylfurfural, and 2-acetylfuran. Cocktail D contained a mixture of 4-vinylguaiacol (5 mg/L) and whiskey lactone (5 mg/L). Five serial dilutions (1:1) of these cocktails were prepared with HPLC grade water. Samples for calibration were made up of 1.00 ± 0.01g of tomato paste, 1.0 mL of cocktail, and 1 uL of thymol (internal standard of 5 g/L). All vials were mixed using a Velp F202A0175 Wizard Vortex Mixer at 3000 rpm for 30 s and analyzed in triplicate by SPME at each dilution. For the smoked tomato paste samples (Trad and PST), 1.00 ± 0.01 g was diluted with 1 mL of water; 1 uL of internal standard was added and the samples were analyzed in triplicate under the same conditions as the standards.

SPME GC-MS was carried out using a DVB/Carboxen/PDMS Stableflex fiber (11mm, SupelCo, Poole, U.K.) and GC-MS was performed on an Agilent 7890-5975C GC-MS equipped with a Zebron ZB-5MSi column (30 m × 0.25 mm i.d. × 1 μm film thickness). Samples were equilibrated at 40 °C for 10 min with intermittent stirring prior to exposing the fiber for 10 min at 40 °C. The fiber was desorbed in the injection port for 20 min and the volatile compounds were analyzed. Helium was the carrier gas at 1.2 mL/min. After desorption, the oven was maintained at 40 °C for 5 min, then raised to 250 °C at 4 °C/ min. Mass spectra were recorded in electron ionization mode at 70 eV and at a source temperature of 230 °C. A scan range of $m/z$ 29–400 with a scan time of 0.69 s was used and the data were controlled and stored by the ChemStation system.

Good linearity was observed for all compounds except 4-vinylguaiacol and whisky lactone which were close to the limit of detection for the method, and their calibrations were based on fewer points ($R^2$ for these were 0.61 and 0.84, respectively). Otherwise, $R^2$ was always greater than 0.9 and generally greater than 0.95. For (Z)-isoeugenol, the calibration curve for (E)-isoeugenol was used. For the unresolved 3- and 4-methylphenols, the calibration curve for 4-methylphenol was used.

**Statistical Analysis**

The data for the volatile analyses were analyzed with XLStat (AddinSoft, Paris, France, 2015.6.01) using one-way analysis of variance (ANOVA), and post-hoc multiple pairwise comparisons were carried out using the Fisher's least significant difference (LSD) test with the significance level set at $p = 0.05$. Two-way ANOVA was used to determine the significance of fat type and smoke technology. For the sensory data, SENPAQ version 3.2 (Qi Statistics, Reading, U.K.) was used to carry out the two-way ANOVA where main effects were tested against the sample by using assessor interaction. Multiple pairwise comparisons were done using the Fisher's LSD at $p = 0.05$. Principal component analysis (PCA) was carried out on the sensory data in XLStat with the volatile data used as supplementary data. Results from the paired preference test were evaluated using the binomial model in V-Power (Jesionka; macro for Microsoft Excel).

# Results

**Flavor Changes in Smoked Tomato Flakes Added to Cream Cheese**

*Volatile Analysis*

Twenty compounds were selected for comparison between samples based on published GC–O data (8), their abundance, and their relevance to cheese and smoke flavor (Table 1). Comparing just the peak areas of the Trad and the PST in full-fat cheese, all the smoke-derived compounds had smaller peak areas in PST compared to those in the Trad, 11 of these being significant at $p < 0.05$ and the remainder at $p < 0.1$. These were all compounds that were observed in smoked oil (12) including furfurals (2), guaiacols (5), eugenols (3), 4-methylphenol, 2-acetylpyrrole, syringol, and cyclotene. In the low-fat cheese, the same trends were observed, except for 2-acetylpyrrole and (Z)-isoeugenol where the differences were either not significant or not consistent between the two cheeses. This decrease in smoke-derived compounds is similar to the decrease observed by Parker et al. (12) where sunflower oil was either Trad smoked or filtered through zeolite. However, in the smoked oils, furfural and 5-methylfurfural did not decrease when the filter was employed.

In the full-fat cheese, the smoking technology had no impact on the compounds that were already present in the unflavored cream cheese (2,3-butanedione, 2-heptanone, 3-hydroxybutanone, and acetic acid); however, in the low-fat cheese, there was a tendency for these compounds to be higher in the PST sample than the Trad sample. This was also the case for 6-methyl-5-heptene-2-one which is a carotenoid-derived compound found in the unsmoked tomato flakes.

# Table 1. Relative Peak Areas of Selected Volatile Compounds Detected in the Headspace of Low-Fat or Full-Fat Cream Cheese Containing 2.5% w/w Trad or PST Ground Tomato Flakes

| Compound | LRI DB5 | LRI Wax | Full-Fat Cheese | | Low-Fat Cheese | | $S^a$ | $S^b$ |
| --- | --- | --- | --- | --- | --- | --- | --- | --- |
| | | | Trad | PST | Trad | PST | | |
| 2,3-Butanedione | 586 | 961 | 26c | 29 | 29 | 33 | * | * |
| 2-Heptanone | 892 | 1167 | 3.3a | 3.4a | 2.5b | 3.0a | ns | ** |
| 3-Hydroxybutanone | 708 | 1283 | 75b | 75b | 80b | 112a | * | ** |
| 6-Methyl-5-hepten-2-one | 986 | 1330 | 9.6b | 12b | 12b | 22a | *** | *** |
| Acetic acid | 602 | 1458 | 25c | 26c | 35b | 46a | ** | *** |
| 2-Furfural | 833 | 1462 | 5.0b | 2.9c | 6.2a | 3.5c | *** | * |
| 5-Methylfurfural | 965 | 1571 | 2.3a | 0.9c | 2.5a | 1.3b | *** | * |
| Benzenacetaldehyde | 1048 | 1640 | 2.1a | 2.5b | 3.4a | 3.9a | ns | *** |
| Cyclotene | 1030 | 1830 | 2.2a | 0.3c | 1.7b | 0.3c | *** | *** |
| Guaiacol | 1094 | 1859 | 18a | 4.7c | 19a | 7.1b | *** | * |
| 4-Methylguaiacol | 1198 | 1956 | 14a | 3.2b | 16a | 6.1b | *** | * |
| 2-Acetylpyrrole | 1061 | 1970 | 4.4a | 3.0b | 3.9a | 4.3a | * | ns |
| 4-Ethylguaiacol | 1286 | 2029 | 9.4a | 1.9b | 11a | 3.9b | *** | * |
| 4-Methylphenol | 1072 | 2090 | 2.2a | 0.6c | 2.3a | 0.9b | *** | * |
| 4-Propylguaiacol | 1375 | 2108 | 1.9a | 0.3b | 1.9a | 0.7b | *** | ns |
| Eugenol | 1365 | 2166 | 0.9 | 0.3 | 0.7 | 0.5 | * | ns |
| 4-Vinylguaiacol | 1322 | 2195 | 0.3b | 0.1b | 1.5a | 0.3b | ** | ** |
| (Z)-Isoeugenol | 1417 | 2254 | 0.5 | 0.1 | 0.3 | 0.3 | * | ns |
| Syringol | 1357 | 2263 | 4.6a | 1.1b | 4.7a | 2.3b | *** | ns |
| (E)-Isoeugenol | 1460 | 2348 | 1.2a | 0.2b | 1.3a | 0.6b | *** | ns |

[a] Significance, obtained from ANOVA, that there is a difference between the Trad and PST samples where ns = no significant difference ($p > 0.05$); *sig is 0.01 < $p$ 0.05; **sig is 0.001 < $p \leq 0.01$; and ***sig is $p \leq 0.001$.   [b] Significance, obtained from ANOVA, that there is a difference between the full-fat and the low-fat cheese samples.   [c] Mean peak areas $\times 10^4$ AU (n=3); means in the same row that are not labelled with the same letters are significantly different ($p = 0.05$).

Although the primary aim of the experiment was not to compare flavor release in full-fat and low-fat cream cheese, the data for the PST cheese showed a significant increase in flavor release for nine compounds in the low-fat cheese compared to the full-fat cheese, with a similar trend for another nine. This is consistent with current understanding of the role of fat content with regard to flavor release. This has mainly been demonstrated in dairy yogurts (14) or ice cream (15), but for hydrophobic aroma compounds with relatively high log $p$ values, it is well established that a decrease in fat content will promote partitioning into the headspace and increase flavor release. However, with the Trad cheese, this effect was greatly diminished, and only three compounds (acetic acid, furfural, and 4-vinylguaiacol) showed a significant increase in the low-fat cheese.

## Sensory Analysis

The sensory data showed that there were significant differences between the cheese samples for 12 of the 25 attributes (Table 2). One key difference was the significant reduction in the bitter taste when the PST was applied, which is perhaps a result of the filter holding back non-volatile bitter compounds in the tar fraction that does not pass through the filter. The compounds associated with smoke were consistently higher in Trad compared to PST, particularly the overall smoky aroma, flavor and aftertaste, the bonfire aroma and flavor, and the diesel flavor. This is consistent with the instrumental volatile data that showed a higher concentration of typical smoky aroma compounds in the Trad samples. On the other hand, the cheesy aroma was significantly higher in the PST samples. In full-fat cheese, the change in smoke technology did not significantly alter the concentration of the compounds associated with the cheese suggesting that these compounds were masked by the high levels of smoke compounds in the full-fat Trad product, which makes them more prominent in the full-fat PST sample. In low-fat cheese, the compounds associated with cheese did increase in the PST product, which is consistent with an increase in aroma in this product. Either or both of these mechanisms could explain the increase in cheesy aroma in the PST products.

When these data are viewed on a PC plot, the correlations are clear (Figure 1). All the smoky aroma compounds are positioned to the far right of PC1: an area associated with the low-fat Trad product which a) contains more smoky compounds than PST, and b) is based on the low-fat cheese which promotes the release of more smoky compounds. PC1 separates the Trad from the PST (with more cheesy and dairy notes associated with PST), which is consistent with less masking from the smoke volatiles. Interestingly, 2,3-butanedione and 3-hydroxybutanone which have buttery creamy and dairy notes were highly correlated with the sour yogurt note as was acetic acid, which contributes the "sour" to this attribute. The sun-dried tomato flavor attribute is associated with 6-methyl-5-hepten-2-one, a carotenoid-derived compound found in the tomato flake.

**Table 2. Mean Panel Scores for Sensory Attributes Found in Low-Fat or Full-Fat Cream Cheese Containing 2.5% w/w Trad or PST Ground Tomato Flakes**

| Sensory Attribute | Full-Fat Cheese | | Low-Fat Cheese | | Sig[a] |
|---|---|---|---|---|---|
| | Trad | PST | Trad | PST | |
| *Aroma* | | | | | |
| Overall smokiness | 33a[b] | 16b | 41a | 23b | *** |
| Bonfire | 25b | 12c | 33a | 15c | *** |
| Ash | 4.4ab | 1.3b | 6.9a | 2.0b | * |
| Cheesy | 19bc | 33a | 17c | 26ab | ** |

**Table 2. (Continued). Mean Panel Scores for Sensory Attributes Found in Low-Fat or Full-Fat Cream Cheese Containing 2.5% w/w Trad or PST Ground Tomato Flakes**

| Sensory Attribute | Full-Fat Cheese | | Low-Fat Cheese | | Sig[a] |
|---|---|---|---|---|---|
| | Trad | PST | Trad | PST | |
| *Taste* | | | | | |
| Sweet | 18 | 19 | 16 | 18 | ns |
| Salty | 19 | 18 | 21 | 20 | ns |
| Umami | 30 | 27 | 30 | 32 | ns |
| Bitter | 7.6ab | 5.3b | 9.7a | 7.7b | * |
| Sour | 13 | 15 | 13 | 15 | ns |
| *Flavor* | | | | | |
| Overall smokiness | 40a | 20c | 46a | 30b | *** |
| Bonfire | 28a | 13b | 36a | 20b | *** |
| Paprika | 14 | 12.8 | 15 | 12.3 | ns |
| Ash | 3.7b | 1.7b | 8.1a | 2.2b | ** |
| Smoked fish | 19a | 8.5b | 20a | 15a | ** |
| Diesel | 4.2a | 0.6b | 4.1a | 0.8b | * |
| Sundried tomato | 16 | 17 | 15 | 20 | ns |
| Dairy | 24 | 29 | 23 | 26 | ns |
| Sour yogurt | 13 | 13 | 13 | 16 | ns |
| Spicy | 13 | 10 | 16 | 13 | ns |
| Smoked bacon | 17 | 11 | 20 | 15 | ns |
| Cheesy | 27ab | 32a | 20b | 28ab | * |
| Balanced | 44 | 46 | 44 | 46 | ns |
| *After-effects* | | | | | |
| Mouthcoating | 34 | 30 | 31 | 31 | ns |
| Warming | 9.4 | 6.6 | 10 | 7.8 | ns |
| Smoked food | 26a | 14b | 30a | 15b | *** |

[a] Significance, obtained from ANOVA, that there is a difference between the mean scores where ns = no significant difference ($p > 0.05$); *sig is $0.01 < p\,0.05$; **sig is $0.001 < p \leq 0.01$; and ***sig is $p \leq 0.001$. [b] Mean panel scores (n = 9 in duplicate); means in the same row not labelled with the same letters are significantly different ($p = 0.05$).

Overall the sensory profiling showed a decrease in smoky notes and a decrease in bitterness when the PST was applied. As a result, more notes from the cheese and the tomato flakes were perceived by the panelists.

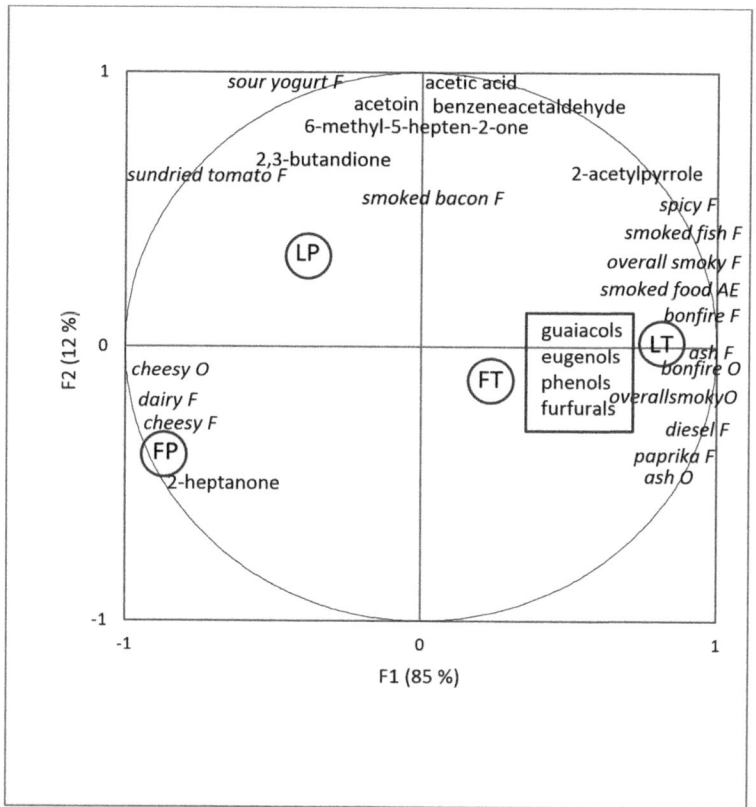

*Figure 1. PC1 vs. PC2 of sensory attributes (italics) for cream cheese with added smoked tomato flakes (F=full-fat, L=low-fat, T=traditional process, P= PST) with volatile data overlaid.*

## Consumer Analysis

Sensory profiling, however, is not hedonic, so we recruited a panel of 115 naïve volunteers to indicate which product they preferred when given a choice of full-fat Trad or full-fat PST. Of these volunteers, 65 preferred the PST product and 50 preferred the Trad (ratio of 57:43). This indicates that there is no adverse effect on flavor when the PST is applied and that there is tendency for the PST to be preferred to Trad at $p = 0.096$. Thus, the removal of PAHs from smoke using PST does not adversely affect the flavor and may also improve it. Other food ingredients are under test.

## Quantitative Comparison in Tomato Paste of Aroma Compounds Generated during Either Trad or PST

However, questions remain as to why the reduction in some of the aroma compounds may improve the flavor and which of these compounds are the key contributors to the less desirable aroma attributes such as ash and diesel. With this in mind, we quantitated the aroma compounds in tomato paste that had been smoked using either Trad or PST. Seventeen compounds were selected based on the literature where GC–O was used to determine odor-active compounds in smoked foods (8). Table 3 shows the concentration of each compound found in the smoked tomato paste. Furfural and 5-methylfurfural were by far the most abundant compounds in the smoked tomato paste, but not necessarily the most odor-active. In order to estimate the odor activity of these compounds, literature thresholds in water were employed, where possible, to calculate odor activity values (OAV, the concentrations in ug/kg divided by the odor thresholds in ug/kg). Of those compounds where thresholds were available, guaiacol had the highest odor activity, which is consistent with data taken from Varlet et al. (8) showing that guaiacol had the highest average intensity by GC–O in smoked salmon, despite very different matrices. According to Table 3, the next most odor-active compounds are likely to be eugenol and 4-methyl-, 4-ethyl-, and 4-vinylguaiacols; in the smoked salmon (8), 4-methylphenol, 2-acetylfuran, and (E)-isoeugenol were the next most intense. No odor thresholds in water were available for (E)-isoeugenol, but 4-methylphenol and 2-acetylfuran were much less odor-active if water thresholds are used. The odor threshold for whiskey lactone was determined in a water-ethanol mix and is therefore likely to be overestimated since the ethanol is likely to reduce the partitioning into the headspace. This would result in an underestimation of the OAV. Schranz et al. (16) have recently reported thresholds in air for many of these compounds. Applying these thresholds to compare OAVs (with arbitrary units) for selected compounds shows that guaiacol is again the most odor-active followed by 4-ethylguaiacol, 4-methylguaiacol, and (Z)-isoeugenol. There are clearly major limitations associated with using literature thresholds from different matrices for calculating OAVs. What is clear though is the fact that the guaiacols, and perhaps also the eugenols, are important contributors to the aroma of the smoked tomato pastes. It is less likely that the furans, the phenols, and the whiskey lactone contribute, but this requires a full sensomics analysis to confirm the role of these compounds in smoky flavor and identify the key differences that make Trad and PST smoke smell slightly different.

**Table 3. Concentration (ug/kg) and Approximated OAV of Aroma Compounds in Trad or PST Tomato Paste**

| Compound | LRI DB5 | Threshold[a] ug/kg | Concentration[b] ug/kg | | OAV | | S[c] |
|---|---|---|---|---|---|---|---|
| | | | Trad | PST | Trad | PST | |
| Guaiacol | 1095 | 3 | 750 | 400 | 250 | 133 | *** |
| Eugenol | 1366 | 6 | 140 | 70 | 23 | 12 | *** |
| 4-Methylguaiacol | 1200 | 90 | 670 | 360 | 7.4 | 4.0 | *** |
| 4-Ethylguaiacol | 1286 | 50 | 230 | 110 | 4.6 | 2.2 | *** |
| 4-Vinylguaiacol | 1323 | 18 | 3.3 | 1.7 | 3.3 | 1.7 | ns |
| 2-Furfural | 836 | 3000 | 3070 | 2460 | 1.0 | 0.8 | *** |
| Cyclotene | 1028 | 300 | 220 | 110 | 0.7 | 0.4 | ns |
| 3/4-Methylphenol | 1073 | 55 | 20 | 10 | 0.4 | 0.2 | *** |
| Syringol | 1358 | 1850 | 380 | 240 | 0.2 | 0.1 | ns |
| Phenol | 978 | 5900 | 650 | 400 | 0.1 | 0.1 | *** |
| Whiskey lactone[d] | 1298 | 790 | 40 | 20 | 0.05 | 0.03 | *** |
| 2-Methylphenol | 1053 | 650 | 30 | 10 | 0.05 | 0.02 | *** |
| 2-Acetylfuran | 913 | 10000 | 110 | 70 | 0.01 | 0.01 | *** |
| 5-Methylfurfural | 967 | na | 3720 | 2890 | | | *** |
| 2,6-Dimethylphenol | 1025 | na | 50 | 20 | | | *** |
| (E)-isoeugenol | 1461 | na | 770 | 360 | | | *** |
| (Z)-isoeugenol | 1418 | na | 380 | 170 | | | *** |

[a] Threshold in water (ug/kg) (17); na = not available. [b] Mean concentration ug/kg (n=3). [c] Probability, obtained from ANOVA, that there is a difference between the Trad and PST samples where ns = no significant difference ($p > 0.05$); and ***significant at $p \leq 0.001$. [d] Odor threshold in water-ethanol mixture (6:4 by vol) (18).

## Overall Conclusion

A new technology (PST) has been developed to remove carcinogenic PAHs from smoke streams used for smoking food ingredients. The aim of this chapter was to determine whether the PST had a detrimental effect on flavor. Tomato flakes were smoked using either Trad or PST, and the smoked tomato flakes were presented in either full-fat or low-fat cream cheese. Using sensory profiling, we demonstrated that in both cases there was a small but significant reduction in bitterness when PST was used to filter the smoke prior to smoking the tomato flakes. There was also a significant reduction in the smoky aroma and flavor attributes, and this was confirmed using instrumental analysis. However, when the two full-fat products were compared in a consumer preference test (n = 115), there was no clear preference for either product. Thus, we conclude that when PST is employed to reduce the concentration of carcinogenic PAHs in products such as smoked tomato flakes (or other spices), it also can reduce some of the bitterness associated with the smoking process. There is a minor impact on the aroma profile, but this did not have an impact on consumer preference.

## References

1. Guillen, M. D.; Manzanos, M. J. Characteristics of Smoke Flavorings Obtained from Mixtures of Oak *Quercus sp.* Wood and Aromatic Plants *Thymus vulgaris* L. and *Salvia lavandulifolia Vahl. Flavour Fragrance J.* **2005**, *20*, 676–685.

2. Guillen, M. D.; Errecalde, M. C. Volatile Components of Raw and Smoked Black Bream *Brama raii* and Rainbow Trout *Oncorhynchus mykiss* Studied by Means of Solid Phase Microextraction and Gas Chromatography/Mass Spectrometry. *J. Sci. Food Agric.* **2002**, *82*, 945–952.

3. Guillen, M. D.; Errecalde, M. C.; Salmeron, J.; Casas, C. Headspace Volatile Components of Smoked Swordfish *Xiphias gladius* and Cod *Gadus morhua* Detected by Means of Solid Phase Microextraction and Gas Chromatography-Mass Spectrometry. *Food Chem.* **2005**, *94*, 151–156.

4. Guillen, M. D.; Ibargoitia, M. L.; Sopelana, P.; Palencia, G.; Fresno, M. Components Detected by Means of Solid-Phase Microextraction and Gas Chromatography/Mass Spectrometry in the Headspace of Artisan Fresh Goat Cheese Smoked by Traditional Methods. *J. Dairy Sci.* **2004**, *87*, 284–299.

5. Guillen, M. D.; Palencia, G.; Ibargoitia, M. L.; Fresno, M.; Sopelana, P. Contamination of Cheese by Polycyclic Aromatic Hydrocarbons in Traditional Smoking. Influence of the Position in the Smokehouse on the Contamination Level of Smoked Cheese. *J. Dairy Sci.* **2011**, *94*, 1679–1690.

6. Majcher, M. A.; Jelen, H. H. Key Odorants of Oscypek, a Traditional Polish Ewe's Milk Cheese. *J. Agric. Food Chem.* **2011**, *59*, 4932–4937.

7. Varlet, V.; Knockaert, C.; Prost, C.; Serot, T. Comparison of Odor-Active Volatile Compounds of Fresh and Smoked Salmon. *J. Agric. Food Chem.* **2006**, *54*, 3391–3401.

8. Varlet, V.; Serot, T.; Cardinal, M.; Knockaert, C.; Prost, C. Olfactometric Determination of the Most Potent Odor-Active Compounds in Salmon Muscle *Salmo salar* Smoked by Using Four Smoke Generation Techniques. *J. Agric. Food Chem.* **2007**, *55*, 4518–4525.

9. Alonge, D. O. Carcinogenic Polycyclic Aromatic Hydrocarbons (PAH) Determined in Nigerian Kundi (Smoke-Dried Meat). *J. Sci. Food Agric.* **1988**, *43*, 167–72.

10. Committee on Diet, Nutrition and Cancer, Assembly of Life Sciences, National Research Council. *Diet, Nutrition and Cancer*; National Academy Press: Washington, DC, 1982.

11. Griffiths, H. D.; Baines, D. A.; Parker-Gray, J. K. *Smoked Food, Method for Smoking Food and Apparatus Therefor*. WO 2015007742 A1, Jan 22, 2015.

12. Parker, J. K.; Lignou, S.; Shankland, K.; Kurwie, P.; Griffiths, H. D.; Baines, D. A. Development of a Zeolite Filter for Removing Polycyclic Aromatic Hydrocarbons (PAHs) from Smoke and Smoked Ingredients While Retaining the Smoky Flavor. *J. Agric. Food Chem.* **2018**, *66*, 2449–2458.

13. Institute of Food Technologists. Sensory Evaluation Division Sensory Evaluation Guide for Testing Food and Beverage Products. *Food Technol.* **1981**, 50–59.

14. Brauss, M. S.; Linforth, R. S. T.; Cayeux, I.; Harvey, B.; Taylor, A. J. Altering the Fat Content Affects Flavor Release in a Model Yogurt System. *J. Agric. Food Chem.* **1999**, *47*, 2055–2059.

15. Ayed, C.; Martins, S. I. F. S.; Williamson, A.-M.; Guichard, E. Understanding Fat, Proteins and Saliva Impact on Aroma Release from Flavoured Ice Creams. *Food Chem.* **2018**, *267*, 132–139.

16. Schranz, M.; Lorber, K.; Klos, K.; Kerschbaumer, J.; Buettner, A. Influence of the Chemical Structure on the Odor Qualities and Odor Thresholds of Guaiacol-Derived Odorants, Part 1: Alkylated, Alkenylated and Methoxylated Derivatives. *Food Chem.* **2017**, *232*, 808–819.

17. Leffingwell, J. C.; Leffingwell, D. GRAS flavor Chemicals—Detection Thresholds. *Perfum. & Flavor.* **1991**, *16*, 1–19.

18. Poisson, L.; Schieberle, P. Characterization of the Key Aroma Compounds in an American Bourbon Whisky by Quantitative Measurements, Aroma Recombination, and Omission Studies. *J. Agric. Food Chem.* **2008**, *56*, 5820–5826.

# Chapter 7

# Development and Performance Characterization of a Lab-Scale Smoke Generator

Yuanyang Zhang,[1] Graham Eyres,[1] Patrick Silcock,[*,1] and Jim Jones[2]

[1]Department of Food Science, University of Otago, P.O. Box 56, Dunedin 9054, New Zealand

[2]School of Engineering and Advanced Technology, Massey University, Palmerston North, 4442, New Zealand

[*]E-mail: pat.silcock@otago.ac.nz.

A lab-scale smoke generator was developed to enable smoke to be generated and collected in a controlled manner to analyze smoke from Mānuka wood, a hardwood species indigenous to New Zealand. The impact of smoke generation parameters, including temperature and atmosphere, on the generation of aroma compounds and polycyclic aromatic hydrocarbons (PAHs) was investigated. Volatile organic compounds (VOCs) were trapped using stir-bar sorptive extraction (SBSE), and analyzed using gas chromatography-mass spectrometry (GC-MS). Mānuka wood smoke was generated at two temperatures (280 °C and 480 °C) under either air or nitrogen. The VOCs in Mānuka smoke varied depending on the smoke generation conditions, demonstrating the possibility of smoke manipulation. Eight PAHs with molecular weight no greater than 202 Da were detected. Higher temperature produced higher levels of PAHs, while the impact of atmosphere composition varied in a compound-specific manner.

## Introduction

Smoke used in food processing is produced from the thermolysis of a variety of biomass types including wood. The thermal decomposition of the three main components in wood (hemicellulose, cellulose, and lignin) yields hundreds of chemical compounds, many of which are important to the flavor, colour, and preservation of food. For instance, phenolic compounds, which are mainly derived from the thermal decomposition of lignin (1), provide flavoring and antimicrobial functions while the carbonyl compounds contribute to color development by reacting with proteins in food (1–3). In New Zealand, the majority of smoked food is produced using smoke generated from what is typically called Mānuka (*Leptospermum scoparium*) wood.

The chemical composition of wood smoke varies due to several factors including smoke generation conditions, collection procedures, species, water content of feedstock, and time (4).

Wood smoke generation parameters reported to have the greatest effect on product yields and composition from wood thermolysis are temperature (heating rate and maximum temperature) and atmosphere (oxygen content and sweeping gas flow rate) (5). Although the literature has detailed the mechanism for how the thermolysis parameters influence the chemical composition of wood smoke (6–8), the chemical composition and sensory profile of Mānuka smoke are not well characterized. In particular, the process for tuning the smoke composition by manipulating generation parameters to achieve a desired intensity and sensory character while minimizing polycyclic aromatic hydrocarbons (PAH) formation, is poorly understood. Generally, volatile organic compounds (VOCs) from thermolysis are largely generated and released at reaction temperatures ranging from 150–400 °C (9), with defined regions of hemicellulose (220–315 °C), cellulose (315–400 °C), and lignin (160–900 °C) decomposition (10).

PAHs are a group of bioaccumulative compounds, some of which are carcinogenic and mutagenic (11). PAHs are mainly formed during biomass pyrolytic reactions (in particular during incomplete combustion) and form at increasing concentrations as the thermolysis temperature increases from 400 °C to 1000 °C (12). PAH content also varies depending on the types of wood species with poplar and hickory producing 35%–55% fewer PAHs than beech wood (13) when smoke is generated by air smoldering. When comparing smoking methods, it has been reported that friction smoking produces fewer PAHs than air smoldering, superheated steam, or hot plate (touch) smoking (8). These studies show that PAH generation has only been studied for a broad class of smoke generation methods and not as a function of the conditions of smoking. This chapter addresses this information gap by describing some preliminary experiments to address the relationship between smoke generation conditions and the VOCs generated. Two system temperatures were selected: 280 °C, where the VOCs arise mostly from hemicellulose and cellulose decomposition, and 480 °C, where most of the lignin has also been decomposed.

Due to the carcinogenic and bioaccumulative properties of PAHs, the European Union enacted regulations in September 2014 that set maximum levels for PAHs in food including smoked meat and smoked fish. Examples of the limits include 5.0 µg/kg and 30.0 µg/kg for benzo(a)pyrene and the sum of benzo(a)pyrene, benz(a)anthracene, benzo(b)fluoranthene, and chrysene, respectively (14). The U.S. Environmental Protection Agency (EPA) also includes these compounds on their list defining 16 priority PAHs according to potential toxicity for human exposure or frequency of occurrence at hazardous waste sites (15). Because of the stringent regulations on these hazardous compounds, it is necessary to understand the mechanism of PAH generation and how to minimize their formation.

Stir-bar sorptive extraction (SBSE) is a solventless sorptive extraction technique for concentrating volatile and non-volatile compounds and has been widely used as a sample pre-treatment since it was first introduced in 1999 (16). Compared to solid-phase microextraction (17), SBSE possesses a higher sensitivity (up to 1000 times greater) due to the higher phase volume, making it a highly suitable technique for analyzing trace and ultra-trace levels of analytes (18, 19). This technique allows adsorption either by direct immersion into liquid or from the gas phase. For example, a study by Kaur et al. successfully characterized the diluted vapor phase of cigarette smoke employing headspace SBSE coupled to gas chromatography–mass spectrometry (GC–MS) (20). Considering the advantages and successful applications of SBSE, direct SBSE sampling of smoke combined with GC–MS analysis was selected to profile the chemical composition of Mānuka smoke in this study.

Our research objective was to design a glass lab–scale smoke generator where Mānuka smoke could be generated under defined conditions of two temperatures and two atmospheres, from which VOCs in smoke could be collected. The impact of thermolysis temperature and atmosphere on the chemical composition of smoke was investigated by direct in-line sampling of the VOCs in smoke with SBSE and subsequent analysis with GC–MS.

## Materials and Methods

### Materials

The Mānuka wood was provided by JB & HA Brosnahan Ltd (Ohope, New Zealand) and ground into a powder using a Stanmore Tabletop Hammer Mill (Glen Creston, London, U.K.). The moisture content of the air-dry Mānuka wood (8.6%) was determined by loss on drying according to the standardized approach outlined by the United States Pharmacopeia <731> Loss on drying (*21*). Water activity was 0.45, which was measured using a water activity meter (AQUALAB 4TE, Meter Group, Pullman, WA, USA). The particle size of the wood powder was determined by sieve analysis using 100 g of wood powder and a geometric series of laboratory test sieves (ENDECOTTS, London, U.K.) with aperture sizes ranging from 100 μm to 850 μm on a test sieve shaker (EFL2 mk3, ENDECOTTS) for 10 min (Figure 1).

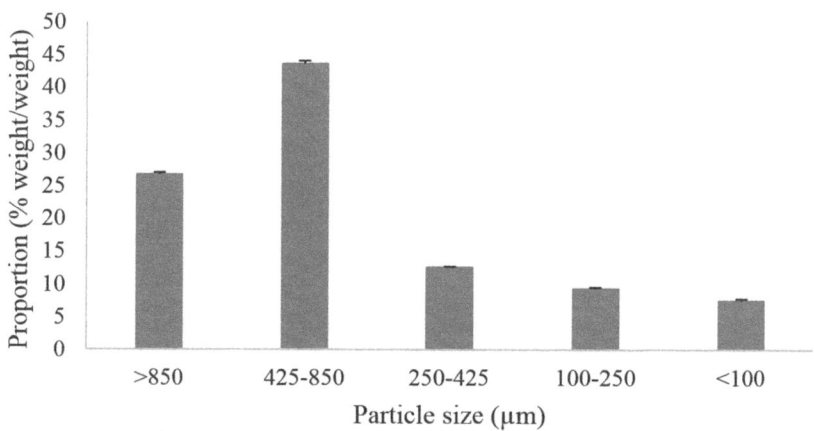

Figure 1. Particle size distribution of Mānuka wood powder as determined by sieve analysis. Particle size limits refer to the aperture sizes of the sieves.

### Methods

*Smoke Generation and Collection*

To generate and collect wood smoke under defined conditions in a controllable manner, a glass lab–scale smoke generator was designed. The system comprised of three parts: a sweep gas flow controller, a smoke generation vessel, and in-line sample collection capability (Figure 2). The sweep gas flow control system helped to control the flow rate and oxygen level. The smoke generation vessel was a custom-made 450 mL glass flat-bottomed flask with four ports that allowed for the introduction of the sweep gas, temperature monitoring, wood powder introduction, and smoke exit and collection, respectively. The VOCs generated from the thermolysis of wood powder were flushed from the smoking vessel and captured in-line using SBSE (Twister; 10 mm; 0.5 mm film thickness; Gerstel, Germany), a magnetic stir bar sealed within a glass tube and coated with polydimethylsiloxane (PDMS) as the extractive phase.

Preliminary experiments were carried out to determine whether the generator could produce smoke under defined conditions allowing for investigation into the influence of generation parameters, including temperature and atmosphere, on the chemical composition of wood smoke. For this study, smoke samples were generated at two temperatures (280 °C and 480 °C) under either air or nitrogen using a flow rate of 50 mL/min. After preheating the smoke generation vessel to one of the set temperatures (280 °C or 480 °C) using a hotplate (MS7-H550-Pro, DLAB Scientific, Beijing, China), Mānuka wood powder (2 g) was added and smoke collection was carried out for 10–40 min in the first experiment to evaluate the effect of exposure time and for a constant 40 min for subsequent experiments.

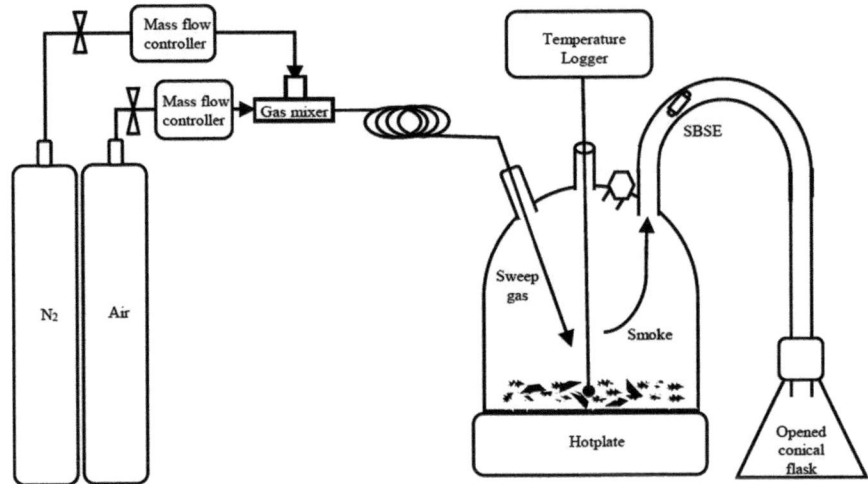

*Figure 2. Schematic of lab-scale smoke generator.*

*Analysis of VOCs in Smoke*

For each experiment, the VOCs in the wood smoke were extracted using SBSE with the PDMS stir bar positioned at the vent of the smoke generation vessel (Figure 2). This stir bar was positioned as close to the vessel as possible but far enough away to ensure that the temperature did not exceed 100 °C. Once the smoke collection was completed, the SBSE bars were rinsed with deionized water and dried with lint-free tissue, then placed in glass desorption tubes (length 60mm, outer diameter 6mm, inner diameter 4mm; Gerstel), and transferred to the multipurpose sampler (MPS, Gerstel) for analysis. The tubes were automatically transferred to the Gerstel thermal desorption unit (TDU), which was initially held at 50 °C for 0.5 min, then increased to 240 °C at 120 °C/min, and followed by a 10-min hold. The thermally desorbed compounds were flushed from the TDU using hydrogen, cryogenically trapped, and then focused at -60 °C with liquid carbon dioxide in the cooled injection system (Gerstel CIS4), which was equipped with an empty baffled glass inlet liner. VOCs were transferred onto the GC column by heating the inlet from -60 °C to 240 °C at 720 °C/min and then held for 10 min in splitless mode.

Separations were performed on a polar SolGel-WAX column (length 30 m, inner diameter 0.25 mm, film thickness 0.25 μm; Trajan Scientific, Australia) with hydrogen as the carrier gas under constant flow (1.6 mL/min). The initial GC oven temperature was 50 °C, which was held for 4 min, then increased to 210 °C at 5 °C/min, then increased again to 240 °C at 10 °C/min, and held for 10 min. Analysis was carried out on an Agilent 7890B GC system equipped with 5977A MSD (Agilent Technologies, Beijing, China) which was operated in electron ionization mode with an electron energy of 70 eV. The temperature of the transfer line to the MS was set to 230 °C and the quadrupole was set to 150 °C. Mass spectra were scanned from 30–300 amu.

*Identification of VOCs in Smoke*

The identification of compounds was primarily achieved based on the compound's mass spectra compared to the National Institute of Standards and Technology (NIST 2014) database supported by retention indices (RIs). The RI of each compound present in the smoke samples was calculated according to the linear RI regression of normal alkanes ($C_7$–$C_{30}$) established under the same GC thermal program. RIs from the literature were used to confirm the identity of compounds by comparing them with experimental RIs from the present study (22).

## Results and Discussion

### Generation and Collection of Smoke

Two temperatures, 280 °C and 480 °C, were selected to evaluate the performance of the smoke generator. Their impact on the thermal conversion process was determined by measuring VOC generation. The smoking vessel was preheated to either 280 °C or 480 °C, after which the wood powder was introduced. The wood powder was at room temperature, so when it is added to the vessel, the thermocouple recorded a temperature decrease of up to 27 °C for the 280 C trial and 89 °C for the 480 °C trial. In the 280 °C trial, it returned to the set point within 3.3 min and for the 480 °C trial, it returned to the set point within 2.5 min. For the remainder of the smoke generation collection time, the thermocouple deviation was no more than ±11 °C and ±4 °C for the two set points, respectively. The decrease in temperature can be attributed to the energy required to evaporate the moisture content of the air-dry wood and to the sensible heating required to raise the temperature to the set point. Here, the measured oxygen levels under air was 22.4%, and the measured oxygen content under nitrogen was 0.0%.

### Comparison of VOC Profiles in Smoke Samples as a Function of SBSE Exposure Times

Other than smoke generation–related parameters such as temperature and atmosphere, the SBSE stir bar exposure time during smoke sampling can also affect the chemical profile of wood smoke. The extraction efficiency for SBSE is determined largely by the partitioning coefficient of target compounds between the adsorbent phase and the sample matrix (23). Therefore, the effect of SBSE stir bar exposure time on the smoke flow was investigated using exposure times of 10, 20, 30, and 40 min.

As shown in Figure 3, the chromatograms of smoke VOCs generated at 280 °C under air and collected after exposure times of 10–40 min are qualitatively similar. To compare the performance at different exposure times, the peak response for each VOC was normalized against its corresponding peak after a 40 min exposure time. In general, the peak response of most VOCs in smoke collected for 20 min was higher than that in samples collected for 10 min, but comparable with the samples collected for 30 and 40 min. For example, the relative peak responses of vanillin (retention time ~37.3 min) after exposure to smoke for 10, 20, 30, and 40 min were 0.9, 1.0, 1.2, and 1.0, respectively. This indicates that a constant concentration had been reached within 20 min. However, it was observed that the response of early peaks that eluted before 24 min decreased as exposure time increased. For example, the relative peak responses of furfural (retention time ~15.8 min) after exposure to smoke for 10, 20, 30, and 40 min were 21, 19, 2, and 1, respectively. This indicates that a sampling time of 20 min is sufficient to maximize the peak response for smoke experiments.

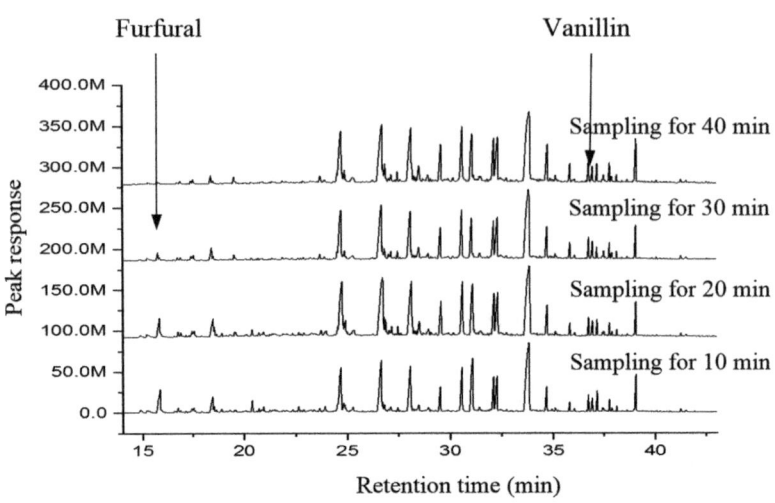

*Figure 3. GC–MS chromatograms (full scan) of Mānuka smoke generated at 280 °C under air with different sampling times.*

*Comparison of VOC Profiles in Smoke Samples Generated under Various Conditions*

The effect of temperature on smoke generated under air at 280 °C and 480 °C is shown in Figure 4. Generally, the VOCs collected from smoke generated at 280 °C and 480 °C were similar from a qualitative point of view in that they shared the same major components, and the VOC with the highest peak response at both temperatures was isoeugenol. However, for some of the VOCs present, their abundance varied considerably. For example, the detected level of furfural (retention time ~16.1 min) and 5-methylfurfural (retention time ~18.7 min) decreased dramatically by 19 and 15 times, respectively, when the temperature was increased from 280 °C to 480 °C. This may be because both compounds are formed from the thermolysis products of hemicellulose with xylose as the precursor (*24*) and the thermal degradation of hemicellulose occurs at relatively lower temperatures compared to cellulose and lignin (*10, 25*). At 480 °C, furfural and 5-methylfurfural may be partially destroyed from secondary decomposition while still within the smoking vessel. However, this result is in conflict with the findings of Shen et al. who investigated the pyrolytic behavior of xylan-based hemicellulose using TG-FTIR and Py-GC-FTIR, finding that high temperature promoted furfural formation (*26*). However, Shen et al. also utilized a higher temperature range (400–690 °C) and a completely different experimental set up. The result and its differences with the results of other researchers do indicate that smoke composition can be highly influenced by system design.

The chromatograms show considerable peak overlapping between 25–35 min, indicating a high degree of chemical complexity. Peak integration and spectral deconvolution indicated that the samples contained more than 300 chemical compounds, making data analysis challenging.

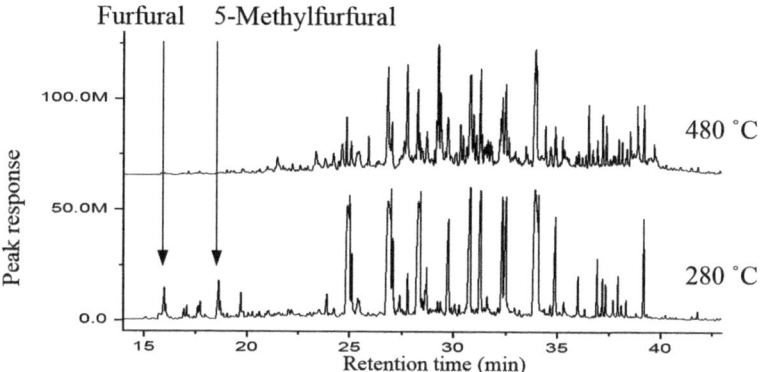

*Figure 4. GC–MS chromatograms (full scan) of Mānuka smoke generated under air at 480 °C and 280 °C (40 min exposure).*

To quantitatively compare the thermolysis products of Mānuka wood generated at 280 °C or 480 °C under nitrogen or air, six representative compounds were selected. These six VOCs, previously identified as odor-active in smoke (*1*), are 2,6-dimethoxy phenol, guaiacol, phenol, m-cresol, acetol, and furfural. Derived from the thermal decomposition of lignin, 2,6-dimethoxy phenol (also known as syringol) has been associated with smoky, phenolic, and balsamic odors. This compound has been found in smoke generated from various types of biomass including oak, rice husk, hickory, pine, cottonwood, cherry tree, bamboo, and cedar (*1, 27, 28*). The highest levels of 2,6-dimethoxy phenol in the current study were observed at 280 °C under both atmospheres; the air atmosphere favored formation more than the nitrogen atmosphere (Figure 5). A similar tendency was observed for guaiacol, which also originates from lignin thermal decomposition. Phenol and m-cresol were previously identified as providing phenolic odors in oak, hickory, and pine smoke (*29–31*). In contrast to the levels of syringol and guaiacol, Figure 5 illustrates that the highest levels of phenol and m-cresol were measured at 480°C under nitrogen followed by levels under air. It has been reported that the formation of non-methoxy phenols increases under higher temperature, whereas the reverse is observed for the formation of methoxy-containing phenols (*32, 33*). Acetol (which has pungent, sweet, and caramellic odors) has been found in the pyrolysis products of rice husk and hickory (*29, 34*). Unlike 2,6-dimethoxy phenol, the highest production of acetol was detected under nitrogen at 280 °C. For furfural, higher levels were detected at the lower temperature regardless of the atmosphere. This is possibly due to the in-line, dynamic sampling approach and the balance between compounds in the gas phase and the extraction phase of the SBSE stir bar.

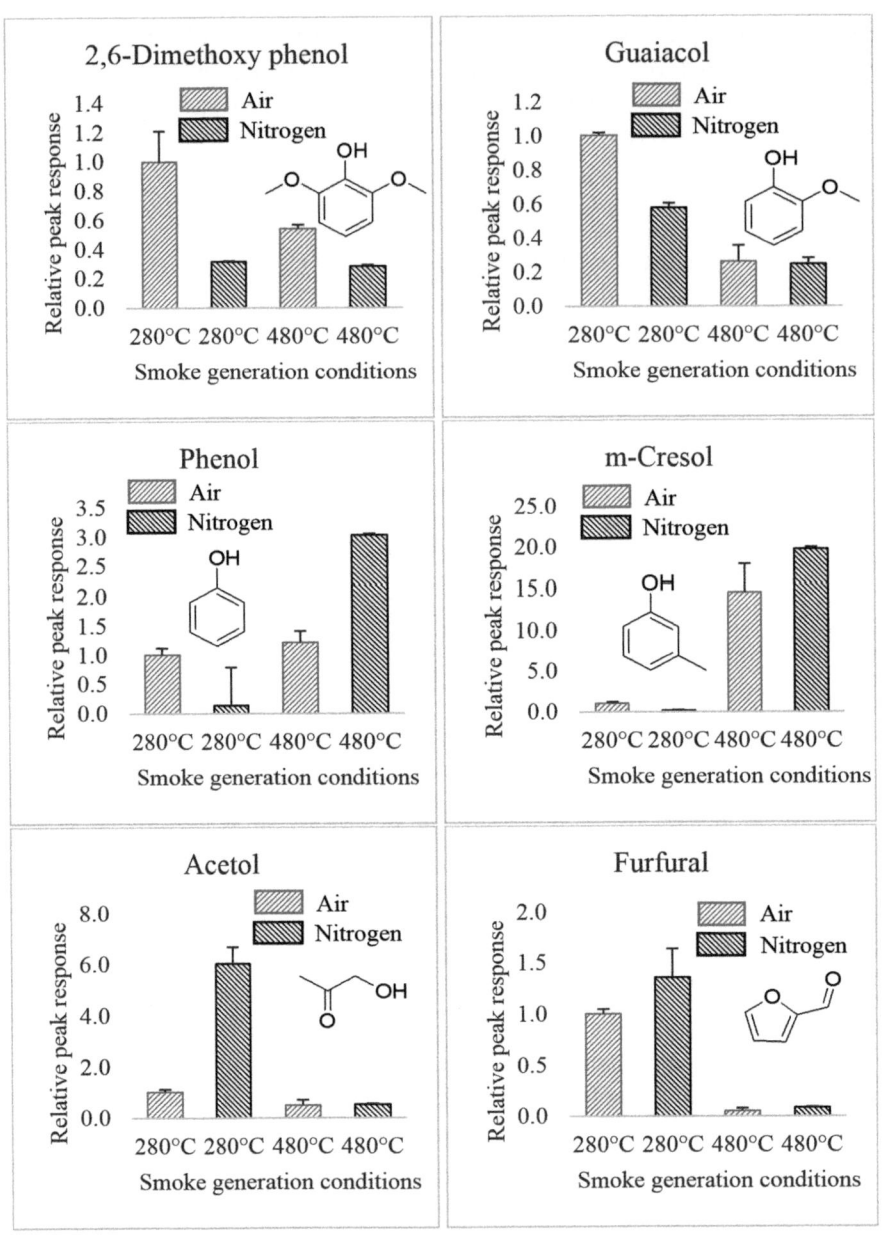

*Figure 5. Comparison of six representative compounds under different smoke generation conditions. The y-axis represents the relative peak response after normalization against the corresponding peak in the smoke sample generated at 280 °C under air.*

*Comparison of PAHs in Smoke Samples Generated under Various Conditions*

The levels of PAHs in smoke were investigated as a function of the smoke generation conditions. Using the gas phase sampling approach, only eight light PAHs (naphthalene, acenaphthylene, acenaphthalene, fluorene, phenanthrene, anthracene fluoranthene, and pyrene) were detectable in the gas phase by SBSE. The molecular weights of the detected compounds were all less than 202 g/mol. Due to their low volatility and high vapor pressure, PAHs with molecular weights greater than 202 g/mol were most likely in the condensed phase prior to contacting the smoke with the SBSE stir bar. It was previously reported that PAHs with four or less aromatic rings are associated with particulate matter and gas phase while PAHs with more than four aromatic rings are mainly only present in condensed particulate phases (35). These results concur with the findings of Fagernäs et al., who investigated the PAH distribution in products produced during slow pyrolysis of birch wood (36).

Figure 6 shows a comparison of naphthalene, acenaphthalene, fluoranthene, and pyrene detected in smoke generated at 280 °C or 480 °C under nitrogen or air. In general, higher levels of all detected PAHs from the smoke were found at 480 °C when under either nitrogen or air. The smoke generation atmosphere had a large effect but in a PAH-specific manner, and it appeared to display a PAH-atmosphere-smoke generation temperature interaction. For fluoranthene and pyrene, smoke generation under an air atmosphere resulted in higher detected levels at 280 °C and 480 °C compared to results from the nitrogen atmosphere. As air causes smoldering in conventional smoke generation, localized temperature hotspots (i.e. embers) are created, thus promoting the formation of these larger PAHs. In contrast, the highest level of naphthalene was detected from smoke generated under nitrogen at 480 °C. For acenaphthalene, the use of air or nitrogen as a smoke generation atmosphere had no effect on detected levels at 280 °C or 480 °C. It should be noted that under the current GC thermal program, phenanthrene and anthracene co-eluted and could not be deconvoluted by either their mass ion as they had the same molecular weight or by RIs due to their similar polarity. In future research, the GC separation method will be refined to resolve this co-elution. In addition, the collection in the smoke aerosol phase on glass fiber (15) will be investigated to determine the concentrations of all PAHs of regulatory concern under different smoke generation conditions.

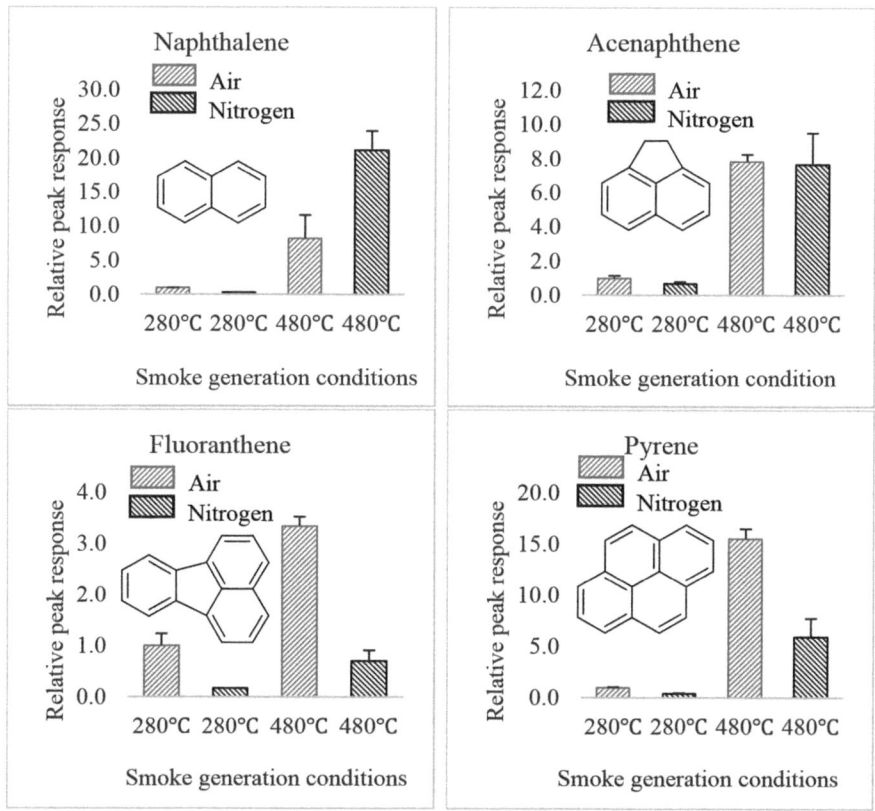

Figure 6. Comparison of levels of selected PAHs under different conditions. The y-axis represents the relative peak response after normalization against the corresponding peak in a smoke sample generated at 280 °C under air.

## Conclusion

A lab–scale smoke generator was designed that allowed for the collection of VOCs from wood smoke generated under defined conditions. The GC–MS analysis of Mānuka wood smoke showed that the formation of VOCs varied depending on the reaction temperature and applied atmosphere. This introduces the possibility of manipulating the smoke generation conditions to achieve specific VOC outcomes with the ultimate goal of modifying smoke flavor. Smoke generation at 480 °C promoted the formation of PAHs, and the use of air or nitrogen resulted in PAH-specific effects. The mechanisms behind VOC generation and the degradation pathways of the wood components under different smoke generation conditions require further research. In future research, GC-olfactometry in combination with GC–MS will be carried out to identify the key odor-active compounds and how smoke generation conditions can affect their production. The use of the glass-smoking vessel under defined conditions in combination with developed GC–MS and GC-olfactometry methodologies will allow for characterization of the Mānuka smoke and, more importantly, allow smoke composition to be tuned to meet consumer, manufacturing, and regulatory requirements.

## Funding Sources

This work was supported by the New Zealand Ministry of Business, Innovation, and Employment as part of the Food Industry Enabling Technologies (FIET) program (contract MAUX1402).

## References

1. Cadwallader, K. R. Wood Smoke Flavor. In *Handbook of Meat, Poultry and Seafood Quality*, 1st ed.; Blackwell: Ames, IA, 2007; pp 201–210.

2. Hollenbeck, C. Contribution of Smoke Flavourings to Processed Meats. In *Flavor of Meat and Meat Products*, 1st ed.; Springer: New York, 1994; pp 199–209.

3. Guillén, M. D.; Errecalde, M. C. Volatile Components of Raw and Smoked Black Bream (*Brama raii*) and Rainbow Trout (*Oncorhynchus mykiss*) Studied by Means of Solid Phase Microextraction and Gas Chromatography/Mass Spectrometry. *J. Sci. Food Agric.* **2002**, *82*, 945–952.

4. Simko, P. Factors Affecting Elimination of Polycyclic Aromatic Hydrocarbons from Smoked Meat Foods and Liquid Smoke Flavorings. *Mol. Nutr. Food Res.* **2005**, *49*, 637–647.

5. Demirbas, A. Effect of Initial Moisture Content on the Yields of Oily Products from Pyrolysis of Biomass. *J. Anal. Appl. Pyrolysis* **2004**, *71*, 803–815.

6. Fretheim, K.; Granum, P.; Vold, E. Influence of Generation Temperature on the Chemical Composition, Antioxidative, and Antimicrobial Effects of Wood Smoke. *J. Food Sci.* **1980**, *45*, 999–1002.

7. Guillén, M. D.; Ibargoitia, M. L. Influence of the Moisture Content on the Composition of the Liquid Smoke Produced in the Pyrolysis Process of *Fagus sylvatica* L. Wood. *J. Agric. Food Chem.* **1999**, *47*, 4126–4136.

8. Pöhlmann, M.; Hitzel, A.; Schwägele, F.; Speer, K.; Jira, W. Influence of Different Smoke Generation Methods on the Contents of Polycyclic Aromatic Hydrocarbons (PAH) and Phenolic Substances in Frankfurter-Type Sausages. *Food Control* **2013**, *34*, 347–355.

9.   Fang, M. X.; Shen, D. K.; Li, Y. X.; Yu, C. J.; Luo, Z. Y.; Cen, K. F. Kinetic Study on Pyrolysis and Combustion of Wood Under Different Oxygen Concentrations by Using TG-FTIR Analysis. *J. Anal. Appl. Pyrolysis* **2006**, *77*, 22–27.

10.  Yang, H.; Yan, R.; Chen, H.; Lee, D. H.; Zheng, C. Characteristics of Hemicellulose, Cellulose and Lignin Pyrolysis. *Fuel* **2007**, *86*, 1781–1788.

11.  Armstrong, B.; Hutchinson, E.; Unwin, J.; Fletcher, T. Lung Cancer Risk after Exposure to Polycyclic Aromatic Hydrocarbons: A Review and Meta-Analysis. *Environ. Health Perspect.* **2004**, *112*, 970–978.

12.  Toth, L.; Blaas, W. Effect of Smoking Technology of the Content of Carcinogenic Hydrocarbons in Smoked Meat Products. *Fleischwirtschaft* **1972**, *52*, 1419.

13.  Hitzel, A.; Pohlmann, M.; Schwagele, F.; Speer, K.; Jira, W. Polycyclic Aromatic Hydrocarbons (PAH) and Phenolic Substances in Meat Products Smoked with Different Types of Wood and Smoking Spices. *Food Chem.* **2013**, *139*, 955–962.

14.  European Commission. Commission Regulation (EU) No. 1327/2014 of 12 December Amending Regulation (EC) No. 1881/2006 as Regards Maximum Levels of Polycyclic Aromatic Hydrocarbons (PAHs) in Traditionally Smoked Meat and Meat Products and Traditionally Smoked Fish and Fishery Products. In *Official Journal of the European Union*, 1st ed.; European Union: Brussels, BE, 2014; pp 13–14.

15.  Sánchez, N. E.; Salafranca, J.; Callejas, A.; Millera, Á.; Bilbao, R.; Alzueta, M. U. Quantification of Polycyclic Aromatic Hydrocarbons (PAHs) Found in Gas and Particle Phases from Pyrolytic Processes Using Gas Chromatography–Mass Spectrometry (GC–MS). *Fuel* **2013**, *107*, 246–253.

16.  Baltussen, E.; Sandra, P.; David, F.; Cramers, C. Stir Bar Sorptive Extraction (SBSE), a Novel Extraction Technique for Aqueous Samples: Theory and Principles. *J. Microcolumn Sep.* **1999**, *11*, 737–747.

17.  Bicchi, C.; Iori, C.; Rubiolo, P.; Sandra, P. Headspace Sorptive Extraction (HSSE), Stir Bar Sorptive Extraction (SBSE), and Solid Phase Microextraction (SPME) Applied to the Analysis of Roasted Arabica Coffee and Coffee Brew. *J. Agric. Food Chem.* **2002**, *50*, 449–459.

18.  Alves, R. F.; Nascimento, A. M. D.; Nogueira, J. M. F. Characterization of the Aroma Profile of Madeira Wine by Sorptive Extraction Techniques. *Anal. Chim. Acta* **2005**, *546*, 11–21.

19.  Nogueira, J. M. F. Stir-Bar Sorptive Extraction: 15 Years Making Sample Preparation More Environment-Friendly. *TrAC, Trends Anal. Chem.* **2015**, *71*, 214–223.

20.  Kaur, N.; Cabral, J. L.; Morin, A.; Waldron, K. C. Headspace Stir Bar Sorptive Extraction–Gas Chromatography/Mass Spectrometry Characterization of the Diluted Vapor Phase of Cigarette Smoke Delivered to an *in vitro* Cell Exposure Chamber. *J. Chromatogr. A* **2011**, *1218*, 324–33.

21.  United States Pharmacopeial Convention. Physical Tests <731> Loss on Drying. In *United States Pharmacopeia,* 40th ed.; United States Pharmacopeial Convention, Inc.: Rockville, MD, 2016; pp 614.

22.  Bianchi, F.; Careri, M.; Mangia, A.; Musci, M. Retention Indices in the Analysis of Food Aroma Volatile Compounds in Temperature-Programmed Gas Chromatography: Database Creation and Evaluation of Precision and Robustness. *J. Sep. Sci.* **2007**, *30*, 563–572.

23.  Qin, Z.; Bragg, L.; Ouyang, G.; Pawliszyn, J. Comparison of Thin-Film Microextraction and Stir Bar Sorptive Extraction for the Analysis of Polycyclic Aromatic Hydrocarbons in Aqueous Samples with Controlled Agitation Conditions. *J. Chromatogr. A* **2008**, *1196–1197*, 89–95.

24. Lu, Y.; Wei, X. Y.; Cao, J. P.; Li, P.; Liu, F. J.; Zhao, Y. P.; Fan, X.; Zhao, W.; Rong, L. C.; Wei, Y. B. Characterization of a Bio-Oil from Pyrolysis of Rice Husk by Detailed Compositional Analysis and Structural Investigation of Lignin. *Bioresour. Technol.* **2012**, *116*, 114–119.

25. Beaumont, O.; Schwob, Y. Influence of Physical and Chemical Parameters on Wood Pyrolysis. *Ind. Eng. Chem. Process Des. Dev.* **1984**, *23*, 637–641.

26. Shen, D.; Gu, S.; Bridgwater, A. V. Study on the Pyrolytic Behaviour of Xylan-Based Hemicellulose Using TG–FTIR and Py–GC–FTIR. *J. Anal. Appl. Pyrolysis* **2010**, *87*, 199–206.

27. Vichi, S.; Santini, C.; Natali, N.; Riponi, C.; Lopez-Tamames, E.; Buxaderas, S. 393 Volatile and Semi-Volatile Components of Oak Wood Chips Analysed by Accelerated Solvent Extraction (ASE) Coupled to Gas Chromatography–Mass spectrometry (GC–MS). *Food Chem.* **2007**, *102*, 1260–1269.

28. Montazeri, N.; Oliveira, A.; Himelbloom, B. H.; Leigh, M. B.; Crapo, C. A. 396 Chemical Characterization of Commercial Liquid Smoke Products. *Food Sci. Nutr.* **2013**, *1*, 102–115.

29. Fiddler, W.; Doerr, R.; Wasserman, A.; Salay, J. 386 Composition of Hickory Sawdust Smoke. Furans and Phenols. *J. Agric. Food. Chem.* **1966**, *14*, 659–662.

30. Mullen, C. A.; Boateng, A. A.; Mihalcik, D. J.; Goldberg, N. M. Catalytic Fast Pyrolysis of White Oak Wood in a Bubbling Fluidized Bed. *Energy Fuels* **2011**, *25*, 5444–5451.

31. Mukarakate, C.; Zhang, X.; Stanton, A. R.; Robichaud, D. J.; Ciesielski, P. N.; Malhotra, K.; Donohoe, B. S.; Gjersing, E.; Evans, R. J.; Heroux, D. S.; Richards, R.; Iisa, K.; Nimlos, M. R. Real-Time Monitoring of the Deactivation of HZSM-5 During Upgrading of Pine Pyrolysis Vapors. *Green Chem.* **2014**, *16*, 1444–1461.

32. Mu, W.; Ben, H.; Ragauskas, A.; Deng, Y. Lignin Pyrolysis Components and Upgrading—Technology Review. *BioEnergy Research* **2013**, *6*, 1183–1204.

33. Lou, R.; Wu, S. B.; Lv, G. J.; Guo, D. L. Pyrolytic Products from Rice Straw and Enzymatic/Mild Acidolysis Lignin (EMAL). *BioResources* **2010**, *5*, 2184–2194.

34. Pino, J. A. 384 Characterisation of Volatile Compounds in a Smoke Flavouring from Rice Husk. *Food Chem.* **2014**, *153*, 81–86.

35. Akyuz, M.; Cabuk, H. Gas-Particle Partitioning and Seasonal Variation of Polycyclic Aromatic Hydrocarbons in the Atmosphere of Zonguldak, Turkey. *Sci. Total Environ.* **2010**, *408*, 5550–5558.

36. Fagernäs, L.; Kuoppala, E.; Simell, P. Polycyclic Aromatic Hydrocarbons in Birch Wood Slow Pyrolysis Products. *Energy Fuels* **2012**, *26*, 6960–6970.

# Chapter 8

# Formation of Desired Smoky Key Odorants in Wheat Beer: A Comparison with the Undesired Toxicologically Relevant Styrene

Valerian Kalb[1] and Michael Granvogl[*,2,3,4]

[1]Lehrstuhl für Lebensmittelchemie und Molekulare Sensorik, Technische Universität München, Wissenschaftszentrum Weihenstephan für Ernährung, Landnutzung und Umwelt, Lise-Meitner-Straße 34, Freising 85354, Germany

[2]Lehrstuhl für Lebensmittelchemie, Technische Universität München, Department für Chemie, Lise-Meitner-Straße 34, D-85354 Freising, Germany

[3]Lehrstuhl für Analytische Lebensmittelchemie, Technische Universität München, Wissenschaftszentrum Weihenstephan für Ernährung, Landnutzung und Umwelt, Maximus-von-Imhof-Forum 2, D-85354 Freising, Germany

[4]Institut für Lebensmittelchemie, Fachgebiet für Lebensmittelchemie und Analytische Chemie (170a), Universität Hohenheim, Fakultät Naturwissenschaften, Garbenstrasse 28, D-70599 Stuttgart, Germany

[*]E-mail: Michael.Granvogl@ch.tum.de.

Wheat beer is a special beer type that is traditionally brewed in Bavaria (southern part of Germany) and Austria. By law, it must be brewed with a wheat malt content of at least 50%. Its characteristic aroma has been described as clovelike and slightly phenolic and is elicited by the presence of 4-vinylphenol and 2-methoxy-4-vinylphenol, two desired decarboxylation products from *p*-coumaric acid and ferulic acid (both cinnamic acid derivatives). Unfortunately, wheat beer also contains an undesired decarboxylation product, namely styrene (vinylbenzene), which is formed via the same pathway from cinnamic acid. Former studies of the brewing process have shown that these compounds are mainly formed by yeast during fermentation from the aforementioned precursors, traditionally referred to as phenolic acids. Despite this knowledge, the current market survey based on 20 commercially available wheat beer samples from different breweries revealed styrene concentrations ranging from 0.22 to 31.6 µg/L, which were similar to amounts reported in previous studies.

# Introduction

Wheat beer is a specialty beer type that is very popular in Bavaria (Germany) and Austria and by law must be brewed with a wheat malt content of at least 50%. The aroma of wheat beer has been described as clovelike and slightly phenolic, evoked by 2-methoxy-4-vinylphenol (2M4VP) with a clovelike and smoky odor impression and 4-vinylphenol (4VP) with an almond-shell-like and phenolic smell (*1*). Wheat beer with relatively low concentrations of these compounds exhibits a less pronounced typical aroma. Furthermore, in addition to these desired volatile vinyl aromatics, the undesired compound styrene (S) is naturally present in typical wheat beers.

Evidence of styrene in beer goes back to the 1970s (*2, 3*), but first attention to its toxicological relevance was not reported until 1996, when the World Health Organization published a tolerable daily intake for drinking water of 7.7 μg/kg body weight per day (*4*). Then, in 2002, styrene was classified by the International Agency of Research on Cancer as possibly carcinogenic to humans (class 2B) based on the results of studies on mice (*5*) and rats (*6*). On the basis of these facts, even if the reported average styrene concentrations in wheat beers will not lead to an excess of the tolerable daily intake, which is enforced for drinking water, mitigation strategies should be applied.

The aforementioned vinyl aromatics have a common biochemical formation pathway, namely the enzymatic decarboxylation of the precursors cinnamic acid (C), *p*-coumaric acid (*p*C), and ferulic acid (F). The conversion of these, traditionally referred to as phenolic acids, occurs mainly in presence of Pof⁺ active top-fermenting yeast strains during fermentation in wheat beer production (*7, 8*). In crops, phenolic acids have an important functional role, as they link arabinoxylan chains to a three-dimensional network. This building block, together with β-glucans, structures the endosperm cell wall of the grains. Thus, barley and wheat are natural sources for these precursor acids.

Several studies by Langos et al. (*1, 9–11*), Langos (*12*), and Schwarz et al. (*8, 13–16*) examined the influence of different process steps of brewing on the release of styrene and the desired vinyl aromatics. It was shown that thermal process steps, like mashing, wort boiling, and pasteurization, have little to no contribution to the overall styrene concentration caused by thermal decarboxylation. Comparative studies on open versus closed fermentation revealed reduced styrene concentrations (−25%) using the former. Styrene is thereby stripped off with the $CO_2$ that is formed, whereas in a closed fermentation, vaporized styrene condenses again without any loss. Nevertheless, open fermentation plays only a minor role in brewing these days.

Known yeast strains exhibit no difference in their selectivity against the phenolic acid precursors. Thus, high concentrations of styrene correlated with high concentrations of 4VP and 2M4VP and vice versa. This is in accordance with the results of Daly et al. (*17*), who classified yeasts into three phenotypes, Pof⁺, Pof⁺ᐟ⁻, and Pof⁻, according to their ability to decarboxylate phenolic acids. Because styrene mitigation strategies within both the thermal-processing steps and fermentation were always accompanied by the loss of the desired vinyl phenolic odorants, further studies focused on the impact of malting conditions on the release of the precursors (*11, 18*). Thereby, the use of "undermodified" malts (malting parameters: steeping degree, 45%; germination temperature, 12 °C; germination time, 5 days) clearly reduces the content of undesired cinnamic acid compared with the contents of the desired phenolic acids, making the malting process a promising tool to lower the styrene content in wheat beer.

Table 1 summarizes past market studies on the concentration of styrene, as well as 4VP and 2M4VP, in commercially available wheat beers. This chapter investigated, in a renewed study on 20 commercially available wheat beers, whether the implementation of the gained knowledge by breweries has led to an improvement of the food safety of wheat beer. Therefore, the aims of the current study were to analyze the precursors *p*-coumaric acid, ferulic acid, and cinnamic acid (using liquid chromatography-tandem mass spectrometry [LC-MS/MS]) as well as the corresponding decarboxylation products 4VP, 2M4VP, and styrene (using comprehensive gas chromatography time-of-flight mass spectrometry [GCxGC-ToF-MS]) based on stable isotope dilution assays. With these data in hand, correlations between the precursors and the decarboxylation products, as well as between the desired vinyl aromatics and styrene, should be calculated.

**Table 1. Published Data on Styrene Concentrations in Commercial Wheat Beers**

| Literature | Fermentation | Pof. act.[a] | S | Concentration [µg/L] of 4VP | 2M4VP | n |
|---|---|---|---|---|---|---|
| Langos et al. 2016 (*10*) | Top | + | 15–33 | 620–1020 | 630–2020 | 6 |
| | Top, r.-alc.[b] | + | 25 | 355 | 795 | 1 |
| Daly et al. 1997 (*17*) | Top | + | 25–31 | — | — | 4 |
| | Top, r.-alc.[b] | + | 9, 25 | — | — | 2 |
| | Top | − | <0.04 | — | — | 3 |
| | Bottom | − | <0.04 | — | — | 7 |
| | Bottom, r.-alc.[b] | − | <0.04 | — | — | 9 |
| Wackerbauer et al. 1982 (*19*) | Top | + | — | 1250 | 2517 | 1 |
| | Bottom | − | — | 10 | 98 | 1 |
| | Special beers | n.a.[c] | — | 555–5204 | 187–251 | 3 |
| Wackerbauer et al. 1982 (*20*) | Top | + | — | 440–2700 | 520–4300 | 21 |
| | Top | n.a.[c] | — | 20 | 189 | 1 |

[a] Pof activity.　[b] Reduced-alcohol.　[c] Not available.

## Experimental Section

### Materials

Twenty different wheat beers from 16 Bavarian (Germany) breweries were purchased from a local store in October 2018. All wheat beers were bottled in 500 mL amber glass bottles and were stored at room temperature. Eighteen of the 20 wheat beers were unfiltered, pale wheat beers with an alcohol content of between 4.9 and 5.8 vol.-%. Additionally, the sample set included two dark wheat beers (both 5.3 vol.-%), two reduced-alcohol (2.8 and 3.3 vol.-%), and two nonalcoholic (<0.5 vol.-%) wheat beers. For comparison with the top-fermented wheat beers, two bottom-fermented lager beers ("Helles") were also analyzed.

The following stable isotopically labeled internal standards were commercially obtained: [$^{13}C_3$]-ferulic acid (99 atom % $^{13}C$), [$^{13}C_3$]-*p*-coumaric acid (99 atom % $^{13}C$), [$^2H_7$]-cinnamic acid (98 atom % $^2H$), and [$^2H_8$]-styrene (98 atom % $^2H$) were purchased from Sigma-Aldrich (Merck, Darmstadt, Germany); [$^2H_3$]-2-methoxy-4-vinylphenol (99.5 atom % $^2H$) was from Toronto Research Chemicals (North York, ON, Canada). [$^2H_5$]-4-Vinylphenol was synthesized according to Jezussek (*21*).

All solvents were of HPLC gradient grade; acetonitrile was from Baker (Sowińskiego, Poland), formic acid from Merck, and ultrapure water from an in-house source.

## Quantitation of Free Phenolic Acids in Wheat Beer by High-Performance Liquid Chromatography-Tandem Mass Spectrometry Based on Stable Isotope Dilution Analysis

Naturally cloudy wheat beer samples were degassed and clarified by filtration (paper filter, 5 H/N, 240 mm, 85 g/m²; Sartorius, Göttingen, Germany) and by an additional ultrasonification step (10 min). Reduced-alcohol and regular wheat beers were used without dilution, lager beers and nonalcoholic wheat beers were diluted 1:5 (v:v) with water prior to further sample preparation steps. Afterward, the stable isotopically labeled internal standards [$^{13}C_3$]-ferulic acid, [$^{13}C_3$]-*p*-coumaric acid, and [$^2H_7$]-cinnamic acid (amounts determined in preliminary experiments) were added and the samples were equilibrated by a multitube vortexer (VWR, Darmstadt) for 15 min. Subsequently, the samples were membrane-filtrated (Minisart RC, hydrophilic, 15 mm; Sartorius), and the phenolic acids were quantitated via high-performance liquid chromatography-tandem mass spectrometry.

High-performance liquid chromatography-tandem mass spectrometry analysis was performed using an UltiMate 3000 HPLC system (Thermo Fisher Scientific; Dionex Softran, Germering, Germany) coupled to a triple quadrupole mass spectrometer (TSQ Vantage, ThermoFisher Scientific, Bremen, Germany). A Kinetex C18 column (100 × 2.1 mm, 2.6 μm, 10.0 nm) (Phenomenex, Aschaffenburg, Germany) was applied for the separation of the precursor acids using the following conditions: solvent A, aqueous formic acid (FA) (0.1%, v/v), solvent B, FA (0.1%, v/v) in acetonitrile; gradient: 0–2 min, 10% B; 2–15 min, from 10 to 90% B; 15–18 min, 90% B; 18–19 min, from 90 to 10% B; 19–30 min, 10% B; flow rate, 0.2 mL/min; injection volume, 10–20 μL; column temperature, 24 °C. The ion source was operated in positive atmospheric pressure chemical ionization mode (APCI⁺) using the following parameters: discharge current, 4.0 μA; vaporizer temperature, 250 °C; sheath gas pressure, 30 arbitrary units; auxiliary gas pressure, 10 arbitrary units; declustering voltage, −10 V; capillary temperature, 300 °C. Selected reaction monitoring (SRM) was used to analyze the transitions from precursor to product ions using experimentally optimized collision energies.

## Quantitation of Volatile Vinyl Aromatics in Wheat Beer by Headspace-Solid Phase Microextraction in Combination with Comprehensive Gas Chromatography-Time-of-Flight-Mass Spectrometry Based on Stable Isotope Dilution Analysis

Wheat beer samples were prepared as mentioned above for the analysis of the precursor acids. Sodium chloride (2 g), tap water (4 mL), and wheat beer (1 mL) were mixed in a headspace vial (20 mL) equipped with a magnetic stir bar. Lager beers (5 mL) were analyzed without a dilution with water. Then, the stable isotopically labeled internal standards [$^2H_3$]-2-methoxy-4-vinylphenol,

[$^{2}H_{5}$]-4-vinylphenol, and [$^{2}H_{8}$]-styrene (amounts determined in preliminary experiments) were added, the headspace vial was immediately sealed (silicon, PTFE septum screw caps), and the sample was stirred for 1 h on a magnetic stirrer at room temperature to adjust the sample headspace equilibrium.

Headspace-solid phase microextraction was performed using a Gerstel MultiPurpose Sampler equipped with a DVB/CAR/PDMS-coated fiber (50/30 μm 2 cm; Supelco; Bellefonte, PA, USA). Prior to sample analysis, a blank run was performed each day to bake out the fiber. The SPME parameters were set as follows: incubation temperature, 50 °C; incubation time, 3.0 minutes; extraction temperature, 50 °C; extraction time, 30 minutes. The adsorbed analytes were thermally desorbed in the injection port of the gas chromatography-time-of-flight-mass spectrometry system at 250 °C for 32 min.

The headspace-solid phase microextraction analysis was performed using a Leco Pegasus 4D gas chromatography-time-of-flight-mass spectrometry instrument (St. Joseph, MI) comprising an Agilent model 7890A GC, a dual-stage quad-jet thermal modulator, and a secondary oven coupled to the mass spectrometer providing unit mass resolution (Waldbronn, Germany). An Agilent Multimode Inlet operated in splitless mode for thermal desorption of the analytes in combination with a Gerstel MultiPurpose Sampler autosampler (Mülheim a.d. Ruhr, Germany) were used. Separation was achieved by a DB-FFAP fused silica column (30 m × 0.25 mm i.d., 0.25 μm film thickness, equipped with a deactivated precolumn (2 m × 0.53 mm i.d.; both J&W, Agilent) in the first dimension and a VF-5-MS column (2 m × 0.15 mm i.d., 0.3 μm film thickness; Varian, Darmstadt) in the second dimension. The column mode was set to constant flow (1.90 mL/min). The first oven temperature program was: 35 °C for 5 min, raised by 6 °C/min to 230 °C, and held for 5 min. The second oven was started at 60 °C for 5 min, then raised at 6 °C/min to 250 °C, and finally kept for 5 min. The modulator offset was set to + 40 °C. Mass spectra were acquired via electron ionization (70 eV) over a mass range $m/z$ 35–300 at a rate of 100 spectra/s. Data were processed using GC Image (Lincoln, NE).

**Statistical Analysis**

Calculation of mean values and standard deviations were performed by Microsoft Office Excel 2007 (Microsoft Corporation, Seattle, WA) and Pearson correlation coefficients by R 3.4.3 (RStudio, Boston, MA).

## Results and Discussion

**Vinyl Aromatic Contents of 20 Commercial Wheat Beers**

Styrene, 4-vinylphenol, and 2-methoxy-4-vinylphenol concentrations were measured in 26 beer samples: 18 pale wheat beers, two dark wheat beers, two reduced-alcohol wheat beers, two nonalcoholic wheat beers, and two bottom-fermented lager beers used as "negative" controls. The results, summarized in Table 2, showed a concentration range for styrene in regular pale and dark wheat beers from 9.8 to 31.6 μg/L, for 4VP from 434 to 1350 μg/L, and for 2M4VP from 620 to 2490 μg/L, respectively. Thereby, wheat beers 3 and 5 differed strongly from the remaining regular wheat beer samples due to their significantly lower vinyl aromatic concentrations, which were 0.22 and 0.70 μg/L for styrene, 65.4 and 55.7 μg/L for 4VP, and 140 and 109 μg/L for 2M4VP, respectively.

**Table 2. Volatile Vinyl Aromatics and Free Phenolic Acids of 18 Pale Wheat Beers (3, 7–9, 11, 14–26), Two Dark Wheat Beers (D; 5, 10), Two Reduced-Alcohol Wheat Beers (AR; 6, 12), Two Nonalcoholic Wheat Beers (AF; 4, 13), and Two Bottom-Fermented Beers (BF; 1, 2) from 17 Different Breweries (Ascending Order of Styrene Concentration)**

| Sample | | Styrene | 4-Vinylphenol | 2-Methoxy-4-vinylphenol | Cinnamic acid | p-Coumaric acid | Ferulic acid |
|---|---|---|---|---|---|---|---|
| | | | | Concentration[a] [μg/l] of | | | |
| 1 | BF | <LoQ[b] | 31.1 | 79.5 | 95.5 | 423 | 1200 |
| 2 | BF | <LoQ | 29.4 | 81.4 | 97.0 | 500 | 1390 |
| 3 | | 0.22 | 65.4 | 140 | 79.6 | 1150 | 2040 |
| 4 | AF | 0.57 | 40.1 | 77.7 | 81.4 | 1010 | 1550 |
| 5 | D | 0.70 | 55.7 | 109 | 101 | 1180 | 2150 |
| 6 | AR | 1.78 | 31.9 | 60.5 | 65.4 | 744 | 1340 |
| 7 | | 9.75 | 829 | 2490 | <LoD[c] | 11.1 | 353 |
| 8 | | 11.4 | 1010 | 1650 | <LoD | 4.87 | 23.6 |
| 9 | | 12.1 | 1220 | 1450 | <LoD | 10.1 | 35.8 |
| 10 | D | 12.1 | 434 | 620 | <LoD | 7.11 | 39.1 |
| 11 | | 12.4 | 816 | 1660 | <LoD | 22.9 | 81.2 |
| 12 | AR | 13.3 | 648 | 900 | <LoD | 6.66 | 28.3 |
| 13 | AF | 14.2 | 506 | 372 | <LoD | 356 | 660 |
| 14 | | 15.0 | 969 | 2070 | 5.97 | 76.7 | 198 |
| 15 | | 16.0 | 560 | 1090 | <LoD | 4.45 | 17.7 |
| 16 | | 16.1 | 1020 | 1390 | <LoD | 3.70 | 15.5 |
| 17 | | 18.3 | 896 | 1420 | 2.69 | 9.50 | 58.7 |
| 18 | | 19.8 | 860 | 1160 | <LoD | 4.65 | 76.0 |
| 19 | | 20.1 | 1050 | 1280 | <LoD | 3.59 | 11.3 |
| 20 | | 20.4 | 803 | 1790 | <LoD | 4.53 | 22.9 |

**Table 2. (Continued). Volatile Vinyl Aromatics and Free Phenolic Acids of 18 Pale Wheat Beers (3, 7–9, 11, 14–26), Two Dark Wheat Beers (D; 5, 10), Two Reduced-Alcohol Wheat Beers (AR; 6, 12), Two Nonalcoholic Wheat Beers (AF; 4, 13), and Two Bottom-Fermented Beers (BF; 1, 2) from 17 Different Breweries (Ascending Order of Styrene Concentration)**

| Sample | Styrene | 4-Vinylphenol | 2-Methoxy-4-vinylphenol | Cinnamic acid | p-Coumaric acid | Ferulic acid |
|---|---|---|---|---|---|---|
| | | | Concentration[a] [μg/l] of | | | |
| 21 | 21.7 | 628 | 1260 | 3.37 | 2.69 | 20.3 |
| 22 | 22.6 | 750 | 1370 | <LoQ[d] | 12.5 | 38.0 |
| 23 | 24.7 | 1290 | 1880 | 2.91 | 24.0 | 69.6 |
| 24 | 27.9 | 1080 | 2120 | <LoD | 4.64 | 26.3 |
| 25 | 28.8 | 1120 | 1830 | <LoD | 4.25 | 22.2 |
| 26 | 31.6 | 1350 | 1900 | <LoD[c] | 4.86 | 19.9 |

[a] All results are mean values of triplicates. Average relative standard deviations are 2.5% for styrene, 1.6% for 4-vinylphenol, 1.9% for 2-methoxy-4-vinylphenol, 6.7% for cinnamic acid, 3.8% for p-coumaric acid, and 4.0% for ferulic acid.   [b] LoQ = 0.10 μg/L.   [c] LoD = 0.75 μg/L.   [d] LoQ = 2.51 μg/L.

According to Wackerbauer et al. (20), wheat beers with <600 μg/L of 2M4VP elicit an atypical to neutral wheat beer aroma (e.g., wheat beers 3 and 5). Within the current sample set, 18 of the 20 wheat beers and one reduced-alcohol wheat beer exhibited a typical wheat beer aroma, as was sensorially assessed during sample preparation (data not shown). This correlates with concentrations >620 μg/L found for 2M4VP in these samples. For "negative" controls, two pale lager beers were analyzed, as the flavor of this beer type does not typically have any of the odor attributes of wheat beer. This is linked to the low amounts of vinyl aromatics, with approximately 30 μg/L of 4VP, 80 μg/L of 2M4VP, and styrene concentrations <LOQ (0.10 μg/L) found here in beers 1 and 2.

**Correlation of the Styrene Content with the Amount of Desired Vinyl Aromatics as well as the Amount of Free Cinnamic Acid in Wheat Beer**

Pearson correlation coefficients were calculated to determine the relationship between styrene, the desired vinyl aromatics, and the free phenolic acids in the analyzed beer samples (Table 3). Within the decarboxylation products, correlation coefficients >0.75 were found, indicating that an increase of styrene is linked to an increase of 4VP and 2M4VP. By comparison, a positive correlation was also found within the set of free phenolic acids, with correlation coefficients >0.88. A comparison of both groups with each other revealed a negative correlation, with correlation coefficients <−0.75; thus, a decrease of free phenolic acids is related to an increase of styrene, 4VP, and 2M4VP. To corroborate these results, the null hypothesis was tested, and a high significance was confirmed by $p$-values $\leq 1.0 \times 10^{-5}$.

**Table 3. Pearson Correlation Coefficients between Vinyl Aromatics and Phenolic Acids in the Analyzed Set of Wheat and Lager Beers**

|        | S       | 4VP     | 2M4VP   | C      | pC     | F |
|--------|---------|---------|---------|--------|--------|---|
| S      | 1       |         |         |        |        |   |
| 4VP    | 0.84[a] | 1       |         |        |        |   |
| 2M4VP  | 0.75[a] | 0.88[a] | 1       |        |        |   |
| C      | −0.80[a]| −0.84[a]| −0.78[a]| 1      |        |   |
| pC     | −0.75[a]| −0.80[a]| −0.77[a]| 0.88[a]| 1      |   |
| F      | −0.81[a]| −0.84[a]| −0.77[a]| 0.94[a]| 0.98[a]| 1 |

[a] $p$-Value $\leq 1.0 \times 10^{-5}$.

The causality of these correlations is linked to the presence of yeast. Wackerbauer et al. (22) proved that yeasts are able to convert $p$-coumaric acid and ferulic acid into their corresponding vinyl aromatics. The decarboxylation ability of yeasts to convert cinnamic acid into styrene was later shown by Goddey et al. (7). Thus, the negative correlation can be explained by the conversion of free phenolic acids, extracted from the malts into the pitching wort, into the corresponding vinyl aromatics by yeasts during fermentation. The positive correlation within the vinyl aromatics highlights an additional fact, namely the lack of selectivity of yeasts against the phenolic acids. This lack in selectivity was also found by Daly et al. (17), who analyzed the release of 2M4VP and styrene from different yeast strains and POF types. Consequently, changing the yeast strain is not an appropriate option to reduce the styrene concentration, as the characteristic odor impression of wheat beer will also be lost. Comparison of the results of wheat beers 3 and 9 (cf. Table 2) highlights that lower concentrations of styrene are accompanied by lower amounts of 4VP and 2M4VP and vice versa (Figure 1). Interestingly, both wheat beers (3 and 9) were from the same brewery and differed only in the yeast strain used for fermentation.

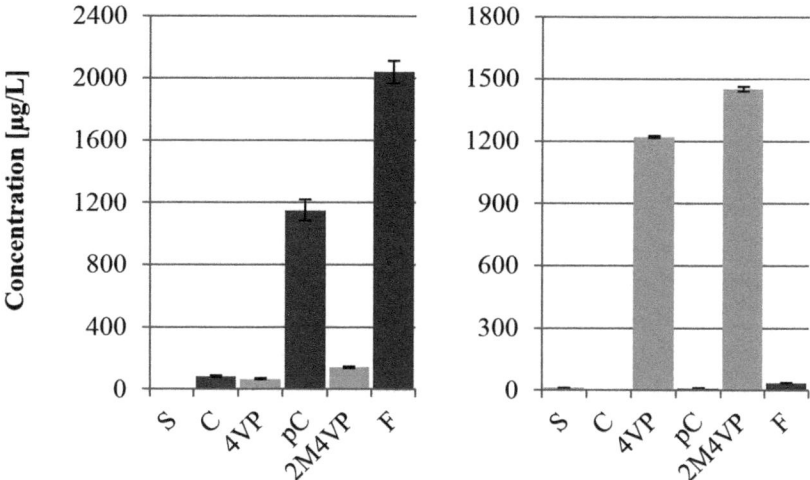

*Figure 1. Wheat beer brewed with a yeast strain with strongly reduced Pof^{+/−} activity (left). Wheat beer brewed with the standard Bavarian wheat beer yeast strain (right).*

To the best of our knowledge, wheat beer breweries in Bavaria usually use the same top-fermenting yeast strain for fermentation, except one brewery that uses a different strain (wheat beers 3 to 6 in Table 2). Therefore, different ratios between styrene, 4VP, and 2M4VP within the remaining set of pale wheat beer samples in Table 2 are influenced by other factors, for example, barley and wheat varieties used (23, 24), malting conditions applied (12), malt ratio used (8, 25), or mashing conditions applied (13, 26). All of them may affect the input of free phenolic acids into the brewing process as well as the overall fermentation management (14, 15, 27). This explains why the calculated Pearson correlation coefficients of a dataset including only pale wheat beers showed no significant correlations.

### Effect of Dealcoholization Process on the Styrene Concentration in Wheat Beers

For the dealcoholization process of beer, either fermentation is stopped when the desired alcohol content is reached (stopped fermentation) or the alcohol is removed after the complete fermentation. For the latter, different methods are applied (e.g., thermal dealcoholization, reducing the alcohol content by evaporation, or reverse osmosis, removing the alcohol via a semipermeable membrane). Schwarz et al. (15) proved a nearly complete conversion of cinnamic acid into styrene after 2 h after starting the fermentation by addition of Pof^{+/−} active yeasts. Therefore, it is expected that a stopped fermentation has no effect on the styrene concentration but does reduce the content of the desired vinyl aromatics as they reach their maximum of release after 3 to 6 days, depending on the fermentation temperature (14). In case of thermal dealcoholization, a clearly reduced amount of all vinyl aromatics is expected, with the lowest content for styrene due to its volatility and its low initial concentration. Regarding the precursors, a reverse effect for the stopped fermentation will occur, leading to high remaining concentrations. For the case of methods that remove the alcohol after fermentation, the amount of free phenolic acids will remain low.

Table 4 compares the vinyl aromatic and free phenolic acid composition of a regular, a reduced-alcohol, and a nonalcoholic wheat beer from the same brewery using the standard yeast strain. The regular wheat beer, with an original gravity of 12.8% and an alcohol content of 5.4%, showed a normal wheat beer pattern of precursors and vinyl aromatics. Regarding the reduced-alcohol wheat beer, the low amount of free phenolic acids indicated that the fermentation was completed. However, the alcohol content of 3.3% was adjusted in this case by a lower original gravity of 7.8% instead of removing the alcohol from the regular wheat beer. With the lower original gravity, less cinnamic acid was introduced into the wort, which explains the lowered styrene concentration compared with the regular wheat beer. The nonalcoholic wheat beer, with an original gravity of 6.5%, had a similar styrene concentration as the reduced-alcohol wheat beer and lower concentrations of 4VP and 2M4VP. Further, it showed high amounts of free $p$-coumaric and ferulic acid, but no cinnamic acid, which is exactly the pattern assumed for a reduced-alcohol wheat beer brewed by stopped fermentation, as a complete conversion of cinnamic acid into styrene, but just a partial conversion of the precursors into 4VP and 2M4VP, based on the reduced fermentation time, was found. The lower concentration of 2M4VP compared with 4VP may lead to the assumption that the reactivity of the decarboxylase is higher for $p$-coumaric acid compared with ferulic acid.

**Table 4. Styrene Content in a Reduced-Alcohol and a Nonalcoholic Wheat Beer Compared with a Regular Wheat Beer from the Same Brewery**

| Type | Concentration[a] [µg/L] of | | | | | |
| --- | --- | --- | --- | --- | --- | --- |
| | S | 4VP | 2M4VP | C | pC | F |
| Wheat beer | 24.7 | 1290 | 1880 | 2.91 | 24.0 | 69.6 |
| Red.-alcohol | 13.3 | 648 | 900 | <LoD[b] | 6.66 | 28.3 |
| Nonalcoholic | 14.2 | 506 | 372 | <LoD | 356 | 660 |

[a] All results are mean values of triplicates.    [b] LoD = 0.75 µg/L.

## Impact of Roasting Conditions on the Styrene Concentration in Wheat Beers

A further comparison of the data revealed higher concentrations of styrene and desired vinyl aromatics in a pale wheat beer compared with a dark wheat beer from the same brewery (Table 5). In both samples, the low free phenolic acid contents point to the fact that a complete fermentation took place. Thus, the lower styrene and desired vinyl aromatic contents in the dark wheat beer must be related to a lower input of the corresponding free phenolic acids by the malt into the wort. Samaras et al. (28) determined the relationship between the amounts of free $p$-coumaric acid and ferulic acid to the roasting degree of barley malts. With increasing kilning temperature, a massive decrease of free phenolic acids was found, which can be explained by the high temperatures used during kilning (220– 229 °C). To prove whether this finding was due to a thermal decarboxylation, the authors exemplarily analyzed the 2M4VP concentration in the respective malts. Thereby, malts kilned with a temperature <140 °C did not contain any 2M4VP, whereas malts kilned above 220 °C exhibited concentrations of 2M4VP of 267 to 439 µg/kg malt per dry mass. As both wheat beers, pale and dark, have almost the same original gravity of approximately 12.8%, a lower input of free phenolic acids as in reduced-alcohol wheat beers is not the explanation for the lower vinyl aromatic contents. As such, the lower amounts of styrene and desired vinyl aromatics in dark wheat beer can be explained by a loss of the precursor during kilning of dark malts, as shown by Samara and co-workers (28). Regarding the vinyl aromatics generated at high kilning temperatures, these results suggested that they are lost within the thermal processing steps due to the fact that they were not transferred into the beer, which is evident from the lower amounts found in dark wheat beer.

**Table 5. Styrene Content in One Pale and One Dark Wheat Beer from the Same Brewery**

| Type | S | 4VP | 2M4VP | C | pC | F |
|------|---|-----|-------|---|----|---|
| | | | Concentration[a] [µg/L] of | | | |
| Pale wheat beer | 20.4 | 803 | 1790 | < LoD[b] | 4.53 | 22.9 |
| Dark wheat beer | 12.1 | 434 | 620 | < LoD | 7.11 | 39.1 |

[a] All results are mean values of triplicates.  [b] LoD = 0.75 µg/L.

### Influence of the Yeast Type on the Styrene Content in Wheat Beers

The ability of yeast strains to convert phenolic acids into their corresponding volatile vinyl aromatics is linked to the presence of the POF1 gene (7). POF stands for phenolic off-flavor because originally these vinyl aromatics were undesired in beers and mostly occurred when wild yeast strains were unknowingly involved in the fermentation process. However, this conversion ability of yeasts to decarboxylate the precursors is now desired, leading to the characteristic aroma of wheat beer.

Daly et al. (17) found that bottom-fermenting yeasts lack the ability to decarboxylate phenolic acids, consequently called Pof[−]. In this regard, top-fermenting yeasts were found to be an inhomogeneous group. Those with a high Pof activity are classified as Pof[+] and those with a very low conversion rate are Pof[+/−]; also, top-fermenting yeasts with no Pof activity have been found. Analysis of three beers fermented with yeast strains of different Pof activities highlighted the influence of this characteristic on the pattern of vinyl aromatics and free phenolic acids (Table 6). In this study, Pof[−] was represented by a bottom-fermented lager beer. According to the absent decarboxylation activity, the free phenolic acids were still high, and nearly no vinyl aromatics were formed after the brewing process. The small amounts can be explained by thermal decarboxylation during the thermal processing steps, like mashing, wort boiling, and pasteurization (10). Pof[+] was represented by a pale wheat beer. Due to the high Pof activity, the decarboxylation of the free phenolic acids took place during fermentation, resulting in high amounts of the desired and undesired vinyl aromatics. Pof[+/−] was represented also by a pale wheat beer; however, this brewery used a traditional in-house yeast strain, which shows only a very low Pof activity, leading to the low vinyl aromatic concentrations and, correspondingly, to the very high remaining precursor amounts. The same yeast strain was also used for fermentation of wheat beers 3 to 6 (cf. Table 2), which explains their different vinyl aromatic and free phenolic acid pattern.

**Table 6. Effect of the Yeast Type on the Styrene Content in Beer**

| Yeast type | S | 4VP | 2M4VP | C | pC | F |
|------------|---|-----|-------|---|----|---|
| | | | Concentration[a] [µg/L] of | | | |
| Pof[−] | <LoQ[b] | 31.1 | 79.5 | 95.5 | 423 | 1200 |
| Pof[+/−] | 0.22 | 65.4 | 140 | 79.6 | 1150 | 2040 |
| Pof[+] | 12.1 | 1220 | 1450 | <LoD[c] | 10.1 | 35.8 |

[a] All results are mean values of triplicates.  [b] LoQ = 0.10 µg/L.  [c] LoD = 0.75 µg/L.

A comparison of the vinyl aromatic concentrations found in this study with the published data summarized in Table 1 indicated that there was no mitigation of styrene in most of the commercially available wheat beers over the past years. Therefore, further cooperation and investigations on the complex topic of styrene reduction in wheat beer brewing should be done.

## References

1.  Langos, D.; Granvogl, M.; Schieberle, P. Characterization of the key aroma compounds in two Bavarian wheat beers by means of the sensomics approach. *J. Agric. Food Chem.* **2013**, *47*, 11303–11311.

2.  Renner, R.; Bartels, H.; Tressl, R. Enrichment and quantitative determination of beer aroma substances by gas extraction. *Monatsschr. Brau.* **1976**, *29*, 478–480.

3.  Tressl, R.; Friese, L.; Fendesack, F.; Koeppler, H. Gas chromatographic-mass spectrometric investigation of hop aroma constituents in beer. *J. Agric. Food Chem.* **1978**, *6*, 1422–1426.

4.  *Guidelines for Drinking-Water Quality: Health Criteria and Other Supporting Information*, 2nd ed.; World Health Organization: Geneva, 1996.

5.  Cruzan, G.; Cushman, J. R.; Andrews, L. S.; Granville, G. C.; Johnson, K. A.; Bevan, C.; Hardy, C. J.; Coombs, D. W.; Mullins, P. A.; Brown, W. R. Chronic toxicity/oncogenicity study of styrene in CD-1 mice by inhalation exposure for 104 weeks. *J. Appl. Toxicol.* **2001**, *21*, 185–198.

6.  Cruzan, G.; Cushman, J. R.; Andrews, L. S.; Granville, G. C.; Johnson, K. A.; Hardy, C. J.; Coombs, D. W.; Mullins, P. A.; Brown, W. R. Chronic toxicity/oncogenicity study of styrene in CD rats by inhalation exposure for 104 weeks. *Toxicol. Sci.* **1998**, *46*, 266–281.

7.  Goddey, A. R.; Tubb, R. S. Genetic and biochemical analysis of the ability of *Saccharomyces cerevisiae* to decarboxylate cinnamic acids. *J. Gen. Microbiol.* **1982**, *128*, 2615–2610.

8.  Schwarz, K.; Methner, F.-J. Styrene concentrations during wheat beer production. *Brew. Sci.* **2011**, *64*, 156–158.

9.  Langos, D.; Granvogl, M.; Meitinger, M.; Schieberle, P. Development of stable isotope dilution assays for the quantitation of free phenolic acids in wheat and barley and malts produced thereof. *Eur. Food Res. Technol.* **2015**, *5*, 637–645.

10. Langos, D.; Granvogl, M. Studies on the simultaneous formation of aroma-active and toxicologically relevant vinyl aromatics from free phenolic acids during wheat beer brewing. *J. Agric. Food Chem.* **2016**, *11*, 2325–2332.

11. Langos, D.; Gastl, M.; Granvogl, M. Reduction of toxicologically relevant styrene in wheat beer using specially produced wheat and barley malts. *Eur. Food Res. Technol.* **2017**, *193*, 558–568.

12. Langos, D. *Optimierung technologischer Parameter zur Minimierung der Bildung von 4-Vinylbenzol (Styrol) beim Brauprozess unter Erhalt des typischen Aromas von Weizenbier (in German).* Ph.D. thesis, Technical University of Munich, Freising, Germany, 2017.

13. Schwarz, K. J.; Boltz, L. I.; Methner, F.-J. Release of phenolic acids and amino acids during mashing dependent on temperature, pH, time and raw materials. *J. Am. Soc. Brew. Chem.* **2012**, *70*, 290–295.

14. Schwarz, K. J.; Stübner, R.; Methner, F.-J. Formation of styrene dependent on fermentation management during wheat beer production. *Food Chem.* **2012**, *4*, 2121–2125.

15. Schwarz, K. J.; Boitz, L. I.; Methner, F.-J. Enzymatic formation of styrene during wheat beer fermentation is dependent on pitching rate and cinnamic acid content. *J. Inst. Brew.* **2012**, *3*, 280–284.

16. Schwarz, K. J.; Boltz, L. I.; Methner, F.-J. Influence of mashing conditions on the release of precursors of phenolic wheat beer aroma. *Brauwelt* **2013**, *153*, 274–279.

17. Daly, B.; Collins, E.; Madigan, D.; Donnelly, D.; Coakley, M.; Ross, P. An investigation into styrene in beer. *Proc. Congr.—Eur. Brew. Conv.* **1997**, *26*, 623–630.

18. Langos, D.; Granvogl, M.; Gastl, M.; Schieberle, P. Influence of malt modifications on the concentrations of free phenolic acids in wheat and barley malts. *Brew. Sci.* **2015**, *68*, 93–101.

19. Wackerbauer, K. Phenolic aromatic substances in beer: Phenolic carboxylic acid and phenols in the raw material and beer preparation. *Brauwelt* **1982**, *15*, 618–620.

20. Wackerbauer, K.; Krämer, P. Bavarian wheat beer - an alternative. Production and composition (in German). *Brauwelt* **1982**, *18*, 758–762.

21. Jezussek, M. M. *Zur Aromabildung beim Kochen von Naturreis (Oryza sativa L.) sowie Blättern von Pandanus amaryllifolius Roxb (in German)*. Ph.D. thesis, Technical University of Munich, Garching, Germany, 2002.

22. Wackerbauer, K.; Kossa, T.; Tressl, R. Phenol formation by yeasts (in German). *Monatsschr. Brauwiss.* **1978**, *31*, 52–55.

23. Vanbeneden, N.; Gils, F.; Delvaux, F.; Delvaux, F. R. Variability in the release of free and bound hydroxycinnamic acids from diverse malted barley (*Hordeum vulgare* L.) cultivars during wort production. *J. Agric. Food Chem.* **2007**, *26*, 11002–11010.

24. Cui, Y.; Wang, A.; Zhang, Z.; Speers, R. A. Enhancing the levels of 4-vinylguaiacol and 4-vinylphenol in pilot-scale top-fermented wheat beers by response surface methodology. *J. Inst. Brew.* **2015**, *1*, 129–136.

25. Coghe, S.; Benoot, K.; Delvaux, F.; Vanderhaegen, B.; Delvaux, F. R. Ferulic acid release and 4-vinylguaiacol formation during brewing and fermentation: indications for feruloyl esterase activity in *Saccharomyces cerevisiae*. *J. Agric. Food Chem.* **2004**, *3*, 602–608.

26. Narziß, L.; Miedaner, H.; Nitzsche, F. The formation of 4-vinylguaiacol during the manufacture of Bavarian wheat beer (in German). *Monatsschr. Brauwiss.* **1990**, *3*, 96–100.

27. Kieninger, H.; Narziß, L.; Hecht, S. Changes of quality determining compounds in the preparation of Bavarian wheat beer (in German). *Monatsschr. Brauwiss.* **1984**, *1*, 9–18.

28. Samaras, T. S.; Camburn, P. A.; Chandra, S. X.; Gordon, M. H.; Ames, J. M. Antioxidant properties of kilned and roasted malts. *J. Agric. Food Chem.* **2005**, *20*, 8068–8074.

# Chapter 9

# On the Importance of Phenol Derivatives for the Peaty Aroma Attribute of Scotch Whiskies from Islay

**V. Mall[*,1] and P. Schieberle[2]**

**[1]Leibniz-Institute for Food Systems Biology at the Technical University Munich, Lise-Meitner-Str. 34, 85354 Freising, Germany**
**[2]Chair for Food Chemistry, Technical University Munich, Lise-Meitner-Str. 34, 85354 Freising, Germany**
**[*]E-mail: v.mall.leibniz-lsb@tum.de.**

Whisky-making has a long tradition in the Scottish territories. Whiskies from the Isle of Islay are known to have a unique smoky and phenolic character, referred to as peatiness. The traditional malt kilning process with peat smoke brings the smoky aroma to the malt and, ultimately, to the whisky itself. After applying the concept of sensomics to two peaty whiskies, the phenol and methoxyphenol derivatives eliciting a smoky or phenolic aroma character were found to be abundant. In an attempt to correlate the concentrations of these smoky aroma compounds with the intensity of the peaty aroma, 10 whiskies from various Islay distilleries were ranked according to their peatiness, and their identified phenol derivatives were quantitated. A combination of sensory and quantitative data showed that although phenolic compounds were responsible for the peaty aroma of the whiskies, their concentrations and odor activity values were not well correlated with the perceived intensity of the smoky, phenolic aroma attribute.

## Introduction

Whisky-making has a long tradition in the Scottish territories. The process only requires three ingredients: water, barley malt, and yeast. After mixing the barley malt with yeast, a double-batch distillation yields the raw spirit, which is aged for at least three years in secondhand oak casks before bottling.

Whiskies from the Isle of Islay have a particular smoky and phenolic note, the so-called peaty aroma, and the intensity of the smoky, phenolic attribute is determined by the different whisky-makers in order to generate their characteristic taste. Islay whiskies that are especially peaty usually come from the southeastern coast of the island, while the distilleries on the East Coast produce less peaty spirits.

The production of Scotch whisky traditionally dries the green malt using peat-fired kilning. The original "fuel" peat is decayed plant material formed over thousands of years and can be found in wetland areas of the Scottish islands. Nowadays, a mixture of peat and anthracite is carefully burned without flaming to produce peat smoke, the so-called peat reek which is used for kilning.

The resulting peatiness of the malt is created by a set of phenolic compounds and may be up to 80 mg per kg of the total phenolic compounds (1, 2). Various derivatives including phenol, o-, m-, and p-cresol; 2-methoxyphenol; and a number of isomers of xylenol have been identified in the peated malt and their concentrations have been correlated with the degree of peating. Additionally, it has been shown that whiskies from Islay (where intense peaty malt is used) contain higher concentrations of phenol compared to whisky made from light peaty malt (1, 2).

Former studies identified phenolic compounds in Scotch whisky (3–5). When found in Scotch and Japanese whisky, 4-allyl-2-methoxyphenol and 4-ethyl-2-methoxyphenol were quantitatively the most important compounds (3). However, later studies (4, 5) included determinations of individual orthonasal odor thresholds in ethanol solutions of 10%–20%, which showed the importance of phenol, 2-, 3-, and 4-methylphenol to the phenolic notes of the whisky. However, the concentration of 4-allyl-2-methoxyphenol was lower than its threshold, and it was consequently considered to be less important to the overall aroma (5).

Using the Sensomics approach (6), the aroma of Bourbon whiskey was investigated and the key aroma compounds were clarified on a molecular basis by Poisson and Schieberle (7–9). When the concept of aroma extract dilution analysis (AEDA), stable isotope dilution assays (SIDAs), and the subsequent determination of odor activity values (OAVs), were applied to the aroma extract of a Bourbon whisky, 26 aroma compounds out of a total of 42 identified odorants were characterized as having an impact on the overall aroma of the whiskey. In particular, high OAVs were determined for ethanol, esters (such as [S]-ethyl 2-methylbutanoate, ethyl hexanoate, ethyl butanoate and ethyl octanoate), 3-methylbutanal, 4-hydroxy-3-methoxybenzaldehyde, (E)-β-damascenone, methylpropanal, and cis-whisky lactone.

When comparing these outcomes with the investigation of a Scotch single malt whisky with a particular peaty aroma, it was determined that the distinctive smoky and peaty aroma of the Scotch whisky could be traced back to the high concentrations and OAVs of several methoxyphenol derivatives including 2-methoxyphenol, 4-allyl-2-methoxyphenol, and 5-methyl-2-methoxyphenol (9).

While it was possible to correlate the concentrations of phenols in peated malt with the degree of peatiness, this correlation was not confirmed for the aroma of the final whisky. Because of this, 10 Scotch whiskies from the island of Islay were used in this study with the aim of finding a correlation between the concentrations of phenol and 2-methoxyphenol derivatives, their OAVs (the ratios of concentration to odor threshold), and the intensity of the respective peaty aroma.

## Materials and Methods

### Whisky Samples

Ten different whiskies originating from various distilleries on Islay, Scotland, were purchased from a local dealer. Samples were from Douglas Laing & Co (Big Peat; BP), Kilchoman (KI), Lagavulin (LV), Ardbeg (AR), Laphroaig (LP), Bowmore (BM), Caol Ila (CI), Bruichladdich (BL), Port Charlotte from Bruichladdich (PC), and Bunnahabhain (BH).

## Chemicals

Reference samples of the following odorants and isotopically labeled internal standards were purchased from commercial sources: phenol; 2-methylphenol; 3-methylphenol; 4-methylphenol; 2,3-dimethylphenol; 2,4-dimethylphenol; 2,5-dimethylphenol; 2,6-dimethylphenol; 3,4-dimethylphenol; 3,5-dimethylphenol; 2-ethylphenol; 3-ethylphenol; 4-ethylphenol; 2-methoxyphenol; 2-methoxy-4-methylphenol; 2-methoxy-5-methylphenol; 4-ethyl-2-methoxyphenol; 2-methoxy-4-propylphenol; 4-allyl-2-methoxyphenol; $[^2H_{6-8}]$-4-methylphenol; $[^{13}C_6]$-phenol (all obtained from Sigma-Aldrich, Taufkirchen, Germany); and $[^2H_4]$-4-ethylphenol (obtained from CDN Isotopes, Pointe-Claire, Canada). The following labeled compounds were synthesized according to the cited literature: $[^2H_3]$-2-methoxy-4-methylphenol and $[^2H_3]$-2-methoxy-5-methylphenol (9); $[^2H_2]$-2-methoxy-4-propylphenol (10); $[^2H_3]$-2-methoxy-phenol (11); and $[^2H_{2-5}]$-4-ethyl-2-methoxyphenol (12). $[^2H_{7-8}]$-2,6-Dimethylphenol was synthesized according to a routine procedure previously used by Czerny and Schieberle (based on unpublished results).

## AEDA and Identification Experiments

An aliquot of the spirit (25 mL) was mixed with an aqueous solution of sodium chloride (50 mL; 1 mol/L) before extraction with diethyl ether (3×50 mL). The combined organic layers were dried over anhydrous sodium sulfate, subjected to solvent-assisted flavor evaporation (SAFE) distillation (13), and then concentrated by means of a Vigreux column and a micro-distillation apparatus to approximately 0.5 mL. The concentrated distillate was subjected to AEDA and the flavor dilution (FD) factors for each single volatile was subsequently determined (6). The original aroma distillate was diluted stepwise using diethyl ether to obtain dilutions of 1:1, 1:2, 1:4, 1:8, 1:16 ... 1:4096 of the original extract. Each dilution was analyzed using high–resolution gas chromatrography-olfactometry (HRGC-O) (with injection volume 1 $\mu$L) until no odor was detectable during GC-O. The odorants were identifed following the previously described protocol (14). In particular, a comparison of retention indices on at least two columns with different stationary phases (DB-FFAP and DB-5), the odor quality, and the mass spectra of the analytes with data of reference compounds were crucial for a successful identification.

## Quantitation of Phenol Derivatives

After spiking a whisky sample with known amounts of the isotopically labeled internal standards, the spirit was subjected to SAFE distillation. The SAFE distillate was extracted using diethyl ether, washed with brine to remove ethanol from the organic layer, and concentrated by means of a Vigreux column followed by micro-distillation to about 0.2 mL. The concentrated distillate was then subjected to HRGC-ion trap MS (chemical ionization; MS-CI) using methanol as the reactant gas. For the quantitation of 2-methoxy-4-propylphenol and 4-ethyl-2-methoxyphenol, a two-dimensional heart cut HRGC-ion trap MS was needed to separate overlapping peaks. Quantitation was done by monitoring the mass traces for the phenol derivatives and their respective labeled standards as described previously (15). Table 1 gives an overview on the 17 quantified phenol derivatives and their respective isotopically labeled internal standards.

**Table 1. Selected Ions, Response Factors and Mass Traces Used in the Quantitation Experiments**

| Odorant | Isotopically Labeled Internal Standard | $R_f{}^a$ | m/z Analyte | m/z Standard |
|---|---|---|---|---|
| phenol | $[^{13}C_6]$-phenol | 0.9099 | 95 | 101 |
| 2-methylphenol | $[^2H_{6\text{-}8}]$-4-methylphenol | 0.7454 | 109 | 115–117 |
| 3-methylphenol | $[^2H_{6\text{-}8}]$-4-methylphenol | 0.9917 | 109 | 115–117 |
| 4-methylphenol | $[^2H_{6\text{-}8}]$-4-methylphenol | 1.0574 | 109 | 115–117 |
| 2,3-dimethylphenol | $[^2H_{7\text{-}8}]$-2,6-dimethylphenol | 0.8621 | 123 | 130+131 |
| 2,4-dimethylphenol | $[^2H_{7\text{-}8}]$-2,6-dimethylphenol | 0.6849 | 123 | 130+131 |
| 2,5-dimethylphenol | $[^2H_{7\text{-}8}]$-2,6-dimethylphenol | 0.5246 | 123 | 130+131 |
| 2,6-dimethylphenol | $[^2H_{7\text{-}8}]$-2,6-dimethylphenol | 0.7995 | 123 | 130+131 |
| 3,4-dimethylphenol | $[^2H_{7\text{-}8}]$-2,6-dimethylphenol | 0.5681 | 123 | 130+131 |
| 3,5-dimethylphenol | $[^2H_{7\text{-}8}]$-2,6-dimethylphenol | 0.6267 | 123 | 130+131 |
| 2-ethyl phenol | $[^{13}C_{3\text{-}4}]$-4-ethylphenol | 1.0478 | 123 | 126+127 |
| 3-ethylphenol | $[^{13}C_{3\text{-}4}]$-4-ethylphenol | 1.0072 | 123 | 126+127 |
| 4-ethylphenol | $[^{13}C_{3\text{-}4}]$-4-ethylphenol | 0.8998 | 123 | 126+127 |
| 2-methoxyphenol | $[^2H_3]$-2-methoxyphenol | 0.7814 | 125 | 128 |
| 2-methoxy-4-methylphenol | $[^2H_3]$-2-methoxy-4-methylphenol | 1.1274 | 139 | 142 |
| 2-methoxy-5-methylphenol | $[^2H_3]$-2-methoxy-5-methylphenol | 0.8767 | 139 | 142 |
| 4-ethyl-2-methoxyphenol | $[^2H_{2\text{-}5}]$-4-ethyl-2-methoxyphenol | 0.8681 | 153 | 155–158 |
| 2-methoxy-4-propylphenol | $[^2H_2]$-2-methoxy-4-propylphenol | 1.0364 | 167 | 169 |
| 4-allyl-2-methoxyphenol | $[^2H_2]$-2-methoxy-4-propylphenol | 0.8782 | 167 | 169 |

$^a$ $R_f$= response factor.

## Calculation of OAVs

OAVs were calculated based of their odor thresholds in 40% ethanol ABV (alcohol by volume). It was important to previously check the absence of other odor-active minor components by means of HRGC-O (*16*).

### Ranking the Peatiness of Islay Scotch Whiskies Using a Balanced Incomplete Block Design

To compare the different Islay Scotch whiskies based on the intensity of the "peaty" odor attribute, all whiskies were presented to the institute's trained sensory panel in a ranking test (17). To assure an equal assessment of all the whiskies, sets of five whiskies were presented according to an incomplete balanced block design (18). A sample of the very peaty whisky (Big Peat) was used as a reference for the distinct smoky and phenolic character of a peaty whisky. The significance of the differences in the ranking test was statistically calculated using the Friedman test.

## Results and Discussion

### Comparative AEDA of Two Peaty Single Malt Whiskies from the Isle of Islay Focusing on Phenol Derivatives

Two peaty single malt Scotch whiskies (Ardbeg and Bowmore) were analyzed using AEDA. A broad spectrum of phenol derivatives identified with high FD factors was found. While 10 phenol derivatives with a phenolic or smoky aroma character were found with FD factors ranging from 16 to 4096 in the AEDA of Bowore whisky, the heavily peated Ardbeg whisky had similar or higher FD factors for these compounds. Additionally, phenol, 2,5-dimethylphenol, 2,4-dimethylphenol, 2-methylphenol, and 3,4-dimethylphenol could only be detected in the AEDA of the more peaty Ardbeg whisky (Table 2). The highest FD factors were determined for 2-methoxyphenol (4096), 2-methoxy-5-methylphenol (1024–4096), and 3-ethylphenol (16–2048).

### Quantitation of Phenol Derivatives

As a next step, the odorants showing smoky and phenolic odors were quantitated by SIDA experiments in 10 whiskies from Islay, and their OAVs were calculated.

While 2,6-dimethylphenol, 3,4-dimethylphenol, 3,5-dimethylphenol, and phenol did not exceed their odor threshold in any whisky, major differences in the OAVs were measured for the other phenol derivatives (Table 3).

The highest OAVs, ranging from 18 (in BH) to 940 (in AR), were found in 3-ethylphenol, followed by 2-methoxyphenol (4.0–360), and 2-methoxy-5-methylphenol (16–380). While the quantitative data confirmed the importance of 2-methoxyphenol and 2-methoxy-5-methylphenol for peaty whiskies, 3-ethylphenol was previously not considered important (9).

In order to correlate the quantitative data with the sensory experiments, all OAVs determined for the phenol derivatives were added up. Following this approach, the Ardbeg whisky reached the highest value of 2077 followed by Kilchoman whisky (1593) and Big Peat whisky (1585). The lowest values were found for the Bowmore (490), the Bruchladdich (190), and the Bunnahabian whisky (56).

**Table 2. FD Factors of Phenol Derivatives in a Comparative AEDA of the Volatiles Isolated from Ardbeg (AR) and Bowmore Whisky (BM)**

| Odorant | Odor Quality[a] | RI[b] | | FD Factor[c] | |
|---|---|---|---|---|---|
| | | FFAP | DB5 | AR | BM |
| 2-methoxyphenol | smoky | 1865 | 1088 | 4096 | 4096 |
| 2-methoxy-5-methylphenol | phenolic, smoky | 1946 | 1184 | 4096 | 1024 |
| 3-ethylphenol | phenolic | 2190 | 1168 | 2048 | 16 |
| 2-methoxy-4-methylphenol | vanilla, smoky | 1963 | 1191 | 512 | 32 |
| 4-methylphenol | horse stable-like | 2083 | 1073 | 256 | 512 |
| 2-ethylphenol | phenolic | 2075 | 1140 | 256 | 64 |
| 2-methoxy-4-propylphenol | smoky, phenolic | 2110 | 1194 | 128 | 512 |
| phenol | phenolic | 2011 | 988 | 128 | <1 |
| 4-ethyl-2-methoxyphenol | vanilla-like, smoky | 2033 | 1277 | 64 | 128 |
| 4-allyl-2-methoxyphenol | smoky, phenolic | 2155 | 1360 | 64 | 64 |
| 3-methylphenol | phenolic | 2089 | 1073 | 64 | 32 |
| 2,5-dimethylphenol | vanilla-like, phenolic | 2078 | 1150 | 32 | <1 |
| 2,4-dimethylphenol | phenolic | 2079 | 1148 | 32 | <1 |
| 2-methylphenol | phenolic | 2008 | 1140 | 4 | <1 |
| 3,4-dimethylphenol | phenolic | 2229 | 1195 | 4 | <1 |

[a] Odor quality detected at the sniffing port at a dilution factor five times above the odor threshold of the reference compound. [b] RI = retention index; determined in comparison to a homologous series of n-alkanes. [c] FD factor = Flavor dilution factor; last dilution of an extract in which an odorant was still detectable.

## Table 3. OAVs of Phenol and Methoxyphenol Derivatives in Islay Whiskies

| Phenol/Methoxyphenol Derivative | Odor Threshold [µg/L][a] | OAV | | | | | | | | | |
|---|---|---|---|---|---|---|---|---|---|---|---|
| | | BP | KI | LV | AR | LP | BM | CI | BL | PC | BH |
| phenol | 15000[c] | <1 | <1 | <1 | <1 | <1 | <1 | <1 | <1 | <1 | <1 |
| 2-methylphenol | 90[b] | 58 | 52 | 33 | 46 | 39 | 10 | 22 | 2.6 | 37 | 1.0 |
| 3-methylphenol | 120[c] | 14 | 15 | 9.8 | 12 | 12 | 3.8 | 6.4 | 1.0 | 12 | <1 |
| 4-methylphenol | 30[b] | 110 | 130 | 81 | 97 | 100 | 28 | 57 | 5.8 | 84 | 2.3 |
| 2,3-dimethylphenol | 67[c] | 4.2 | 3.8 | 2.9 | 4 | 4 | 1.0 | 1.7 | <1 | 2.6 | <1 |
| 2,4-dimethylphenol | 320[c] | 5.8 | 5.7 | 4.2 | 6.2 | 5.3 | 1.3 | 2.8 | <1 | 4.1 | <1 |
| 2,5-dimethylphenol | 490[c] | 1.3 | 1.2 | 1.0 | 1.2 | 1.1 | <1 | 1.0 | <1 | 1.5 | <1 |
| 2,6-dimethylphenol | 510[c] | <1 | <1 | <1 | <1 | <1 | <1 | <1 | <1 | <1 | <1 |
| 3,4-dimethylphenol | 4400[c] | <1 | <1 | <1 | <1 | <1 | <1 | <1 | <1 | <1 | <1 |
| 3,5-dimethylphenol | 2500[c] | 1.0 | 1.0 | <1 | 1.0 | 1.0 | <1 | <1 | <1 | <1 | <1 |
| 2-ethylphenol | 83[c] | 3 | 3.4 | 6.4 | 10 | 7.6 | 2.0 | 1.2 | <1 | 2.7 | <1 |
| 3-ethylphenol | 0.57[c] | 550 | 640 | 380 | 940 | 490 | 200 | 240 | 66 | 420 | 18 |
| 4-ethylphenol | 170[b] | 12 | 14 | 7.5 | 16 | 11 | 3.7 | 5.6 | 1.3 | 7.8 | 1.0 |
| 2-methoxyphenol | 9.2[b] | 360 | 250 | 180 | 280 | 220 | 77 | 100 | 16 | 110 | 4.0 |
| 2-methoxy-4-methylphenol | 230[b] | 3.5 | 4.6 | 2.7 | 7.4 | 3.8 | 2.0 | 1.7 | <1 | 4 | <1 |
| 2-methoxy-5-methylphenol | 0.32[b] | 280 | 270 | 190 | 380 | 240 | 84 | 78 | 16 | 232 | 17 |
| 4-ethyl-2-methoxyphenol | 6.9[b] | 140 | 150 | 84 | 200 | 130 | 9.3 | 49 | 13 | 110 | 5.7 |
| 4-allyl-2-methoxyphenol | 7.1[b] | 9.8 | 13 | 6.1 | 20 | 16 | 4.4 | 8.6 | 3.8 | 14 | 5.4 |
| 2-methoxy-4-propylphenol | 1.9[b] | 31 | 41 | 22 | 52 | 34 | 10 | 12 | 3.0 | 47 | 2.0 |
| sum of OAVs | | 1584 | 1595 | 1011 | 2073 | 1315 | 437 | 587 | 129 | 1089 | 56 |

[a] Odor threshold concentration in 40% ethanol ABV.  [b] As previously described by Poisson (8).  [c] Newly determined in 40 % ethanol ABV following the previously described procedure (13).

113

### Results of the Ranking Test Using an Incomplete Balanced Block Design

On the basis of a sensory panel evaluation, Big Peat had the highest peatiness value with 43, followed by a group of five whiskies with similar rankings, namely Kilchoman, Lagavulin, Ardbeg, Laphroaig, and Bowmore whiskies. Bunnahabian whisky was determined to have the lowest vaule (Figure 1).

The statistical analysis of this data resulted in a least significant difference (LSD) of 12.4 in the ranking sum. Figure 1 shows the significant differentiable of the whisky samples in the ranking test with $\alpha=0.05$. Big Peat, the whisky with the highest ranking, was significantly differentiated from all whiskies except Kilchoman and Lagvulin. The group of whiskies with ranking sums between 30–33 (Kilchoman, Lagavulin, Ardbeg, and Laphroaig) was significantly differentiable from Port Charlotte and Bunnahabian whisky, while Bowmore with a ranking sum of 29 was only significantly differentiable from Bunnahabian whisky with the lowest ranking sum of 15 (Figure 1).

| W[a] | | BP | KI | LV | AR | LP | BM | CI | BL | PC | BH |
|------|----|----|----|----|----|----|----|----|----|----|----|
| R[b] | | 43 | 33 | 31 | 30 | 30 | 29 | 21 | 21 | 17 | 15 |
| BP | 43 | 0 | 10 | 12 | 13 | 13 | 14 | 22 | 22 | 26 | 28 |
| KI | 33 | 10 | 0 | 2 | 3 | 3 | 4 | 12 | 12 | 16 | 18 |
| LV | 31 | 12 | 2 | 0 | 1 | 1 | 2 | 10 | 10 | 14 | 16 |
| AR | 30 | 13 | 3 | 1 | 0 | 0 | 1 | 9 | 9 | 13 | 15 |
| LP | 30 | 13 | 3 | 1 | 0 | 0 | 1 | 9 | 9 | 13 | 15 |
| BM | 29 | 14 | 4 | 2 | 1 | 1 | 0 | 8 | 8 | 12 | 14 |
| CI | 21 | 22 | 12 | 10 | 9 | 9 | 8 | 0 | 0 | 4 | 6 |
| BL | 21 | 22 | 12 | 10 | 9 | 9 | 8 | 0 | 0 | 4 | 6 |
| PC | 17 | 26 | 16 | 14 | 13 | 13 | 12 | 4 | 4 | 0 | 2 |
| BH | 15 | 28 | 18 | 16 | 15 | 15 | 14 | 6 | 6 | 2 | 0 |

[a] W= whisky; [b] R = ranking sum.

*Figure 1. Determined ranking sums and differences in ranking sums with $\alpha=0.05$ and LSD=12.4.*

### Correlation of Quantitative and Sensory Analyses

Tthe sum of the OAVs were subsequently compared to the ranking sums (Figure 2). However, the whisky with the highest phenol content from the Ardbeg distillery was not distinguishable from the Bowmore whisky with medium-high phenol concentrations and OAVs. Additionally, the Port Charlotte whisky with a sum of OAVs of 1092 was rated as second-to-last place in terms of "peatiness," while a whisky with a similar OAV sum of 1008 from the Lagavulin distillery was rated among the whiskies with the most intense peaty odor. Moreover, Big Peat was ranked as the most intense peaty whisky, despite its OAV sum not being the highest within the group of Islay whiskies.

These calculations show that it is not simply the total concentrations of phenol derivatives that determine the intensity of the peaty aroma impression. While the Ardbeg whisky contained higher amounts of phenols and methoxyphenols than the Bowmore whisky, it also contained higher amounts of esters and other important aroma compounds like 4-hydroxy-3-methoxybenzaldehyde, *cis*-whisky lactone, or (*E*)-*β*-damascenone (data not provided). These other impact aroma compounds seem to modify the peaty aroma impression (i.e., may downregulate the intensity of the peatiness). Following this concept, the whisky from Lagavulin may contain fewer amounts of fruity esters or other aroma compounds than the Port Charlotte whisky, and Big Peat may contain fewer esters or other aroma compounds than the other whiskies. In order to verify this proposal and to understand the chemistry of peatiness of Scotch whiskies, the esters and other key aroma compounds should be quantified and considered for correlation to the peatiness of whiskies in additional studies.

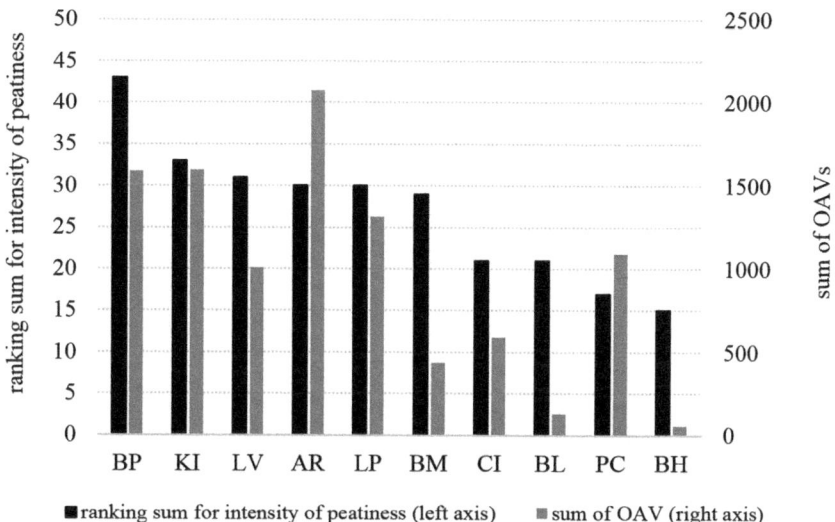

*Figure 2. Contrast of ranking sums for intensity of peatiness (black) and sum of OAVs of phenol derivatives (gray) of Islay whiskies.*

## Conclusions

The application of the sensomics approach to two peaty single malt whiskies from Islay confirmed the general importance of phenol derivatives, particularly for the phenolic and smoky aroma of these whiskies. As confirmed by quantitation experiments on 10 Islay whiskies, high OAVs for phenol derivatives were calculated. The highest OAVs were found for 3-ethylphenol and 2-methoxy-5-methylphenol. However, an approach to correlate the sum of OAVs of phenol derivatives to the overall intensity of the peaty aroma note did not show a direct correlation. Thus, it can be proposed that further key aroma compounds of such whiskies (i.e., esters or 4-hydroxy-3-methoxybenzaldehyde) may play an important role in modifying the overall aroma mitigating the peaty impression.

## References

1.  Bathgate, G.N.; Cook, R. Malting of Barley for Scotch Whisky. In *The Science and Technologies of Whiskies*; Piggott, J.R., Sharp, R., Duncan, R. E. B., Eds.; Longman Scientific & Technical: Harlow, 1989.

2.  Bathgate, G. N.; Taylor, A. G. The Qualitative and Quantitative Measurement of Peat Smoke on Distiller's Malt. *J. Inst. Brew.* **1997**, *83*, 163–168.

3.  Nishimura, K.; Masuda, M. Minor Constituents of Whisky Fusel Oils. I. Basic, Phenolic and Lactonic Compounds. *J. Food Sci.* **1971**, *36*, 819–822.

4. Swan, J. S.; Burtles, S. M. The Development of Flavor in Potable Spirits. *Chem. Soc. Rev.* **1978**, 7, 201–223.

5. Jounela-Eriksson, P.; Lehtonen, M. Phenols in the Aroma of Distilled Beverages. In *Quality of Food and Beverages: Chemistry and Technology*, 2nd; Chamralambous, G.; Proceedings of the Symposium of the International Flavor Conference, Athens, Greece, July 20–24, 1981; Ed.; Elvesier: Amsterdam, 1981, pp 167–181.

6. Schieberle, P.; Hofmann, T. Mapping the Combinatorial Code of Food Flavors by Means of Molecular Sensory Science Approach. In *Food Flavors–Chemical, Sensory and Technological Properties*; Jelen, H., Ed.; CRC Press: Boca Raton, FL, 2011; pp 411–437.

7. Poisson, L.; Schieberle, P. Characterization of the Most Odor-Active Compounds in an American Bourbon Whisky by Application of the Aroma Extract Dilution Analysis. *J. Agric. Food Chem.* **2008**, 56, 5813–5819.

8. Poisson, L.; Schieberle, P. Characterization of the Key Aroma Compounds in an American Bourbon Whisky by Quantitative Measurements, Aroma Recombination and Omission Studies. *J. Agric. Food Chem.* **2008**, 56, 5820–5826.

9. Poisson, L. *Charactization of Key Aroma Compounds in American and Scotch Single Malt Whisky (in German)*. Ph.D. Thesis, Technical University of Munich, 2003.

10. Schmitt, R. *On the Role of Ingredients as Sources of Key Aroma Compounds in Crumb Chocolate (in German)*. Ph.D. Thesis, Technical University of Munich, 2005.

11. Guth, H.; Grosch, W. Odorants of Extrusion Products of Oat Meal–Changes During Storage. *Z. Lebensm Unters Forsch.* **1993**, 196, 22–28.

12. Semmelroch, P.; Laskawy, G.; Blank, I.; Grosch, W. Determination of Potent Odorants in Roasted Coffee by Stable Isotope Dilution Analysis. *Flavor Fragr. J.* **1995**, 10, 1–7.

13. Engel, W.; Bahr, W.; Schieberle, P. Solvent Assisted Flavor Evaporation–A New and Versatile Technique for the Careful and Direct Isolation of Aroma Compounds from Complex Food Matrices. *Eur. Food Res. Technol.* **1999**, 209, 237–241.

14. Molyneux, R.J.; Schieberle, P. Compound Identification: A Journal of Agricultural and Food Chemistry Perspective. *J. Agric. Food Chem.* **2007**, 55, 4625–4629.

15. Mall, V.; Schieberle, P. Evaluation of Key Aroma Compounds in Processed Prawns (Whiteleg Shrimp) by Quantitation and Aroma Recombination Experiments. *J. Agric. Food Chem.* **2017**, 65, 2776–2783.

16. Czerny, M.; Christlbauer, M.; Christlbauer, M.; Fischer, A.; Granvogl, M.; Hammer, M.; Hartl, C.; Hernandez, N.; Schieberle, P. Re-Investigation on dour Thresholds of Key Food Aroma Compounds and Development of an Aroma Language Based on Odour Qualities of Defined Aqueous Odorant Solutions. *Eur. Food Res. Technol.* **2008**, 228, 265–273.

17. *ISO 8587:2006: Sensory Analysis – Methodology – Ranking*, https://www.iso.org/standard/36172.html (accessed Mar 17, 2019).

18. *ISO 29842:2011: Sensory Analysis – Methodology – Balanced Incomplete Block Designs*. https://www.iso.org/standard/45702.html (accessed Mar 17, 2019).

# Chapter 10

# Rapid Quantitation of Phenolic Compounds in Islay Single Malt Scotch Whiskies by Direct Injection Mass Spectrometry

Jonathan Beauchamp,[*] Sonja Biberacher, and Shang Gao

**Department of Sensory Analytics,
Fraunhofer Institute of Process Engineering and Packaging IVV,
Giggenhauser Str. 35, 85354 Freising, Germany**

[*]E-mail: jonathan.beauchamp@ivv.fraunhofer.de.

Scotch whiskies from the Isle of Islay are renowned for their peaty and smoky aroma characteristics, which are imparted by phenolic compounds that are transferred from peat smoke to the malt during the kilning process. Conventional analysis of aroma compounds in beverages such as whisky is carried out by gas chromatography coupled to mass spectrometry (GC-MS), which delivers a comprehensive qualitative and quantitative picture of constituent compounds. Yet, GC-MS is time-consuming when quantifying aqueous-phase concentrations from liquid matrices. An alternative analytical approach to detecting constituent aroma compounds in (alcoholic) drinks based on direct injection mass spectrometry (DIMS) is presented here in a proof-of-concept study. Specifically, proton transfer reaction time-of-flight mass spectrometry (PTR-TOFMS) was used in combination with a liquid calibration unit (LCU) to quantify phenolic compounds in a selection of Islay whiskies in comparison to non-peaty whiskies. The LCU was used in a unconventional manner: Sample aliquots of aqueous whisky solutions were vaporized by the LCU to allow for direct analysis of their constituent volatile aroma compounds in the gas-phase by PTR-TOFMS with subsequent conversion to aqueous-phase concentrations in the sample matrix. This chapter presents this concept as a novel approach for the targeted analysis of specific volatile constituents of alcoholic beverages using a feasibility study focusing on the quantitation of phenol, methylphenols (cresols), and 2-methoxyphenol (guaiacol) in selected whiskies. This technique holds promise for the rapid liquid-phase quantitation of volatile aroma compounds via their direct detection in the gas-phase with minimal sample workup and high sample throughput yields.

# Introduction

Whiskies from the Scottish island of Islay are renowned—and appreciated by whisky connoisseurs—for their peaty and smoky characteristics. This distinguishing flavor is attributed to phenolic compounds that are present in the whisky. Historically, distilleries on the Isle of Islay used peat as a readily-available fuel source for firing the kiln to dry the malt prior to mashing, fermentation, distillation, and maturation. The combustion of this rich organic material generates a complex mixture of compounds including phenolic compounds like phenol, 4-ethylphenol, methylphenols (cresols), 2-methoxyphenol (guaiacol), and 4-ethyl-2-methoxyphenol (4-ethylguaiacol). These compounds impart the whisky with a peaty character (1). Contemporary distilleries have switched to using oil or gas for the kiln but implement a secondary peat furnace to generate the peaty flavor that is characteristic of Islay whiskies (2).

Much research on the aroma of whisky—including Islay whiskies—has been conducted over the years. In addition to sensory descriptive analysis, instrumental analyses have offered insights into the chemical compounds that are responsible for their characteristic flavor attributes. The latter typically utilizes gas chromatography (GC), either with olfactometric detection (GC–O), mass spectrometric detection (GC-MS), or a combination of both (GC-MS/O) (3). Such analyses can be enhanced by using aroma extract dilution analysis (AEDA) to establish the relative contributions of individual odor-active compounds to the overall aroma of a sample (4). For example, this comprehensive approach has been used to identify key aroma compounds in an American Bourbon whiskey (Kentucky Straight Bourbon) (5). Taking such analyses one step further, the elucidation of the key aroma compounds made in this manner can be verified using a sensomics approach, whereby the identified compounds are mixed in a water-ethanolic solution at specific ratios in relation to their concentration in the sample and their odor activity values (OAVs) (4) in such a way to generate a recombinant that exhibits an aroma that is indistinguishable from the original sample; this has been successfully achieved for the aforementioned whiskey (6), as well as for other distilled spirits such as brandy (7) and rum (8).

GC-MS/O analysis relies on the successful extraction of (odor-active) volatiles from the sample under investigation. The previously described AEDA approach requires liquid (solvent) extraction and enrichment, with subsequent dilution steps, whereby extracts are subsequently injected onto the GC column. Alternatively, aroma compounds present in beverages may be captured onto an adsorbent material, as performed in solid-phase microextraction (SPME) (9, 10) or stir-bar sorptive extraction (SBSE) (11, 12), and are then transferred to the GC column by thermal desorption. SPME is primarily used to extract gas-phase aroma compounds present in the headspace of a sample, whereas SBSE is mostly used for extraction in the liquid-phase, although gas- and liquid-phase sampling is possible using both methods.

A major challenge in analyzing the aroma constituents of beverage samples is their quantitation within the original matrix. Methods based on headspace extraction, such as SPME, allow for their quantitation in the gas phase above the drink. However, a conversion to the liquid-phase concentration in the matrix is challenging as the headspace concentration depends on liquid-to-gas-phase partitioning that is driven by many factors ranging from temperature and pH to matrix composition to the physicochemical properties of the individual compounds of interest (e.g., polarity and volatility). Previous studies have demonstrated that liquid-phase quantitation via gas-phase SPME headspace sampling is possible, but this requires prior determination of calibration curves for each target compound using (isotopically labeled) reference standards (13, 14), and, critically, the conditions during calibration must be identical to those of the sample during analysis due to the aforementioned parameters that drive aroma release. In contrast, liquid extraction techniques, such as SBSE, have the advantage of sampling directly in the liquid phase, thereby providing quantitative data in direct relation to the matrix.

In the present work, a new approach using direct injection mass spectrometry (DIMS) is presented as an alternative method to characterize aroma compounds in liquid matrices. DIMS techniques are generally based on chemical ionization mass spectrometry (CIMS) that allow continuous and direct ionization and, thereby, rapid detection and quantitation of volatile organic compound (VOC) constituents in gas samples (15). DIMS methods include atmospheric pressure chemical ionization (APCI-MS) (16), selected ion flow tube (SIFT-MS) (17), and proton-transfer-reaction mass spectrometry (PTR-MS) (15). In this work, PTR-MS, equipped with a time-of-flight mass spectrometer (PTR-TOF-MS) (18), was used in conjunction with a liquid calibration unit (LCU) (19), to analyze a series of Islay whiskies and two non-Islay whiskies for comparison, to quantify selected phenolic compounds. Diluted aqueous solutions of the individual whisky samples were used in combination with the LCU to directly vaporize the constituent phenolic compounds into the gas phase for immediate detection and quantitation by PTR-MS with subsequent calculation of the original liquid-phase concentration based on flow rates established by the LCU, the properties of the target compounds, and their dilution in solution.

## Experimental Procedures

### Whisky and Reference Samples

Four single malt Scotch whiskies from the Isle of Islay were analyzed against a non-Islay single malt Scotch whisky and a Japanese coffee grain whisky for comparison: Details of the samples are presented in Table 1. Phenol (99% purity; Fluka, Steinheim, Germany), p-cresol, and guaiacol (both 99% purity; Alrich, Steinheim, Germany) were used as references to calibrate the PTR-MS.

Whisky samples were diluted in distilled water to establish 5% v/v stock solutions; this was necessary to reduce the amount of ethanol and thereby avoid a depletion of the PTR-MS primary ion signal, which is a well-known phenomenon when analyzing alcoholic beverages using PTR-MS (20).

### Table 1. Overview of Whisky Samples

| Sample code | Description | ABV[a] [%] | Origin |
|---|---|---|---|
| LAPH QC | Laphroaig Quarter Cask | 48 | Islay |
| LAPH PX | Laphroaig PX Cask | 48 | Islay |
| BUNN | Bunnahabhain, 8 years | 43 | Islay |
| LAG | Lagavulin, 16 years | 43 | Islay |
| ARRAN | The Arran Single Malt, 10 years | 46 | Highlands |
| NIKKA | Nikka Coffee Grain Whisky | 45 | Japan |

[a] ABV = alcohol by volume.

### Instrumentation

A PTR-MS equipped with a time-of-flight (TOF) mass spectrometer, specifically a PTR-TOF 8000 (IONICON Analytik GmbH, Innsbruck, Austria) was used for analyzing the whisky samples.

The instrument was operated with the following drift tube parameters: voltage of 600 V (extraction voltage, 35 V); temperature of 60 °C; and pressure of 2.2 mbar. This established an electric field strength (E) to buffer gas number density (N) ratio (E/N) of 138 Td. The 1/16" OD, 0.04" ID PEEK sampling (inlet) line was held at 80 °C and sampled gas at a flow rate of ~28 mL/min (mimimum flow achievable).

An advanced model LCU, a LCU-a (IONICON Analytik, Innsbruck, Austria) was used in the present experiments for two purposes: namely to calibrate the PTR-MS for phenol, p-cresol, and guaiacol; and to analyze the whisky samples by direct injection. In brief, the LCU pumps a liquid solution into a nebulizer that is housed in an oven (held at 100 °C) and flushed with nitrogen (or zero-air) at high pressure (ca. 3 bar) and a flow rate of 1000 mL/min. The nebulizer vaporizes the solution and its volatile contituents with the ensuing water vapor and volatiles dispersed within the evaporation chamber and carried via the carrier gas toward the PTR-MS sampling line. The quantity of the sample entering the chamber can be adjusted by varying the pumping rate. A second micro pump is used to flush water into the evaporation chamber in order to compensate for variations in sample gas humidity that would otherwise occur when the flow rate of the sample solution is varied. This is achieved by maintaining a constant total flow rate between both micro pumps (50 μL/min in the present tests). Knowledge of the flow rates, as well as the properties of the compound(s) under investigation (such as molar mass and density) allows for direct calculation of the gas-phase concentrations that are generated.

The calibrations of phenol, p-cresol, and guaiacol focused on their protonated parent ions (MH$^+$), namely $m/z$ 95.051, $m/z$ 109.065, and $m/z$ 125.060, respectively. A calibration plot of the signal intensities versus the gas-phase concentrations yielded sensitivity factors for these compounds of 185, 29, and 43 cps/ppb, respectively. These sensitivities were subsequently used to calculate the gas-phase concentrations of these compounds in the whisky samples. It should be noted that since potential interferences on those $m/z$ from other compounds present in the whisky samples cannot be ruled out, these reported values are upper limits. Further, the PTR-TOFMS instrument was calibrated for the isomer p-cresol; data reported on cresol should be taken as the sum of the cresol isomers since these cannot be separately analyzed in PTR-TOFMS.

**Whisky Analysis**

Whisky samples were analyzed in a similar manner to the reference compound solutions. The flow rate of the sample into the nebulizer was varied to establish different gas-phase concentrations in the evaporation chamber of the LCU. Despite diluting the whiskies with water, ethanol nevertheless remained present at high concentrations, leading to a depletion of the H$_3$O$^+$ primary ion signal when the sample pumping rate was 50 μL/min. To counteract this problem, the pumping rate of the sample was reduced (but increased for the water pumping rate) until no depletion of the primary ion signal was observed; this was achieved at a sample pumping rate of 10 μL/min and a corresponding water pumping rate of 40 μL/min. It should be noted that the minimal differences in ethanol content (% ABV; see Table 1) between the samples had only a negligible affect on the primary ion signal; thus, it was deemed unnecessary to correct for these in the present study.

Signal intensities of phenol at $m/z$ 95.051, cresol at $m/z$ 109.065, and guaiacol at $m/z$ 125.060 during sample analysis were converted to gas-phase concentrations using the sensitivities for the previously reported compounds. The gas-phase concentrations were then converted to liquid-phase concentrations using the LCU conversion software tool (IONICON Analytik, Innsbruck, Austria) and values for the molar mass (94.11, 108.13, and 124.14 g/mol) and density (1.07, 1.035, 1.11 g/

mL) of phenol, *p*-cresol, and guaiacol, respectively. Finally, the values were adjusted according to the dilution factor of the stock solution to return aqueous-phase concentrations in the samples in units of ng/μL.

## Results and Discussion

The presence of phenol, cresol, and guaiacol was detected in all of the Islay whisky samples but not in either of the two non-Islay samples that were analyzed for comparison (signals were below the limits of detection for those compounds). The liquid-phase concentrations of the three compounds for the whiskies are reported in Table 2. For phenol, these concentrations had a range of 4.9–9.4 ng/μL. These values are comparable to literature reports of phenol quantitation using the SBSE method (range of 3.7–4.8 ng/μL for two unspecified Scotch whiskies) (*12*); because of this, the method presented here delivers plausible values within the range of those concentrations determined via established methods using GC-MS.

A comparison of the samples shows that the highest phenol content was present in the LAPH PX sample, followed by LAG, LAPH QC, and finally BUNN. By comparison, cresol and guaiacol were both highest in LAPH QC, but with LAPH PX exhibiting the second highest concentration of guaiacol, whereas LAG contained the next highest level of cresol, albeit at only slightly higher concentrations compared to LAPH PX. Overall, the concentrations of cresol were similar in all of the samples; by comparison, guaiacol varied by a factor of three across the four samples (Table 2).

**Table 2. Concentrations of Phenol, Cresol, and Guaiacol in Whisky Samples**

| | Concentration in liquid | | | | | |
| --- | --- | --- | --- | --- | --- | --- |
| Sample code[a] | ng/μL | | | Relative to maximum | | |
| | Phenol | Cresol[b] | Guaiacol | Phenol | Cresol[b] | Guaiacol |
| LAPH PX | 9.4 | 47.3 | 17.2 | 1.00 | 0.98 | 0.88 |
| LAG | 7.7 | 47.6 | 6.6 | 0.81 | 0.99 | 0.34 |
| LAPH QC | 5.7 | 48.3 | 19.5 | 0.60 | 1.00 | 1.00 |
| BUNN | 4.9 | 43.9 | 12.1 | 0.52 | 0.91 | 0.62 |

[a] Sample codes: LAPH PX, Laphroaig PX Cask; LAG, Lagavulin; LAPH QC, Laphroaig Quarter Cask; BUNN, Bunnahabhain.   [b] The isomers of cresol (*p*-, *m*-, and *o*-cresol) are not distinguishable by PTR-TOFMS; values therefore represent the sum of all isomers.

Based on anecdotal evidence, it is interesting to note that LAG exhibited a noticeably lower intensity of the smoky, peaty aroma characteristic than the LAPH and BUNN whiskies, which could reflect the guaiacol data presented in this chapter. Although no systematic sensory analysis was performed for the present study, smoky, peaty notes are not characteristic of the ARRAN and NIKKA whiskies; this is also reflected in the analytical data, whereby none of these three compounds was detected above the instrumental detection limits in either of these non-peaty whiskies.

In general, the PTR-TOFMS mass spectra of all whisky samples were rich in analytical features with signals at most *m/z* within the detection range for the present analyses, often at high intensity and in many cases with double or triple peaks, indicating the presence of isobaric compounds. The mass spectrum of LAPH QC is plotted in Figure 1 as the entire mass spectrum and two zoomed-in features. In particular, it is apparent that the signal at $m/z$ 109, at which the cresols are detected ($m/z$ 109.065), has several peaks.

When analyzing alcoholic beverages, especially those with a high ethanolic content such as spirits, the presence of ethanol can interfere with the analysis by sequestering the hydronium primary ions, thereby reducing the upper limit of the linear detection range of PTR-MS (typically on the order of low ppm concentrations). This "problem" can be overcome by utilizing protonated ethanol as the primary reagent ion for subsequent reactions with volatile constituents in the sample. Although this makes the ensuing mass spectra more complex and challenging to interpret compared to protonation via hydronium, by operating the PTR-MS under modified conditions, samples can be analyzed undiluted, which has been successfully demonstrated for mass spectral fingerprinting (21) and in vivo analysis (22) of wines.

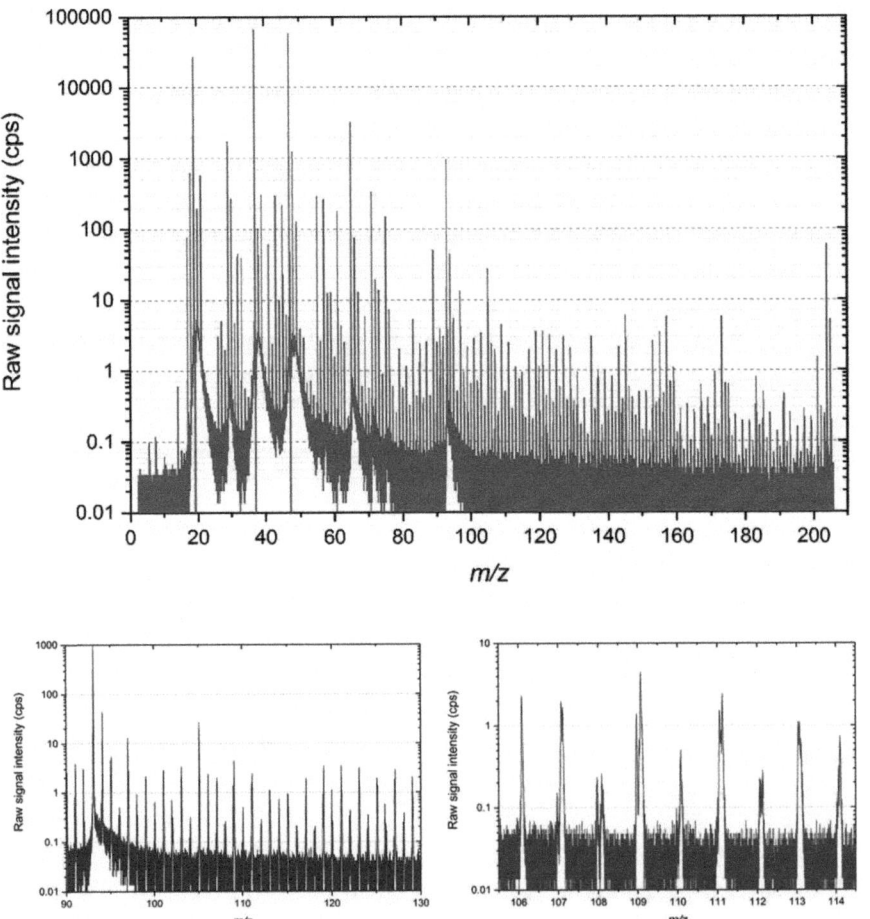

Figure 1. Mass spectra of LAPH QC for the entire m/z range (top, center), at closer ranges of m/z 90–130 (bottom, left), and m/z 106–114 (bottom, right).

As an alternative to exploiting the high production of protonated ethanol, this "interference" phenomenon can be circumvented by using a fastGC interface that separates the volatiles of a sample in time prior to their detection by PTR-MS (23, 24). Ethanol present in a sample elutes from the column at an early stage of the GC temperature program and is therefore no longer present in the PTR-MS reaction chamber when the targeted compounds elute at a later time. This means that the sample can be analyzed at lower dilutions (or undiluted), which increases the detection limit of the system. Additionally, the use of a fastGC makes it possible to separate and identify isomeric compounds in some cases. The fastGC was not employed in the present study but is mentioned here in the context of improving this analytical approach for future studies featuring complex samples containing ethanol.

## Conclusions

The novel use of a liquid calibration unit to volatilize constituent aroma compounds from a liquid matrix—whiskies, in the present case—and detect these in the gas phase using PTR-MS offers an additional analytical approach to aqueous-phase quantitation that obviates the need for sample workup and extract enrichment. In particular, once the system has been calibrated, samples can be analyzed at high throughput. The method presented in this chapter is of particular use for targeted approaches where the compounds of interest are known—and known to be present in the matrix—*a priori*. In the case of complex matrices, as is the case for whisky (and other distilled spirits), complementary analysis using comprehensive GC-MS for elucidating the identities of constituent compounds is beneficial.

## References

1. Lee, K. Y. M.; Paterson, A.; Piggott, J. R.; Richardson, G. D. Origins of Flavour in Whiskies and a Revised Flavour Wheel: A Review. *J. Inst. Brew.* **2001**, *107*, 287–313.
2. Bathgate, G. N.; Taylor, A. G. The Qualitative and Quantitative Measurement of Peat Smoke on Distiller's Malt. *J. Inst. Brew.* **1977**, *83*, 163–168.
3. Chin, S. T.; Eyres, G. T.; Marriott, P. J. Gas Chromatography-Mass Spectrometry in Odorant Analysis. In *Springer Handbook of Odor*; Buettner, A., Ed.; Springer International Publishing: Cham, Switzerland, 2017; pp 47–48.
4. Grosch, W. Determination of Potent Odourants in Foods by Aroma Extract Dilution Analysis (AEDA) and Calculation of Odour Activity Values (OAVs). *Flavour Frag. J.* **1994**, *9*, 147–158.
5. Poisson, L.; Schieberle, P. Characterization of the Most Odor-Active Compounds in an American Bourbon Whisky by Application of the Aroma Extract Dilution Analysis. *J. Agric. Food Chem.* **2008**, *56*, 5813–5819.
6. Poisson, L.; Schieberle, P. Characterization of the Key Aroma Compounds in an American Bourbon Whisky by Quantitative Measurements, Aroma Recombination, and Omission Studies. *J. Agric. Food Chem.* **2008**, *56*, 5820–5826.
7. Willner, B.; Granvogl, M.; Schieberle, P. Characterization of the Key Aroma Compounds in Bartlett Pear Brandies by Means of the Sensomics Concept. *J. Agric. Food Chem.* **2013**, *61*, 9583–9593.
8. Franitza, L.; Granvogl, M.; Schieberle, P. Characterization of the Key Aroma Compounds in Two Commercial Rums by Means of the Sensomics Approach. *J. Agric. Food Chem.* **2016**, *64*, 637–645.
9. Pawliszyn, J. Theory of Solid-Phase Microextraction. *J. Chromatogr. Sci.* **2000**, *38*, 270–278.
10. Fitzgerald, G.; James, K. J.; MacNamara, K.; Stack, M. A. Characterisation of Whiskeys Using Solid-Phase Microextraction with Gas Chromatography–Mass Spectrometry. *J. Chromatogr. A* **2000**, *896*, 351–359.

11. Baltussen, E.; Sandra, P.; David, F.; Cramers, C. Stir Bar Sorptive Extraction (SBSE), a Novel Extraction Technique for Aqueous Samples: Theory and Principles. *J. Microcolumn Sep.* **1999**, *11*, 737–747.

12. Nie, Y.; Kleine-Benne, E. Determining phenolic compounds in whisky using direct large volume injection and stir bar sorptive extraction. *AppNote* [Online], 2012. http://www.gerstel.co.uk/pdf/p-gc-an-2012-02.pdf (accessed Apr 28, 2019).

13. Siebert, T. E.; Smyth, H. E.; Capone, D. L.; Neuwöhner, C.; Pardon, K. H.; Skouroumounis, G. K.; Herderich, M. J.; Sefton, M. A.; Pollnitz, A. P. Stable Isotope Dilution Analysis of Wine Fermentation Products by HS-SPME-GC-MS. *Anal. Bioanal. Chem.* **2005**, *381*, 937–947.

14. Kang, W.; Xu, Y.; Qin, L.; Wang, Y. Effects of Different β-D-Glycosidases on Bound Aroma Compounds in Muscat Grape Determined by HS-SPME and GC-MS. *J. Inst. Brew.* **2010**, *116*, 70–77.

15. Beauchamp, J.; Zardin, E. In *Springer Handbook of Odor*; Buettner, A., Ed.; Springer Nature: Cham, Switzerland, 2017; pp 355-408.

16. Taylor, A. J.; Linforth, R. S. T.; Harvey, B. A.; Blake, A. Atmospheric Pressure Chemical Ionisation Mass Spectrometry for in Vivo Analysis of Volatile Flavour Release. *Food Chem.* **2000**, *71*, 327–338.

17. Smith, D.; Španěl, P. The novel Selected-Ion Flow Tube Approach to Trace Gas Analysis of Air and Breath. *Rapid Commun. Mass Spectrom.* **1996**, *10*, 1183–1198.

18. Zardin, E.; Tyapkova, O.; Buettner, A.; Beauchamp, J. Performance Assessment of Proton-Transfer-Reaction Time-of-Flight Mass Spectrometry (PTR-TOF-MS) for Analysis of Isobaric Compounds in Food-Flavour Applications. *LWT-Food Sci. Technol.* **2014**, *56*, 153–160.

19. Fischer, L.; Klinger, A.; Herbig, J.; Winkler, K.; Gutmann, R.; Hansel, A. In *The LCU: Versatile Trace Gas Calibration*, Proceedings of the 6th International Conference on Proton Transfer Reaction Mass Spectrometry and its Applications, Obergurgl, Austria; Hansel, A., Dunkl, J., Eds.; Innsbruck University Press: Obergurgl, Austria, 2013; pp 192–195.

20. Fiches, G.; Déléris, I.; Saint-Eve, A.; Pollet, B.; Brunerie, P.; Souchon, I. Modifying PTR-MS Operating Conditions for Quantitative Headspace Analysis of Hydro-Alcoholic Beverages. 1. Variation of the Mean Collision Energy to Control Ionization Processes Occurring During PTR-MS Analyses of 10–40% (v/v) Ethanol–Water Solutions. *Int. J. Mass Spectrom.* **2013**, *356*, 41–45.

21. Boscaini, E.; Mikoviny, T.; Wisthaler, A.; Hartungen, E. v.; Märk, T. D. Characterization of wine with ptr-ms. *Int. J. Mass Spectrom.* **2004**, *239*, 215–219.

22. Sémon, E.; Arvisenet, G.; Guichard, E.; Le Quéré, J. L. Modified Proton Transfer Reaction Mass Spectrometry (PTR-MS) Operating Conditions for in Vitro and in Vivo Analysis of Wine Aroma. *J. Mass Spectrom.* **2018**, *53*, 65–77.

23. Romano, A.; Fischer, L.; Herbig, J.; Campbell-Sills, H.; Coulon, J.; Lucas, P.; Cappellin, L.; Biasioli, F. Wine Analysis by FastGC Proton-Transfer Reaction-Time-of-Flight-Mass Spectrometry. *Int. J. Mass Spectrom.* **2014**, *369*, 81–86.

24. Beauchamp, J.; Herbig, J. In *The Chemical Sensory Informatics Of Food: Measurement, Analysis, Integration*; Guthrie, B., Beauchamp, J., Buettner, A., Lavine, B. K., Eds.; American Chemical Society: Washington, DC, 2015; Vol. 1191, pp 235–251.

# Chapter 11

# Overview of Distilled Spirits

Michael C. Qian,*,1 Paul Hughes,1 and Keith Cadwallader2

1Department of Food Science and Technology, Oregon State University, Corvallis, Oregon 97330, United States

2Department of Food Science and Human Nutrition, University of Illinois-Urbana-Champaigne, Urbana, Illinois, United States

*E-mail: michael.qian@oregonstate.edu.

A spirit is a distilled beverage with alcohol content greater than 30% (v/v). Depending on the type of spirit, the raw fermentation materials can be grains, fruits, vegetables, or other carbohydrate-rich ingredients. Following the fermentation process, the alcohols and other volatiles are distilled to concentrate the alcohol. The distillates are typically aged and blended before consumption. Depending on the raw materials used for fermentation, distillation, and the final aging process, the volatile profiles can vary tremendously. This chapter will provide an overview of the production and flavor characteristics of different spirits.

## Distilled Spirits Based on Neutral Alcohol

While the properties of many local and global spirit categories reflect their initial raw materials (either implicitly or explicitly), there are many large volume examples of spirits for which the alcohol used in production is expected, or even required, to be neutral. From a legal viewpoint, the Alcohol and Tobacco Tax and Trade Bureau (TTB) define neutral spirits as those distilled to at least 95% ABV and at least 40% ABV if bottled. This has the advantage of providing a "blank canvas" alcohol base to the distiller or producer, which can be marketed essentially as is (such as vodka or Everclear-type products) or flavored or colored during or after distillation.

Known as a neutral grain spirit (NGS) or grain neutral spirit (GNS), the alcoholic base of one of these spirits can, in principle, be made from any suitable agricultural commodity, with the implication that the subsequent distillation of any fermented alcohol is sufficiently rigorous to remove any residual flavors. This is in contrast to brandies and whiskeys where such residual flavors are expected or even legally mandated. Nevertheless, there can be subtle differences in the flavor qualities of neutral alcohols and industrial producers will often prepare and utilize various grades (such as vodka quality, organic, Kosher, and gluten-free).

It is important to appreciate the substantial role that large-scale industrial producers of neutral alcohol have on the spirits industry. As might be expected, efficient production of neutral alcohol with minimized losses requires significant rigor in the manufacturing plant (and is reliant on continuous distillation). This helps to explain the current situation in the distilled spirits industry, where most producers of neutral-based spirits rely on third-party producers to source their alcohol base. There are smaller-scale, batch neutral spirit producers, although the processes from these producers tend to be less efficient, with economies of scale making the cost of producing neutral alcohol more expensive on a smaller scale.

## Outline of Neutral Spirit Production

As long as the final spirit is at least 95% ABV, the agricultural source of the fermentable carbohydrates is generally unspecified, except in certain cases where there may be an insistence on using specific raw materials for specific products. The processing of the raw materials depends on the format of the carbohydrates present. For raw materials containing simple sugars (such as sugar cane, beets, fruits), extracts can be fermented as is. For starch-bearing raw materials (such as cereals and potatoes), it is required that the raw materials be comminuted and the starches then be saccharified in order to release a fermentable substrate. Efficiencies in fermentation can be achieved by applying approaches such as continuous fermentation, although bear in mind that one of the economic benefits of continuous processes is planned, but infrequent, shut-downs, which can make an unreliable continuous operation a production liability.

Once fermented, the alcohol (typically 8–12% ABV) must be recovered. At the time of writing, this is done by distillation. A generalized process (Figure 1) is based on five distinct stages:

1) Stripping the alcohol from the finished fermentation
2) Rectification of the stripped alcohol (at least 95% ABV)
3) Hydroselective distillation
4) Rectification
5) Optional demethylization

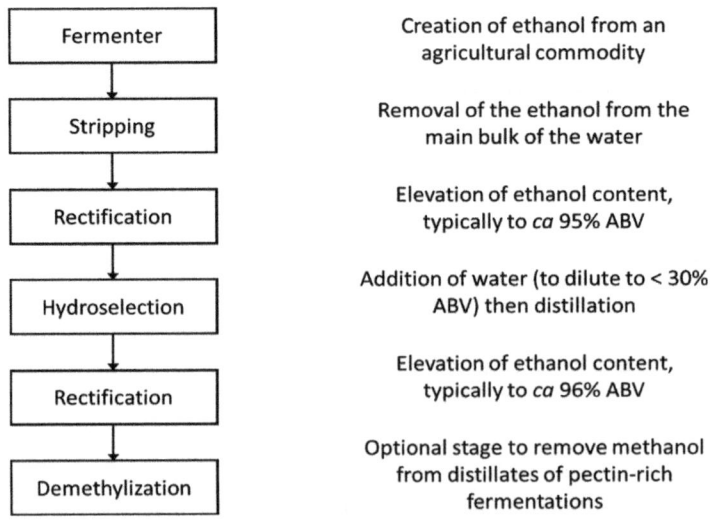

*Figure 1. General manufacturing process of spirits.*

In the first stage, the objective is to remove alcohol from its water matrix as effectively as possible. Generally a recovery of greater than 99% is expected. The feed from a stripping column is then fed into a rectifying column where the collected alcohol is approximately 95% ABV. This concentration can be significantly less, although generally the lower the % ABV from the rectifier, the more congeneric and potentially flavor-active materials are present. While this is a clean spirit compared to what is produced from a two- or three-pot distillation operation, it is still not considered to be neutral. The third stage of neutral alcohol production is the so-called hydroselective distillation. Here, perversely, clean, taint-free water is added back to the rectified spirit to reduce the alcohol content. This diluted spirit is then redistilled and the resulting distillate is then further -rectified to produce the final neutral spirit.

The reason for this hydroselection step is that the small amount of residual volatiles present in the distillate from step 2 conveys modest but detectable levels of flavor activity. By adding water, these volatiles, which are almost all less polar than ethanol itself, become relatively more volatile than ethanol and, therefore, are more easily removed by distillation.

The final optional step is demethylization. This is required only when the raw materials contain significant levels of pectin, which in turn yield unacceptably high levels of toxic methanol if not removed. For neutral spirit production, this is pertinent particularly when potatoes are used, although more generally, grapes, stone fruits, and agave also yield significant amounts of methanol in the distillate. Unlike most volatiles in spirit distillates, methanol is more polar than ethanol and, thus, is more easily removed from mixtures of the two when the ethanol concentration is higher (i.e. after the second rectification).

## Vodka

Translating as "little water," it is overly simplistic to expect that vodka is produced simply by diluting a neutral alcohol source, although in principle this can be an approach to its production. There has been much discussion as to where vodka originated from, with Russia, Poland and Scandinavia all making various claims. In any case, vodka does have some eminent scientific history; Dmitry Mendeleev (of periodic table fame) completed his doctoral studies on vodka production in the mid-1860s and asserted that vodka at around 40% ABV was ideal for consumption, a figure that is close to the sales strength of many vodkas today.

Usually vodka is produced by the fermentation of saccharified cereals or potatoes and subsequent distillation. These fermentations can be batch or continuous, with usual economies more attributable to continuous processes. The quality of alcohol used in vodka production is usually of the highest grade available, although many vodka producers will redistill their sourced neutral alcohol, often subjecting it to multiple passes through charcoal filters to further reduce the levels of residual terpenoids and esters. With the advent of a "super-premium" vodka category, there has been a greater focus on the purity or cleanliness of the product. However, some brands do retain an element of the raw materials flavor from which the alcohol originated. One example is Ciroc, produced from a spirit made in turn from grape fermentations that features a pronounced grape flavor.

## Gin

Gin is one of the most well-known botanically based spirits globally. Originating from the "Low Countries" (present day Belgium and the Netherlands), a combination of juniper and alcohol was once considered to be a panacea with the alcohol enhancing the bioavailability of juniper when extracts were consumed. It seems that the Lowlanders took very kindly to their medicine to the extent that it stimulated the government at the time to begin imposing taxes on the new alcohol. This medication would evolve into genever.

With William III (Prince of Orange from the Netherlands) ascending to the British throne with his wife, Mary II (daughter of the exiled James II), in the late 1600s and English soldiers fighting in the Anglo-Dutch wars in the mid 1600s, the English became familiar with genever with its popularity growing in the first half of the 18th century. At the height of gin consumption in 1733, Londoners were consuming 60 L per capita, with an estimated 25% of all Londoners being incapacitated by the alcohol at any one time. The most famous image of the dangers of gin at that time came from William Hogarth's Gin Lane engraving. By the late 18th century, legislation, including punitive taxes, helped to curb the excesses of gin production and consumption.

Fast-forward to the 20th century and gin saw increasing popularity in mixed drinks. The gin and tonic, gin and orange, and gin and "it" (Italian vermouth) all contributed to the rise of a more reputable gin industry. Over the past 10–15 years, there has been a huge increase in the number of gins available to the public, largely driven by the growing craft industry. There is a substantial amount of flexibility for those wishing to make gin: As long as it tastes perceptibly of juniper and is at least 40% ABV (37.5% ABV in the EU), essentially any flavoring can be used in the production. Thus, producers have a broad palette of botanicals available to them, and there is virtually no lead time from production to sale, making the production of matured spirits rather painful for new businesses in terms of cash flow.

There are two broad classes of gin: compounded gin and (re)distilled gin. The distinction is that the latter must contain flavors that are only imparted to the spirit by distillation, with the redistilled category indicating a second botanical distillation. The former can contain flavors that are infused or, more commonly, a combination of distilled and infused flavors. This can be a useful strategy when thermally labile flavors are demanded, such as rose or cucumber. To introduce these flavors, the botanicals can be soaked in an alcohol-water mixture (macerated) and then distilled. Alternatively, the botanicals can be suspended in a basket above a boiler containing alcohol and water, with the vapor stripping the flavors from the botanicals before being condensed and collected as a spirit. Combinations of the two can be used, adding more flexibility for distillers looking to design a unique product. This collected spirit is generally well above typical sales strength (at 65–75% ABV) and thus must be diluted prior to bottling.

The major botanical used for gin production is juniper berries (or, more accurately, cones). In the EU, a specific species of juniper (*Juniperus communis*) is specified but there are no juniper species restrictions in the U.S. (although it is prudent to avoid toxic species). Juniper berries convey a range of flavor notes to the spirit including herbal, citrus, woody, peppery, and spice. The major volatile compounds in juniper berries are $\alpha$-pinene (I), $\beta$-pinene (II), $\beta$-myrcene (III), $\alpha$-copaene (IV), $\beta$-caryophyllene (V), germacrene D (VI), $\gamma$-cadinene (VII), and $\delta$-cadinene (VIII).

Many juniper samples can be overly resinous, which can overpower some gins; because of this, it is prudent to evaluate juniper samples prior to committing them to any final product. After juniper, coriander seeds are a very common gin botanical with citrus and peppery notes predominating. Some peppery gins have higher levels of coriander seeds in their botanical blend. Other botanicals can be broadly classified as citrus peels, roots, seeds or pods, leaves, and barks. Citrus peels and cardamom seeds are particularly potent sources of flavors, with distinctions being made between sweet and bitter orange peels, although lime and pomelo peels are also used. The use of certain roots (especially orris and angelica roots) contribute to the flavor, but they are also thought to add structure to the gin. The implication is that these roots help to retain the oils in the gin, much in the same way that orris is used in perfumes to help retain the aromas for longer. In any case, their omission often has significant impact on the sensory performance of the resulting gins.

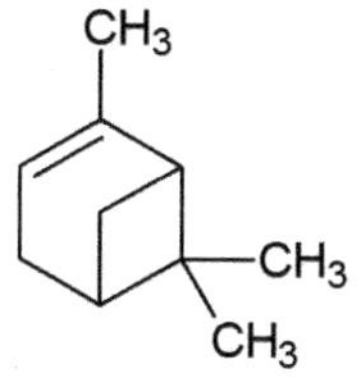

(I)

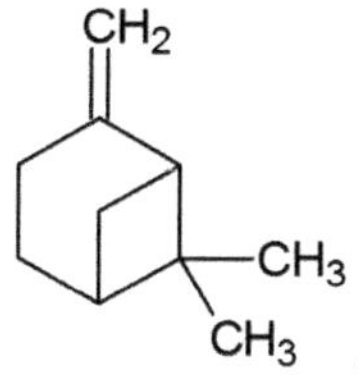

(II)

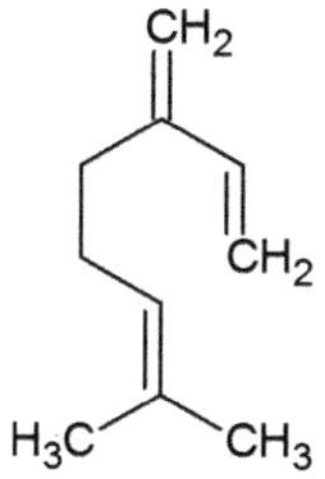

(III)

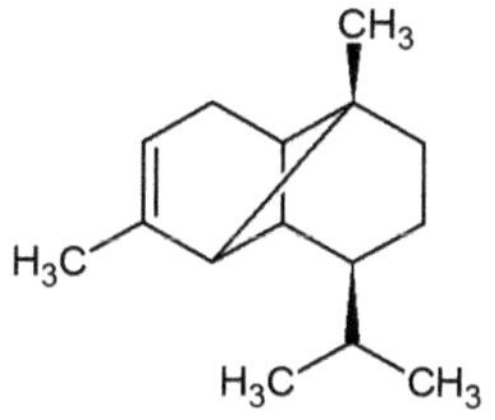

(IV)

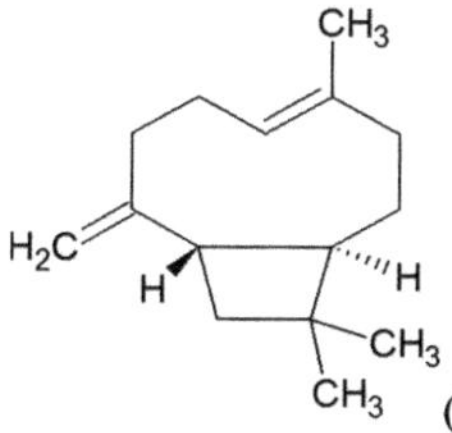

(V)

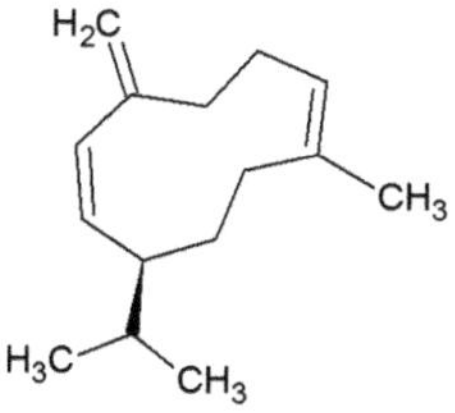

(VI)

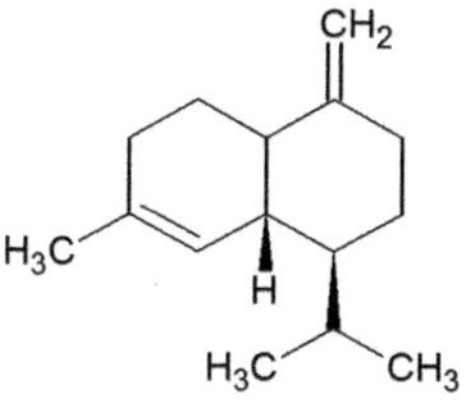

(VII)

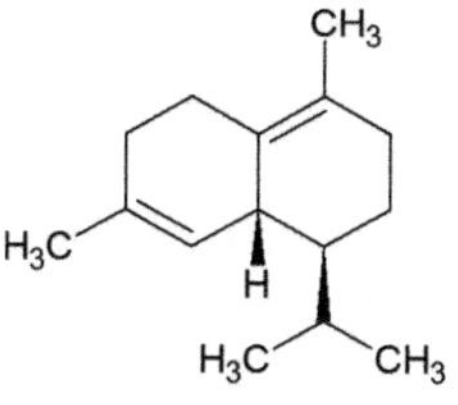

(VIII)

An example recipe (Table 1) shows that the levels of botanicals used for the production of gin are rather low, usually around 10 g/L 50% ABV still charge. Using too much plant material will result in haze formation or louching ("loo-shing") when diluting the gin to sales strength for bottling. A common practice is to deliberately use high levels of botanicals for the initial distillation and then diluting first with a neutral alcohol and then with water. In this way, so-called multi-shot gin can be produced multiplying up the production volume from a still by an order of magnitude. However, moving a product from single- to multi-shot production is not usually a linear scaling exercise and some trial and error can be required to ensure adequate flavor matching.

**Table 1. A Typical Gin Recipe Where Juniper and Coriander Seeds are the Dominant Botanicals Gravimetrically**

| Botanical | Formula | Dose (g/L of a 50% ABV Charge) |
|---|---|---|
| Juniper berries | $x$ | 4.00 |
| Coriander seeds | $x / 2$ | 2.00 |
| Cassia andcinnamon bark | $x / 10$ | 0.40 |
| Angelica root | $x / 10$ | 0.40 |
| Liquorice root | $x / 10$ | 0.40 |
| Bitter almonds | $x / 10$ | 0.40 |
| Lemon peel | $x / 100$ | 0.04 |
| Orange peel | $x / 100$ | 0.04 |
| Cardamom seeds | $x / 100$ | 0.04 |
| Orris root | $x / 100$ | 0.04 |

## Absinthe

If gin has a dubious history, then absinthe is positively scurrilous. It is a highly alcoholic spirit with some production similarities to gin. Absinthe originated in French-speaking Switzerland in the late 1700s and was subject to widespread prohibition in 1915, with that ban being lifted piecemeal in the late 1990s. The rationale for the ban was the apparent psychotropic properties of greater wormwood (*Artemisia absinthium*) or, rather thujone, a major component of the wormwood essential oil, that exists as $(-)$-$\alpha$-thujone (IX) and $(+)$-$\beta$-thujone (X), a monoterpene ketone with a menthol odor. This presumably led to absinthe becoming known as *la fee verte* (the green fairy). The evidence for this psychotropic effect is tenuous at best, although thujone was recognized to be a convulsant in the mid-19th century, leading to widespread public opinion that absinthe was a more dangerous form of alcohol than other spirits. Today, absinthe is restricted to less than 35 mg/L of thujone in the EU, while in the U.S., this limit is 10 mg/L.

The production of absinthe is a multi-step process (Figure 2). Botanicals are introduced into the absinthe spirit via maceration and subsequent distillation. Historically, the alcohol source was wine spirit, but today many absinthes rely on neutral spirits. If the absinthe is to be colored, the resulting distillate is macerated with a further tranche of botanicals. Typically, the initial botanicals are greater wormwood, anise, or fennel, although many other botanicals can be used. These botanicals are macerated at high alcohol concentrations, typically 85% ABV. The subsequent distillation of this macerate usually requires dilution. This is necessary to take advantage of hydroselection as used in the production of neutral spirit, which allows for more efficient transfer of botanical oils into the distillate. In our laboratory, we have observed oil yield increases even when diluting macerates to 25% ABV prior to distillation. The disadvantage of such high dilutions is that the capacity of the still is reduced.

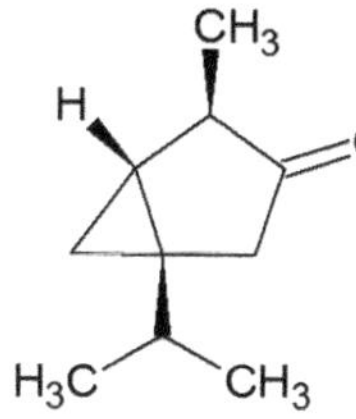

(IX)

(−)-α-thujone

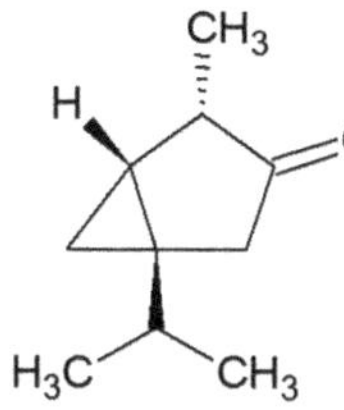

(X)

(+)-β-thujone

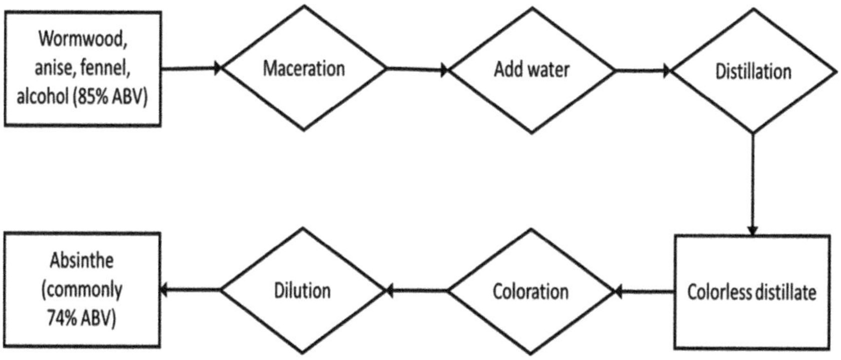

Figure 2. Scheme outlining the typical absinthe production process.

Once the colorless distillate is collected, it can be diluted to sales strength to be sold as a white absinthe or macerated further with other botanicals. In particular, this second maceration with wormwood conveys a characteristically bitter flavor to the final product. Traditional absinthes are sold at elevated alcohol contents (up to around 74% ABV) and are usually emerald green due to the presence of chlorophyll. The high alcohol content helps to keep the oils in solution as absinthe (in comparison to gin) typically uses around 10–20 times the amount of botanicals in the still. This explains why dilution of absinthe readily results in a milky emulsion that is remarkably stable, a phenomenon known as louching. In contrast to gin, absinthe louching is desirable or even demanded. The traditional way of dispensing absinthe, by dripping ice-cold water through a sugar cube into the absinthe, encourages emulsion formation in three ways: by increasing the polarity of the spirit by adding sugar, by increasing the polarity by adding water, and by cooling caused by the ice-cold water.

131

Today, absinthe is enjoying a modest resurgence and is an ingredient in various cocktails (such as the Sazerac), and several craft distillers are producing their own renditions of this historic and wrongly maligned spirit.

## Liqueurs

Many, but not all, liqueurs are based on a neutral alcohol base. Generally, they can range from 15 to more than 40% ABV and contain at least 10% (w/v) sugar. The alcohol though must be fermented and any additives should be of agricultural origin, with "natural" flavoring. All that being said, it is clear that this is a wide-ranging category (Table 2). Production usually follows one of two broad methods:

1) Mixing a macerated flavor base with a sweetened spirit or
2) Maceration of a flavor base with a spirit, followed by distillation and mixing with a sweetened spirit.

Because sugar is non-volatile and will not carry over into the distillate as is, all liqueurs are compounded. Some liqueurs are designed to be consumed as is or as part of a mixed drink, while others are only used as part of a mixed drink.

The greatest shelf-life challenges in the global distilled spirits category are usually attributed to liqueurs. Emulsion liqueurs, such as those based on cream, can be prone to phase separation although many producers of these spirits now understand what is required compositionally and physically to keep their products relatively stable. Advocaat, traditionally based on egg yolks and brandy, can pose a slight microbiological threat, especially if unpasteurized yolks are used.

The sugar requirement for liqueurs can also be problematic. With increasing alcohol concentration, the solubility of sugars generally decreases and they may crystallize out of the solution during the shelf-life of the product. For instance, selecting sugar syrups instead of crystallized sugars can help stave off this separation.

There are many liqueurs that have enjoyed a long life cycle in the market, but some, whether they are brand extensions or novel products, may only be on the market for a short time. For instance, while Bailey's Original cream liqueur is globally successful, there have been brand extensions released in the market for the end-of-year holiday period, as liqueur sales tend to increase in the months of November and December.

## Brandies

In its broadest sense, "brandy," derived from the medieval Dutch term *brandewijn* (literally "burned wine"), refers to any spirit produced by the distillation of a fruit-derived fermentation or the fermentation and distillation of pomace, the wet solid residue of fruits after pressing for their juice. Examples of fruit brandies include slivovitz (eastern European plum brandy), rajika (Serbian fruit distillates), and palinka (central European fruit brandies), while marc and grappa are well-known pomace distillates. Another term, *eau-de-vie*, is used to refer to unmatured fruit distillates, although in English-speaking countries, this term is reserved for spirits made from fruits other than grapes.

The TTB defines brandies as, "spirits distilled from the fermented juice, mash or wine of fruit or from its residue at less than 95% alcohol by volume (190 Proof) having the taste, aroma and characteristics generally attributed to brandy and bottled at not less than 40% alcohol by volume (80 Proof)" (*1*).

**Table 2. Examples of Common Liqueurs, Illustrating the Various Alcohol and Flavor Sources Used**

| Flavor<br>Spirit | Fruit | Botanical | Cocoa, Coffee and Tea | Nut and Emulsion |
|---|---|---|---|---|
| Neutral | Crème de cassis, cherry liqueur, limoncello | Jägermeister, galliano, sambuca | Crème de cacao | Amaretto, Davis Walnut |
| Brandy | Cherry brandy, winter pimm's no. 3 | Advocaat | | |
| Gin | Sloe gin | | | |
| Rum | Malibu | | Kahlúa, Tia Maria | |
| Shochu or soju | Podoju, maeshilju, plum wine | Ginseng, yagyongju | | |
| Whisky or whiskey | Southern comfort | Drambuie, glayva | | Bailey's irish cream |
| Vodka | Cherry vodka, sloe vodka | Bärenfang, krupnikas | Alchemia Czekoladowa | Crème likier mleczny |

This general class definition is further resolved into types, such as pisco (Peruvian grape brandy), applejack (fruit brandy made from apples), and region-specific products such as calvados (an apple brandy that satisfies the French definitions pertaining to its production), cognac, and armagnac; the two latter are grape brandies from their respective areas in southwest France. The European regulations are generally more detailed and are well-defined in Regulation (EC) No 110/2008 (2).

By far the most well-known brandies internationally are produced from grapes. According to EU definitions, brandy or weinbrand is a spirit drink derived from a wine spirit. This must be produced exclusively by distilling the wine (fortified or not with other wine distillates) at less than 86% ABV, with specifications for a minimum amount of volatiles present (1.25 g/L of pure alcohol) and a maximum level of methanol of 2 g/L of pure alcohol. Maturation must last at least one year and the alcohol level should be between 36% and 50% ABV.

Similar to other distilled spirits, the origin of brandy is linked with the development of distillation methods, which only began to be applied to alcohol production in the late 14th century. Originally, the wine distillation was performed to provide a preservative for different wines as adding distillates back to wine reduced the risk of microbial spoilage. Additionally, the production of fortified wines was seen as a more efficient method of transportation with water being added back prior to sale. It was also suggested that this was a way to reduce levels of taxation, as taxes were initially based on volume rather than the strength of the alcohol.

In any case, the resulting fortified wines and distillates were stored in oak casks as was typical for many liquids since the Roman Empire; it soon became apparent, as with other products like whiskies, that storage in oak improved the flavor of the resulting liquid. The development of the grape brandy industry really grew from the area where grapes were produced for viniculture. By the end of the 19th century, western European markets and their overseas colonies were dominated by brandies from France and Spain, while eastern Europe was supplied by brandies from Bulgaria, Georgian, and the Crimean peninsula. These latter brandies were considered to be excellent products and competed well with western European products, although the French brandy industry was still reeling from the devastation of their vines from the introduction of the aphid phylloxera in the mid-to-late 19th century.

While there are many versions of grape brandy worldwide, cognac, armagnac, and jerez brandies are some of the best-known. Cognac brandy is produced from grapes grown and fermented in the Cognac region of France, just southeast of La Rochelle near the Atlantic coast. Across this region, the soil is characterized by its chalk content and this chalk has a material impact on the quality of the resulting brandy produced. As a result, cognacs are usually further defined from grandes champagne as the highest quality to bon bois and bois ordinaire for lower quality. This is always a useful aspect to check for when purchasing cognac.

Cognacs are produced from the wine of various grape varieties, most notably Ugni blanc, Colombard, and Folle blanche. In any case, harvesting in late summer requires that the resulting white wine must be distilled before March 31 of the following year. Cognac must be produced in a legally-defined alembic still of the Charentais style. In this still, wine is introduced into a simple pot which is direct-fired. The vapor travels through a narrow swan-neck into a pipe running through a wine heater (a buffer tank for introducing wine into the still) before being condensed and collected. The first distillate, known as *le brouillis*, is redistilled with cuts from the previous distillation and the hearts are retained for maturation. This spirit must be aged for at least two years in oak casks made from oak grown in either the French Limousin or Tron çais regions.

In practice, many cognacs are aged for much longer than the minimum two years and the spirit is assessed continually, removed, transferred, and blended depending on such assessments. Eventually some are transferred into demi-johns (or "demi-jeannes") to essentially eliminate the impact of oxygen on the spirit. The final blended cognacs are broadly classified according to age. For example, a CS or three-star brandy is in its third or fourth year of aging. Very special old pale (VSOP) is in its fifth or sixth year of aging and products termed Napoleon, XO, Extra, or Hors d'age are older.

Armagnacs are a distinct category of French brandies with historical records documenting their existence as predating that of cognacs. They are produced in the southwest of France in Gascony, just to the north of the Pyrenees Mountains. Unlike cognacs, they can be prepared either by batch or continuous distillation. Traditionally, preheated wine, made from Ugni blanc and Baco blanc grape varieties, is introduced into the top of a short distillation column, and vapor from wine collected at the bottom of the column provides an alcoholic stream that strips an alcohol-rich vapor that is condensed and collected for maturation in oak casks produced in the Limousin or Monlezun regions of France. The strength of the distillate is typically a relatively modest 58% ABV, which in part accounts for the robustness of matured armagnacs.

Outside of France, Jerez, an Andalusian region in southern Spain, became an important area for producing what is known as brandy de Jerez. Originally produced to make alcohol for non-potable uses by the occupying Islamic Moors and then latterly to fortify local wines, brandies from Jerez have since become popular in their own right. Much like cognacs and armagnacs, there are specific stipulations regarding the manufacture of brandies de Jerez. They can be produced using either batch (pot) or continuous (column) distillation and must be matured in casks no larger than 500 L.

An unusual feature of brandies de Jerez is their mode of maturation. Casks are stacked up into a so-called solera. The spirit in the lowest part of the stack is the oldest, and the spirit is withdrawn from these casks for sale. While not completely emptied, there is space in these casks, which is filled with spirits from the layer of casks above. This continues until the top layer is reached, where the unoccupied volume is filled with a newly made spirit. It is therefore difficult to assign anything more than an average age to brandies de Jerez, and Spanish legislation defines three categories for these brandies based on average age: brandy de Jerez solera (average age of 1 year), brandy de Jerez solera reserve (average age of 3 years), and brandy de Jerez solera gran reserve (average age of 10 years). Similarly to other brandies, the concentration of volatiles in brandies de Jerez increases with age.

The brandy category has developed across the world in traditional markets such as Europe and "New World" markets such as the U.S. While fruits can be used to make a wide array of brandies, the truly international products are invariably made from a wine spirit.

## Whiskey

Whiskey (sometimes spelled whisky) has been an economically important product for centuries (3). It is regarded as the world's oldest distilled beverage and is thought to have originated in Irish monasteries in the 12th century and later adopted in Scotland during the 14th and 15th centuries. Whiskey production came to the Americas in the 16th century and became especially important in the late 17th century with the arrival of Scottish and Irish immigrants. Whiskey is now produced and consumed in many countries throughout the world with Scotland, the United States, Ireland, Canada and Japan accounting for most of the global world production.

Whiskey can be generally defined as an alcoholic distillate derived from yeast fermented cereal grains including barley, corn, rye or wheat. Like other distilled spirits, it is composed primarily of water and alcohol, but its unique flavor is derived from low levels of volatile and non-volatile congeners, which refer to any substances that are not ethanol- or water-derived from the fermented grains or the container (e.g. oak barrel) in which the whiskey is aged.

Whiskeys are defined and strictly regulated on the basis of their origin and the materials and methods used in their production. For example, American whiskeys differ from other types of whiskeys because of the country of origin (must be produced, matured, and bottled in the United States) and some of the unique steps used in their production, foremost being that American whiskeys must be aged in new, charred, American oak barrels. There are some other ways in which the various whiskeys may differ, such as in the distillation methods used in their production and the final proof of the spirit. There may also be some sub-classification of whiskeys, such as bourbon vs Tennessee American whiskeys. Furthermore, American whiskeys can be further subdivided on the basis of the main grain used in the mash bill (e.g. rye whiskeys must made from 51% rye and corn whiskeys from 80% corn). Scotch may be subdivided into lowland and highland whiskeys, which differ not only in origin, but also in terms of production practices. In addition, some Scotches are produced from "peated" barley (barley that has been dried using heat from burning peat), which imparts a unique smoky and phenolic flavor to the whiskey.

While it is clear that these various types of whiskeys can be defined by their origin and production practices, they all share a common ancestry and general flavor chemistry derived through fermentation of grains, distillation, and aging in oak. It is the subtle flavor differences among the various types of whiskeys that make this distilled beverage interesting and so popular throughout the world.

## Rum

The first distilled liquor to be made in quantity in the New World was rum, not whiskey. This is due mainly to the readily available supply of sugar cane and molasses produced in the West Indies and exported to the American colonies, ca. mid 16th century. Rum is defined by the US Code of Regulations (4) as an alcoholic distillate made from fermented sugar cane juice, syrup, molasses, or other sugar cane byproducts. It must be produced at less than 190° proof (95% ABV) in such a manner that the spirit possesses the taste, aroma, and other characteristics generally attributed to rum, and it must be bottled at not less than 80° proof (40% ABV). This definition is quite limited when compared with the other previously mentioned spirits. Therefore, aside from having to be produced from cane sugar or its byproducts, rums can be made using a variety of methods and aged in any type of container. Furthermore, the age statement is not standardized and is generally set by the country of origin. As a result, rum is a highly varied spirit, especially in terms of its flavor profile.

Rum is a complex spirit, possessing a range of flavor characteristics due to the use of a variety of different raw materials and production practices (5, 6). Rums can be categorized based on the raw materials, the fermentation and distillation, and the aging practice used in their production. When shopping, one can find white, gold, aged, dark (or black), over-proof, flavored, and spiced rums. In the French Caribbean, a distinction is made between rums produced from cane juice (*rhum agricole*) and those made from molasses (*rhum industriel*). Additionally, there are light, heavy, and traditional rums. Heavy rums are highly aromatic and are produced by prolonged fermentation of both yeast and bacteria, which produce a more complex flavor profile, and distillation is done using pot stills. Some Jamiacan rums typically utilize this style. On the other hand, light rums have low flavor intensity and are produced by continuous fermentation, while traditional rums have medium flavor intensities.

Much like whiskey, aging is one of the most important practices affecting the flavor of rum. However, the aging practices with rums are not prescribed or regulated. Rums can be aged in any type of material, but previously used oak barrels are most common. As expected, the rums take on some of the character of the spirit originally aged in the barrel (e.g. from barrels previously used for whiskey, sherry, wine, or cognac). Aging can be done for a short period (e.g. for white rums, in which the color is later removed using activated carbon) or for longer periods depending on the complexity of the flavor desired by the manufacturer. In addition, the conditions and method of aging are important. Some high-end rums are aged using a solera system similar to what is used for aging of sherry.

## Baijiu

Baijiu is a traditional Chinese liquor fermented from grain. Baijiu typically has a high alcohol content of 50–55% (ABV), although lower alcohol (38% v/v) products have gained some popularity in recent years. Unlike whiskey and other spirits, which are normally consumed with some sort of dilution (on the rocks or in cocktails), baijiu is normally drunk directly from bottle without any dilution. Baijiu is typically served at room temperature with meals.

Alcohol consumption is deeply rooted in Chinese civilization. The earliest evidence of alcohol fermentation was discovered from the pottery from a neolithic settlement (between 7000 and 5800 B.C.) in Jiahu, the central plain of ancient China near the Yellow River (7). Chemical analysis proved that the unique alcohol beverage during the Neolithic period was fermented from rice, honey, grapes, and hawthorn. Alcohol fermentation and consumption from grain has been seen in all the ancient dynasties in China. Although the fermentation process was invented much earlier, the production of baijiu began during the era of the Song (960–1270 A.D.) and Yuan (1271–1368 A.D.) dynasties when the distillation process was first introduced. The of fermentation and distillation of baijiu were later passed onto the later dynasties.

Baijiu has become a beloved drink throughout China. It is the second most popular alcoholic beverage in China next to beer, with a total production of 12 billion L in 2017, which accounted for 17% of total alcoholic beverages produced in China. Baijiu is the most important alcoholic beverage to the economy and government tax income of China. In 2017, the sales of baijiu generated RMB 565.4 billion (U.S. $88.5 billion) in revenue, accounting for 61% of the total revenue from all alcoholic beverages in China.

### Ingredients of Baijiu

There are more than a hundred different types of baijiu produced from over 10,000 distillers in China using different raw materials, fermentation conditions, and distillation techniques. Baijiu is typically fermented from sorghum or a mixture of sorghum, wheat, corn, rice, and sticky rice (8, 9), although other starchy raw materials can be used depending on the availability and economics. The unique part of baijiu manufacturing is the solid-state fermentation and solid-state distillation. In solid-state fermentation, the saccharification of starch and alcoholic fermentation of sugar occur simultaneously catalyzed by "Qu." Qu is a mixture of yeasts, molds, and bacteria cultured on grains. Qu can be divided into Daqu (big Qu) (Figure3a), Xiaoqu (small Qu) (Figure 3b), and Fuqu, and each is used in different processes.

 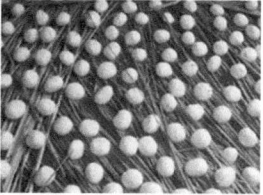

*Figure 3. Typical shape of Qu. 3a, Daqu; 3b, Xiaoqu.*

Daqu is used for the fermentation of sorghum-based grain. Daqu is typically made from wheat or a mixture of wheat, wheat bran, barley, and peas. The raw materials are first milled, mixed with warm water, and pressed into a brick-shaped block. This block is then incubated in a controlled environment for several months to let the yeasts, molds, and bacteria to grow. The Qu is then cooled and grinded before being used for fermentation. The condition under which the Daqu is made is the one of the most guarded secrets for manufacturers. Based on the maximum temperature at which the Daqu is incubated, the types of Daqu can be further classified into low temperature Daqu (less than 45°C), moderate temperature Daqu (45–60°C), and high temperature Daqu (greater than 60°C). Daqu is rich in yeast, bacteria, fungi, and other microorganisms (8). In addition, complex enzyme systems are accumulated in the finished Daqu (10). All of these microorganisms and enzyme systems are essential for the production of rich baijiu aroma. Most baijiu are fermented using Daqu.

In contrast, Xiaoqu is used for the fermentation of rice-based grain in semi-solid fermentation (11). Xiaoqu is made from rice and glutinous rice. The rice is soaked in water then made into a small ball or paste. The ball is then coated with old Qu powder to allow inoculation. The Qu is then moved to a controlled room for microbial growth. Compared to Daqu, the microbial population in Xiaoqu is much less complex. Only a small amount of baijiu is made from Xiaoqu.

Fuqu is used for a few types of baijiu. Fuqu is the saccharification starter of Aspergillus that is inoculated in wheat bran. Yeasts need to be added to the starter during the alcoholic fermentation.

**Manufacturing Process of Baijiu**

There is no standard procedure to make baijiu. The fermentation procedures and conditions can vary from one plant to another. A general process of strong-aroma type of baijiu is illustrated in Figure 4.

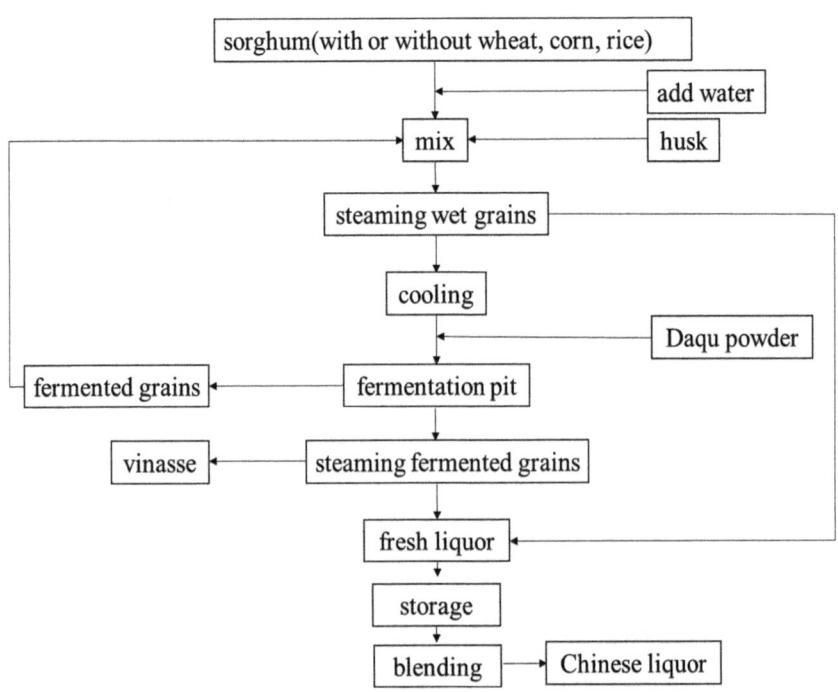

Figure 4. Baijiu manufacturing process (strong-aroma type).

The steps in the baijiu fermentation process are described as follows:

1) Raw ingredient formulation: Sorghum, corn, wheat, and rice are grinded and mixed according to specification. Water is the added at the right ratio. Rice husks are also added to improve the porosity of the fermentation mixture. Rice hulls are sometimes pre-steamed to remove undesired aroma and flavor.

2) Cooking: Steam is typically used to cook the grains to gelatinize the starch granules. After cooking, the grains are cooled down to the right temperature.

3) Fermentation: The cooked grains are mixed with Qu powder and then loaded into a fermentation pit. The fermentation pit is typically a cube with the insides coated in a layer of fermentation mud made of clay, spent-grain, bean cake powder, and fermentation bacteria. The fermentation is carried out in a solid state for up to two months at a controlled temperature under anaerobic condition (Figure 5).

4). Distillation: After fermentation, the liquor is distilled out with steam under a solid state. (Normally, the cooking process and liquor distillation process are combined in industrial manufacturing into one process, in which the former usually takes 30 min and the later takes 20–30 min). The forerun (head) and the after-run (tails) are collected separately, and the middle-run is saved as fresh distillate (Figure 6).

5). Aging: The fresh distillate is harsh and contains harmful aldehydes and other undesirable aroma compounds and must be aged for flavor development and stabilization. All baijiu are aged in a china jar for six months or more to develop the bouquet aroma. Some premium baijiu are aged three to five years in the jar (Figure 7).

6). Blending: Water is added to adjust the alcohol strength from 38% to 65%. A different fraction or different aging process for the liquor can be used to create the signature flavor of different baijiu.

*Figure 5. Baijiu fermentation pit.*

*Figure 6. Solid state distillation apparatus.*

*Figure 7. China jar used for baijiu aging.*

**Flavor of Baijiu**

The aroma profile of a baijiu is complex and greatly influenced by its raw ingredients, type of Qu, fermentation conditions, distillation, aging, and many other manufacturing conditions. Based on aroma characteristics, Chinese liquors are generally classified into 11 aroma types: strong aroma, light aroma, soy sauce aroma, sweet and honey aroma, roasted sesame-like, *chi* aroma, complex, herb-like, *Feng* aroma, *Laobaigan* aroma, and *Te* aroma types (*12*). Of these, the most popular aroma types are strong aroma, light aroma, soy sauce aroma, and sesame aroma. Regardless of the aroma type, all baijiu types have strong fruity, pineapple-like, and banana-like aromas, especially for the strong-aroma type, although other aroma signatures differentiate the aroma type during classification.

In addition to alcohols, the major volatile compounds in baijiu are esters and short-chained fatty acids. Aldehydes, ketones, acetals, and heterocyclic compounds also play important roles in different aroma types of baijiu. These compounds largely come from the fermentation, distillation, and aging processes, and the quantitative differences produce the unique aroma of different baijiu.

The strong-aroma style of baijiu accounts for about 70% of total liquor production. Within this category, *Wuliangye, Jiannanchun,* and *Yanghe Daqu* are among the most famous brands. Fan and Qian first systematically reported the aroma-active compounds in strong-aroma type baijiu using gas chromatography–olfactometry (*13, 14*). Using aroma extract dilution analysis, extensive fractionation, and GC–olfactometry, Fan and Qian (*13, 14*) have identified more than 70 odor-active compounds in Yanghe Daqu liquor. The results show that the Yanghe Daqu liquor aroma is mainly produced by fatty acid esters and short-chained fatty acids. Ethyl hexanoate is the dominating ester and also the most important aroma contributor. Ethyl butanoate, ethyl pentanoate, ethyl octanoate, and 3-methylethyl hexanoate are important aroma compounds in the liquor. In addition to these esters, short-chained fatty acids such as hexanoic, acetic, and butanoic acids are also important. Esters were mostly formed by esterifying alcohols with fatty acids during the fermentation and aging processes. Daqu has both high hydrolase and esterase activities (*15*). Esters and short-chained fatty acids are also important to *Wuliangye* and *Jiannanchun* liquors (*16*). *Wuliangye* and *Jiannanchun* liquors have other aroma-active compounds that contribute to soy sauce and roasted aromas. The flavor differences are the results of differences in the raw ingredients used in the fermentation, fermentation conditions, distillation practices, and aging processes (*16*).

Maotai is definitely the most well-known baijiu and is considered to be the premium national liquor. It is produced in the Guizhou province of China and has a unique soy-sauce aroma. Maotai liquor is fermented from sorghum with Daqu powder made from wheat. The fermentation process is different from that of strong-aroma type liquor, and eight cycles of fermentation and seven runs of distillation are involved (Figure 8).

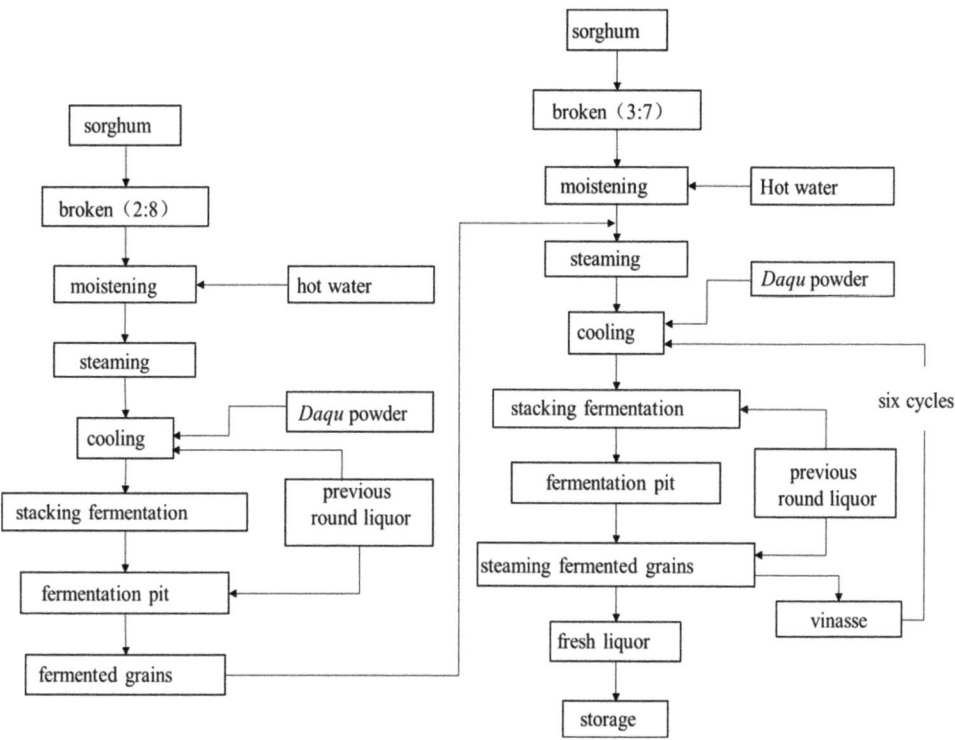

*Figure 8. Typical flow chart for Maotai liquor production.*

The aroma compounds of Maotai liquor have been of great interest to flavor chemists and baijiu researchers. Using fractionation, GC–olfactometry, and quantitative analysis, Qian and others have identified the aroma compounds in Maotai and other soy-sauce aroma type baijiu (*12, 17, 18*). In addition to esters and acids, the research has demonstrated that alkylpyrazines are important to the characteristic soy sauce and roasted aromas (*12*). Some of the alkylpyrazines identified in Maotai include 2,3,5,6-tetramethylpyrazine (XI), 2,3-dimethyl-5-ethylpyrazine (XII), 2,3,5-trimethyl-6-ethylpyrazine (XIII), and 2,3,5-trimethylpyrazine (XIV). Pyrazines contributed to roasted and baked aromas. Pyrazines are generated from Maillard reactions or metabolic activities of microorganisms (*19, 20*). It has been demonstrated that 2,3,5,6-tetramethylpyrazine (XI) is generated from the reaction of 3-hydroxy-2-butanone from glucose by *Bacillus subtilis* with ammonium phosphate (*21–23*). Some pyrazines were produced by the Maillard reaction between saccharide and amino residues in the liquor production process (*24*).

In addition to pyrazines, other important aroma compounds identified in Maotai liquor include ethyl hexanoate, hexanoic acid, 3-methylbutanoic acid, 3-methylbutanol, ethyl 2-phenylacetate, 1,2-dimethoxy-3-methylbenzene, 2-phenylethyl acetate, ethyl 3-phenylpropanoate, 4-methyl-2-methoxyphenol, γ-decalactone, vanillin, γ-nonalactone, Z-whiskylactone, 2,5-dimethyl-4-hydroxy-3(2H)-furanone, 4,5-dimethyl-3-hydroxy-2,5-dihydrofuran-2-one (sotolon), and geosmin (*12, 17, 18*). The aroma compositions of soy-sauce aroma liquor are much more complex than strong, light, and other aroma types of baijiu (*12, 17, 18*).

H3C  N  CH3
H3C  N  CH3  (XI)

N  CH3
H3C  N  CH3  (XII)

CH3
N  CH3
H3C  N  CH3  (XIII)

N  CH3
H3C  N  CH3  (XIV)

## Consuming Distilled Spirits

How to consume distilled spirits is a potentially contentious issue, with widely varying opinions from consumers and experts alike. Perhaps the aspect that evokes most discussion is the use of water and ice. For instance, there are many that eschew the addition of water to Scotch whisky, but in fact this addition helps to break up the alcohol "clusters" within the spirit, which in turn helps to release flavor. A viable alternative is to allow the evaporation of neat whiskey or cognac in the glass, which evaporates slowly with a resulting enhanced flavor release as the proportion of water in the glass gradually increases.

The use of ice and its format is also widely discussed. The cooling effect of ice is substantial due to its latent heat of fusion (at around 334 kJ/kg), whereas the specific heat capacity of water is around 4.2 kJ/kg· K. Thus, the heat required to melt 1 kg of ice is around 10 times greater than that required to raise water temperature from 0 to 8 °C. It is the melting, not the warming of water, from ice that produces the critical cooling effect. This calls into question the efficacy of cooling stones that experience no melting in the glass and generally have lower heat capacities than water. As a result, their impact on liquid cooling is minimal. Of course, diluting a drink due to melting ice is a legitimate concern with two approaches used to minimize this. Firstly, there is using a large surface area of ice (e.g., large cubes or a single cube or sphere) to slow down the melting (and cooling). Alternatively, any mixologist will know that by creating a mixed drink over ice and then straining out the excess ice, you can restrict the amount of dilution. A related issue is the addition of mixers to spirits, in as much as mixers are added for taste preference and therefore should be the choice of the consumer, no matter what the opinions of others might be.

The creation and recreation of cocktails has become increasingly popular in recent years as consumers seek out new experiences. It can be argued that the spirit components of a cocktail need not be of the highest quality—is it wise to use top-shelf spirits? Again this is a moot point, but it is arguably up to the consumer to decide what they like. There is an old adage that whiskey and coke is fine, as it is the best way to drink coke. I suggest that this is a position taken by some to ameliorate their lack of empathy with the choices of others.

## Acknowledgments

The authors thank Dr. Shuang Chen of Jiangnan University for providing baijiu production flow charts and photos.

## References

1. *The Beverage Alcohol Manual, Chapter 4: Class and Type Designation*. https://www.ttb.gov/spirits/bam/chapter4.pdf (accessed May 29, 2019).
2. European Parliament and Council. Regulation (EC) No 110/2008 of the European Parliament and of the Council of 15 January 2008 on the Definition, Description, Presentation, Labelling and the Protection of Geographical Indications of Spirit Drinks and Repealing Council Regulation (EEC) No 1576/89. *Off. J. Eur. Union* **2008**, *39*, 16–54.
3. Russell, I.; Stewart, G. *Whisky: Technology, Production and Marketing*; Elsevier: Amsterdam, 2014.
4. *US Code of Regulations* (Part 5.22, Title 27, 2015).
5. Foss, R., *Rum: A Global History*; Reaktion Books: London, 2012.
6. Bamforth, C. W.; Ward, R. E. *The Oxford Handbook of Food Fermentations*; Oxford Handbooks: Oxford, 2014.
7. McGovern, P. E.; Zhang, J.; Tang, J.; Zhang, Z.; Hall, G. R.; Moreau, R. A.; Nuñez, A.; Butrym, E. D.; Richards, M. P.; Wang, C. S. Fermented Beverages of Pre- and Proto-Historic China. *Proc. Natl. Acad. Sci. U. S. A.* **2004**, *101*, 17593–17598.
8. Shen, Y. *Manual of Chinese Liquor Manufactures Technology*; Light Industry Publishing House of China: Beijing, 1996.
9. Fan, W.; Teng, K. Brewing Technology of Yanghe Daqu Liquor. *Niangjiu* **2001**, *28*, 36–37.
10. Fan, W.; Xu, Y. Research Progress of Enzyme in Daqu. *Niangjiu* **2000**, *27*, 35–40.
11. Wang, H. H. Development and/or Reclamation of Bioresources with Solid State Fermentation. *Proc. Natl. Sci. Council* **1999**, *23*, 45–61.
12. Fan, W. L.; Xu, Y.; Qian, M. C. Identification of Aroma Compounds in Chinese "Moutai" and "Langjiu" Liquors by Normal Phase Liquid Chromatography Fractionation Followed by Gas Chromatography/ Olfactometry. In *Flavor Chemistry of Wine and Other Alcoholic Beverages*; Qian, M. C., Shellhammer, T. H., Eds.; ACS Symposium Series 1104; American Chemical Society: Washington, DC, 2012; pp 303–338.
13. Fan, W.; Qian, M. C. Headspace Solid Phase Microextraction and Gas Chromatography-Olfactometry Dilution Analysis of Young and Aged Chinese "Yanghe Daqu" Liquors. *J. Agric. Food Chem.* **2005**, *53*, 7931–7938.
14. Fan, W.; Qian, M. C. Identification of Aroma Compounds in Chinese "Yanghe Daqu" Liquor by Normal Phase Chromatography Fractionation Followed by Gas Chromatography/ Olfactometry. *Flavour Fragrance J.* **2006**, *21*, 333–342.
15. Fan, W.; Xu, Y.; Lu, H.; Dao, Y. Study on Esterifying Power and Rate of Breaking-Up Ester from Daqu F Chinese Strong Flavored Liquor. *Niangjiu* **2003**, *30*, 10–12.

16. Fan, W.; Qian, M. C. Characterization of Aroma Compounds of Chinese "Wuliangye" and "Jiannanchun" Liquors by Aroma Extract Dilution Analysis. *J. Agric. Food Chem.* **2006**, *54*, 2695–2704.

17. Qian, M. C.; Burbank, H. M.; Wang, Y. Preseparation Techniques in Aroma Analysis. *Sens.-Dir. Flavor Anal.* **2007**, 111–154.

18. Fan, W.; Xu, Y.; Zhang, Y. Characterization of Pyrazines in Some Chinese Liquors and Their Approximate Concentrations. *J. Agric. Food Chem.* **2007**, *55*, 9956–9962.

19. Rizzi, G. P. Formation of Pyrazine from Acyloin Precursors Under Mild Conditions. *J. Agric. Food Chem.* **1988**, *36*, 349–352.

20. Rizzi, G. P. Biosynthesis of Aroma Compounds Containing Nitrogen. In *Heteroatomic Aroma Compounds*; Reineccius, G. A., Reineccius, T. A., Eds.; ACS Symposium Series 826; American Chemical Society: Washington, DC, 2002; pp 132–149.

21. Zhu, B.; Xu, Y.; Fan, W. Tetramethylpyrazine Production by Fermentative Conversion of Endogenous Precursor from Glucose by *Bacillus sp. J. Biosci. Bioeng.* **2009**, *108*, S122.

22. Zhu, B.; Xu, Y.; Fan, W. L. High-Yield Fermentative Preparation of Tetramethylpyrazine by *Bacillus sp.* Using an Endogenous Precursor Approach. *J. Ind. Microbiol. Biotechnol.* **2010**, *37*, 179–186.

23. Zhu, B.; Xu, Y. A Feeding Strategy for Tetramethylpyrazine Production by *Bacillus subtilis* Based on the Stimulating Effect of Ammonium Phosphate. *Bioprocess Biosyst. Eng.* **2010**, *33*, 953–959.

24. Scarpellino, R.; Soukup, R. J. Key Flavors from Heat Reactions of Food Ingredients. In *Flavor Science: Sensible Principles and Techniques*; Acree, T. E., Teranishi, R., Eds.; ACS Professional Reference Book; American Chemical Society: Washington, DC, 1993; pp 309–335.

# Chapter 12

# Current Practice and Future Trends of Aroma and Flavor Research in Chinese Baijiu

Wenlai Fan,[*,1] Yan Xu,[1] and Michael Qian[2]

[1]Key Laboratory of Industrial Biotechnology, Ministry of Education, Laboratory of Brewing Microbiology and Applied Enzymology, Center of Brewing Science and Enzyme Technology, School of Biotechnology, Jiangnan University, Wuxi, Jiangsu, China 214122

[2]Department of Food Science and Technology, Oregon State University, Corvallis, Oregon 97330, United States

[*]E-mail: Wenlai.Fan@163.com.

Chinese baijiu (liquor) is one of the oldest distillates in the world and can be classified into 11 categories according to their aroma characteristics. Volatile compounds are extremely important to the flavor profile of baijiu. However, not all of these compounds are important to the aroma and flavor. The aroma and flavor compounds in baijiu was first studied using gas chromatography–olfactometry (GC–O) in 2004 by Qian and Fan. Since then, more than 300 aroma compounds of 10 different aroma type baijius were detected by GC–O coupled with gas chromatography–mass spectrometer (GC–MS), and key aromas of 5 aroma type baijius were confirmed using GC–O, OAV, recombinate, and omission tests. Esters were very important aroma compounds; these include ethyl hexanoate, ethyl butanoate, and ethyl acetate. Ethyl hexanoate was the key aroma compound in strong aroma and flavor type baijiu. Fatty acids were important aromas for Chinese baijiu. Some compounds are particularly important for certains types of baijiu. $\beta$-Damascenone is a positive aroma compound for light aroma type of baijiu, whereas geosmin contributes a negative characteristic aroma for light aroma and flavor type baijiu. However, geosmin was a key aroma contributor for laobaiganxiang-aroma type baijiu. (E)-2-Enals are very important aroma compounds for chixiang-aroma type baijiu. Sulfur-containing compounds were very important aroma compounds. Many of the volatile compounds also affect the bitter taste and astringent mouthfeel. The knowledge obtained in the past decade about aromas and flavors of baijius has provided a direction for baijiu quality improvement and new technology implementation in industry.

## Introduction

Chinese liquor, baijiu, is one of the oldest distillates in the world. Compared with other spirits, such as vodka, whisky, and brandy, baijiu has an higher ethanol content (normally 40–55% by volume). The annual production of baijiu is approximately 13 million kL in 2017.

Baijiu can be classified into 11 categories according to their aroma characteristics: strong (nongxiang), light (qingxiang), soy-sauce (jiangxiang), sweet-honey (mixiang), complex (jianxiang), roasted-sesame-like (zhimaxiang), Chinese herb-like (yaoxiang), fenxiang, laobaiganxiang, chixiang, and texiang aroma and flavor type liquors (1, 2). The famous Moutai and Langjiu liquors belong to the soy-sauce aroma style liquor, while Fenjiu and Baofeng liquors belong to light aroma style liquor and Wuliangye, Yanghe Daqu, and Louzhouliaojiao liquors belong to strong aroma style liquor. Of these, the strong aroma type liquor accounts for about 60% of total liquor production in China (3).

Baijiu is typically fermented from grains with daqu in a solid state or xiaoqu in a semi-solid state. Daqu and xiaoqu are crude enzymes and starters used for baijiu fermentation. Daqu is typically made from wheat fermented at high temperatures of 60–70 °C for 7–10 days of fermentation (named high-temperature daqu), wheat fermented at high temperatures of 55–60 °C for 7–10 days of fermentation (named moderate-high-temperature daqu), a mixture of wheat, barley, and peas fermented at 50–55 °C for around 10 days (named moderate-temperature daqu), or a mixture of barley and peas fermented at 45–50 °C (named low-temperature daqu) (4). The raw materials for making daqu are cracked, mixed with water, and pressed into a brick-shaped block, which is then incubated under controlled conditions for approximately 30 days and naturally air-dried for 3–6 months before being used to make Chinese liquor (5). Xiaoqu is fermented from milled rice for five to seven days.

The grain used for baijiu manufacturing is sorghum or a mixture of sorghum, wheat, rice, sticky rice, and corn. The grain is ground and cooked, mixed with daqu powder, and then fermented for 30–60 days. After fermentation, the liquor is distilled out with steam. The distillate is then aged for more than two years to develop the bouquet aroma.

Volatile compounds are only trace compounds in baijiu (approximately 2–5 g/L), but they are extremely important for the aroma profile of baijiu. The important aroma- and flavor-active compounds in baijiu are esters, acids, terpenes, aldehydes, alcohols, acetals, ketones, furans, and nitrogen- and sulfur-containing compounds. However, the presence of different volatile compounds can contribute differently to the flavor profile of baijiu. Therefore, the studies of active-aroma compounds in baijiu are useful for estimating the contribution of volatile compounds to the flavor of Chinese liquor. The aim of this review is to discuss the relative importance of volatile compounds to the aroma and flavor of baijius.

## Volatile Compounds Identified in Baijiu

Volatile compounds of Chinese baijiu were isolated and identified using paper chromatography on the "Moutai Test" organized by the Ministry of Light Industry in China in 1963 (6). From then on, paper chromatography and gas chromatography (GC) (especially the latter) were widely applied to trace volatiles analyses of baijiu. By 2007, more than 300 volatiles were detected using GC–flame ionization detector (FID), GC–nitrogen-phosphorus detector (NPD) (7), GC–flame photometric detector (FPD) (8), and GC–mass spectrometry (MS) (9, 10). In 2013, 698 volatiles were identified using liquid-liquid extraction (LLE) and fractionation method coupled by GC–MS, including 167 esters, 67 alcohols, 33 aldehydes, 48 ketones, 18 acetals, 34 fatty acids, 69 terpenes, 111 benzoic compounds, 26 volatile phenols, 21 sulfur-containing compounds, 35 furanic compounds, 31 pyrazines, 11 other nitrogen-containing heterocyclic compounds, 11 lactones, 8 alkanes and alkenes, and 8 other compounds (11, 12).

With application of more sophisticated sampling and new instrumental analysis technology, more volatile compounds in baijiu were detected and identified with liquid extraction-basic fraction and headspace (HS)–solid phase microextraction (SPME) (13, 14). Pyrazines were identified in baijiu, including pyrazine, 2-methylpyrazine, 2,3-dimethylpyrazine, 2,5-dimethylpyrazine, 2,6-dimethylpyrazine, 2-ethylpyrazine, 2-ethyl-3-methylpyrazine, 2-ethyl-5-methylpyrazine, 2-ethyl-6-methylpyrazine, 2,3,5-trimethylpyrazine, 2,6-diethylpyrazine, 2,3-dimethyl-5-ethylpyrazine, 2,5-dimethyl-3-ethylpyrazine, 3,5-dimethyl-2-ethylpyrazine, 2,3,5,6-tetramethylpyrazine, 3,5-diethyl-2-methylpyrazine, 2,3,5-trimethyl-6-ethylpyrazine, 2,5-dimethyl-3-isobutylpyrazine, 2-methyl-6-vinylpyrazine, 2-acetyl-3-methylpyrazine, 2-butyl-3,5-dimethylpyrazine, 2-acetyl-6-methylpyrazine, 2-methyl-6-[(Z)-1-propenyl]pyrazine (tentatively identified), 2-acetyl-3,5-dimethylpyrazine (tentatively identified), 2,5-dimethyl-5-pentylpyrazine (tentatively identified), and 2,3-dimethyl-5-[(Z)-1-propenyl]pyrazine (1, 4, 5, 11–13, 15, 16).

About 20 sulfur-containing compounds were identified using HS–SPME coupled with GC–FPD (17) or GC–PFPD (18), including carbonyl sulfide, hydrogen sulfide, carbon disulfide, methanethiol, ethanethiol, 2-furfruylthiol, dimethyl sulfide (DMS), dimethyl disulfide (DMDS), dimethyl trisulfide (DMTS), dimethyl tetrasulfide, diethyl disulfide (DEDS), S-methyl thioacetate (MeSAc), S-ethyl thioacetate (EtSAc), ethyl (methylthio)acetate, ethyl 3-methylsulfanylpropanoate, 3-methylsulfanylpropanal, 3-methylsulfanylpropan-1-ol , 3-methylthiophene, thiazole, and benzothiazole (17, 18).

A total of 53 carbonyl compounds from baijiu were identified and quantified based on simultaneous extraction and derivatization with o-(2,3,4,5,6-pentafluorobenzyl)hydroxylamine (PFBHA) using HS–SPME coupled with GC–MS-selected ion monitoring mode (SIM) in 2014 (19). The carbonyl compounds of baijiu identified by PFBHA derivatization included (1, 2, 4, 5, 11, 12, 15, 16, 19–21):

1) 15 saturated fatty aldehydes: formaldehyde, acetaldehyde, propanal, butanal, pentanal, hexanal, heptanal, octanal, nonanal, decanal, undecanal, dodecanal, 2-methylpropanal, 2-methylbutanal, and 3-methylbutanal;

2) 14 unsaturated fatty aldehydes: (E)-pentenal, (E)-hexenal, (E)-heptenal, (E)-octenal, (E)-nonenal, (E)-decenal, (E)-undecenal, (E)-dodecenal, (E,E)-2,4-hexadienal, (E,E)-2,4-heptadienal, (Z,E)-2,6-nonadienal, (E,E)-2,4-octadienal, (E,E)-2,4-nonadienal, and (E,E)-2,4-decadienal;

3) 9 methyl ketones: propanone, 2-butanone, 2-pentanone, 2-hexanone, 2-heptanone, 2-octanone, 2-nonanone, 2-decanone, 2-undecanone;

4) 6 aromatic carbonyl compounds: benzaldehyde, phenylacetaldehyde, acetophenone, 2-phenyl-2-butenal, 4-methoxybenzaldehyde, cinnamaldehyde; and

5) 9 others: 2-acetylfuran, 2-acetyl-5-methylfuran, furfural, 5-methylfurfural, 3-hydroxy-2-butanone, 2,3-butandione, 2-acetylpyridine, geranylacetone, and 3-methylsulfanylpropanal.

Currently, over 1000 volatile compounds comprising numerous chemical classes are present in baijiu, including esters, alcohols, terpenes, and sulfur compounds, and the concentrations of the individual components range from several g/L (e.g., ethyl acetate, ethyl hexanoate, and ethyl lactate) to less than a few µg/L (e.g., geosmin).

# Aroma Compounds Threshold in Chinese Baijiu

Currently, a total of 143 volatile aroma compound thresholds in a 46% by volume ethanol–water solution were determined by ASTM method (22), including 31 esters, 15 alcohols, 15 aldehydes, 2 acetals, 5 ketones, 14 fatty acids, 9 pyrazines, 16 aromatic compounds, 7 furans, 14 phenols, 4 γ-lactones, 6 sulfur compounds, and 5 terpenes. Of these, diethyl azelate (diethyl nonane-1,9-dioate) had the highest threshold at 1280 mg/L (2) and 1-octen-3-one had the lowest threshold at 0.067 µg/L (Table 1) (23).

### Table 1. Thresholds of Aroma Compounds Detected in Chinese Baijiu

| Aroma Compounds | Threshold (µg/L) | Aroma Compounds | Threshold (µg/L) |
|---|---|---|---|
| *Esters* | | | |
| Ethyl acetate | 32,600 (24) | Ethyl 2-methylpropanoate | 57.5 (24) |
| Ethyl propanoate | 19,000 (24) | Ethyl 2-methylbutanoate | 18.0 (2) |
| Ethyl butanoate | 81.50 | Ethyl 3-methylbutanoate | 6.89 (24, 25) |
| Ethyl pentanoate | 26.8 (24) | 2-Methylpropyl acetate | 922 (24) |
| Ethyl hexanoate | 55.3 (24) | 3-Methylbutyl acetate | 93.9 (24) |
| Ethyl heptanoate | 13,200 (24) | Isopentyl butanoate | 915 (26) |
| Ethyl octanoate | 12.9 (24) | 2-Methylpropyl hexanoate | 5350 (27) |
| Ethyl nonanoate | 3150 (24) | 3-Methylbutyl hexanoate | 1400 (2) |
| Ethyl decanoate | 1120 (24) | Ethyl lactate | 128,000 (24) |
| Ethyl undecanoate | 1000 (28) | Ethyl 2-hydroxyhexanoate | 51,400 (26) |
| Ethyl dodecanoat | 400 (28) | Diethyl butanedioate | 353,000 (24) |
| Hexyl acetate | 5560 (26) | Diethyl pimelate | 396,000 (2) |
| Propyl hexanoate | 12,800 (26) | Diethyl suberate | 641,000 (2) |
| Butyl hexanoate | 678 (26) | Diethyl azelate | 1,280,000 (2) |
| Pentyl hexanoate | 13,800 (27) | Ethyl *trans*-4-decenoate | 112.3 (27) |
| Hexyl hexanoate | 1890 (26) | | |
| *Alcohols* | | | |
| 1-Propanol | 54,000 (26) | 3-Methylbutanol | 179,000 (24) |
| 1-Butanol | 2730 (24) | 2-Butanol | 50,000 (26) |
| 1-Pentanol | 37,400 (26) | 2-Pentanol | 194,000 (26) |
| 1-Hexanol | 5370 (24) | 2-Heptanol | 1430 (24) |
| 1-Heptanol | 26,600 (2) | 3-Octanol | 393 (26) |
| 1-Octanol | 1100 (26) | 1-Octen-3-ol | 6.12 (24, 25) |
| 1-Nonanol | 806 (26) | *trans*-2-Octen-1-ol | 1008 (25) |
| 2-Methylpropanol | 28,300 (24) | | |

## Table 1. (Continued). Thresholds of Aroma Compounds Detected in Chinese Baijiu

| Aroma Compounds | Threshold (μg/L) | Aroma Compounds | Threshold (μg/L) |
|---|---|---|---|
| *Aldehydes* | | | |
| Acetaldehyde | 1200 (26) | 2-Methylpropanal | 1300 (26) |
| Butanal | 2902 (25) | 3-Methylbutanal | 17 (26) |
| Pentanal | 725 (25) | (E)-2-Octaenal | 15.1 (2) |
| Hexanal | 25.5 (24) | trans-2-Nonenal | 50.5 (2) |
| Heptanal | 410 (25) | (E)-2-Decenal | 12.1 (2) |
| Octanal | 39.6 (25) | (E)-2-Undecenal | 240 (2) |
| Nonanal | 122 (24) | (E,E)-2,4-Decadienal | 7.71 (2) |
| Decanal | 70.8 (24) | | |
| *Acetals* | | | |
| 1,1-Diethoxyethane | 2090 (24) | 1,1,3-Triethoxypropane | 3700 (2) |
| *Ketones* | | | |
| 2-Nonanone | 483 (26) | 1-Octen-3-one | 0.067 (23) |
| 2-Decanone | 483 (2) | 3-Hydroxy-2-butanone | 259 (2) |
| 2-Undecanone | 7.00 (28) | | |
| *Fatty Acids* | | | |
| Acetic acid | 160,000 (24) | Nonanoic acid | 3560 (26) |
| Propanoic acid | 18,200 (26) | Decanoic acid | 13,700 (26) |
| Butanoic acid | 964 (24) | Dodecanoic acid | 9154 (25) |
| Pentanoic acid | 389 (24) | 2-Methylpropanoic acid | 1580 (24) |
| Hexanoic acid | 2520 (24) | 2-Methylbutanoic acid | 5932 (25) |
| Heptanoic acid | 13,300 (26) | 3-Methylbutanoic acid | 1050 (24) |
| Octanoic acid | 2700 (26) | 4-Methylpentanoic acid | 144 (26) |
| *Pyrazines* | | | |
| 2-Methylpyrazine | 121,900 (25) | 2,3-Diethylpyrazine | 172 (26) |
| 2,3-Dimethylpyrazine | 10,820 (25) | 2-Ethyl-5-methylpyrazine | 91,940 (27) |
| 2,5-Dimethylpyrazine | 3202 (25) | 2,3,5-Trimethylpyrazine | 730 (26) |
| 2,6-Dimethylpyrazine | 791 (26) | 2,3,5,6-Tetramethylpyrazine | 80,100 (26) |
| 2-Ethylpyrazine | 21,810 (25) | | |
| *Aromatic Compounds* | | | |
| Styrene | 1400 (2) | Acetophenone | 256 (26) |
| Naphthalene | 159 (2) | 4-(4-Methoxyphenyl)-2-butanone | 5566 (25) |

## Table 1. (Continued). Thresholds of Aroma Compounds Detected in Chinese Baijiu

| Aroma Compounds | Threshold (μg/L) | Aroma Compounds | Threshold (μg/L) |
|---|---|---|---|
| Benzaldehyde | 4200 (24) | Ethyl benzoate | 1430 (24) |
| Phenylacetaldehyde | 262 (24) | Ethyl 2-phenylacetate | 407 (24) |
| trans-Cinnamaldehyde | 4800 (2) | Ethyl 3-phenylpropanoate | 125 (24) |
| 2-Phenyl-2 butanal | 472 (25) | 2-Phenylethyl acetate | 909 (24) |
| Benzyl alcohol | 40,900 (2) | 2-Phenylethyl butyrate | 961 (26) |
| 2-Phenylethanol | 28,900 (24) | 2-Phenylethyl hexanoate | 94 (26) |
| *Furans* | | | |
| Furfural | 44,000 (24) | 2-Furanmethanol | 54,700 (26) |
| 2-Acetylfuran | 58,500 (25) | Ethyl 2-furoate | 132,000 (27) |
| 5-Methylfurfural | 466,000 (26) | Hexyl 2-furoate | 24,200 (27) |
| 2-Acetyl-5-methylfuran | 40,900 (24) | | |
| *Phenols* | | | |
| Phenol | 18,900 (24) | 4-Propylguaiacol | 220 (23) |
| 4-Methylphenol | 167 (24) | 4-Vinylguaiacol | 209 (25) |
| 4-Ethylphenol | 123 (26) | Eugenol | 21.24 (25) |
| 4-Vinylphenol | 39,000 (23) | Isoeugenol | 22.54 (25) |
| Guaiacol | 13.41 (25) | Vanillin | 438 (25) |
| 4-Methylguaiacol | 315 (24) | Ethyl vanillate | 3358 (25) |
| 4-Ethylguaiacol | 123 (24) | Acetylvanillin | 5588 (25) |
| γ-Lactones | | | |
| γ-Octalactone | 2820 (2) | γ-Decalactone | 10.87 (25) |
| γ-Nonalactone | 90.7 (24) | γ-Dodecalactone | 60.68 (25) |
| *Sulfur Compounds* | | | |
| DMDS | 9.13 (25) | 3-(Methylthio)-propanal | 7.12 (2) |
| DMTS | 0.36 (25, 26) | 3-(Methylthio)-1-propanol | 2110 (2) |
| Methyl mercaptan | 2.21 (27) | Ethyl 3-(methylthio)-propanoate | 3080 (27) |
| *Terpenes* | | | |
| Geosmin | 0.11 (25, 29) | a-Terpinenol | 1960 (27) |
| β-Damascenone | 0.12 (24) | Geranyl acetate | 636 (25) |
| a-Cedrene | 11,940 (27) | | |

# Important Aroma Compounds in Baijiu

The study on aroma compounds of baijiu started in 1963 (6). In that year, ethyl hexanoate was detected and identified by paper chromatography, and it was found to be the key aroma compounds of the raw baijiu distilled from the fermented grains located in the bottom of fermentor (30), which was named strong aroma type baijiu in 1979.

The systematic investigation of aroma and flavor compounds in baijiu was first approached by Fan and Qian by using GC–olfactometry (GC–O)-MS in 2005 (21). Currently, important aroma compounds of 10 aroma type baijius were detected by GC–O coupled with GC–MS, and key aromas of 5 aroma types of baijiu were evaluated by GC–O, odor activity value (OAV), recombinate, and omission test.

## Strong Aroma Type Baijiu

The strong aroma type baijiu is made from sorghum or a mixture of sorghum, wheat, rice, sticky rice, and corn. The moderate-high-temperature daqu, which is naturally fermented using ground wheat or wheat, barley, pea, and water, are saccharification and fermentation agents. The fermentation period of daqu is about 30 days (31, 32).

The crushed sorghum is mixed with jiupei (5–5.5 times of sorghum weight), which is fermented grains, and paddy hull (14–25%). The mixture is distilled with steam. The distilled jiupei is added to hot water, cooled by forced air, mixed with daqu powder (20–25% by sorghum weight), and fermented 60 days in solid state, sometimes up to 90 or 120 days in a special fermentor made of clay, the side of which is coated with a layer of fermentation mud made of clay, spent grains, bean cake powder and fermentation bacteria (*Clostridium* sp.) (31, 32). After fermentation, the fermented grains (jiupei) are mixed with creaked sorghum and rice hull and distilled in a special distillation (zongtong) by steam. This producing process is known as a continuous-cycle process (xuchafa process). In each run, some spent grains (jiuzao) are moved (33).

The raw baijiu are aged in a pottery jar for more than two years to develop the bouquet aroma. The finished product has strong fruity, sweet, and alcoholic aromas (34).

The main producing regions of strong aroma type baijiu are located in the Sichuan, Jiangsu, and Anhui provinces of China. The famous brands of this aroma type include Wuliangye, Yanghe Daqu, Louzhouliaojiao, and Gujinggong.

Ethyl hexanoate, ethyl butanoate, ethyl acetate, and ethyl lactate were dominating esters in strong aroma type baijiu (nongxiangxing baijiu). Ethyl hexanoate was the key aroma compound (32, 35, 36), with a concentration range from 1.2 g/L to 2.8 g/L in high alcohol baijiu and from 0.70 g/L to 1.2 g/L in low alcohol baijiu (37).

The short-chain to medium-chain fatty ethyl esters, especially ethyl hexanoate, are extremely important aroma compounds of strong aroma type baijiu. Using HS–SPME-aroma extract dilution analysis (AEDA), it was found that ethyl hexanoate, ethyl heptanoate, ethyl benzoate, and butyl hexanoate were very important aromas in both young and aged Yanghe Daqu, one of most typical strong aroma type of baijiu (21). Using LLE coupled with normal phase chromatography fractionation and GC–Osme study, ethyl hexanoate and ethyl butanoate were determined to be most important esters (5). In 2006, another two of the most typical strong aroma type baijiu (Wuliangye and Jiannanchun) were researched using the AEDA technique. On the basis of flavor dilution (FD) values, the most important aroma compounds in both liquors could be ethyl butanoate, ethyl pentanoate, ethyl hexanoate ethyl octanoate, butyl hexanoate, ethyl 3-methylbutanoate, hexanoic acid, and 1,1-diethoxy-3-methylbutane (4). Further research using OAV suggests that the top four aroma compounds were ethyl hexanoate (OAV of 33,454), ethyl butanoate (OAV of 1815), ethyl pentanoate (OAV of 1633), and ethyl octanoate (OAV of 881) (Table 2) (*38*).

Ethyl hexanoate has a high concentration in strong aroma type baijiu (*38*). The concentration of ethyl hexanoate was 1.85 g/L, its threshold was 55.3 μg/L, and its OAV (33,454) was highest among all aroma compounds using GC–O (Table 2). The results of recombination and omission experiments of 24 aroma compounds showed ethyl hexanoate was the key aroma compound (Table 3).

The short-chain to medium-chain fatty acids are also very important aroma compounds for strong aroma type baijiu. Many researchers have demonstrated the fatty acids, especially hexanoic, butanoic, 3-methylbutanoic, and pentanoic acids, were important aroma compounds (*4, 5, 32*). The OAVs of butanoic and hexanoic acids were 191 and 146, respectively (Table 2).

**Light Aroma Type Baijiu**

Sorghum is the raw material for making light aroma type baijiu. Sorghum is cracked, added to water, and cooked with steam. The low-temperature daqu powder, which naturally ferments from barley (60% by weight) and peas (40% by weight) for 30 days, is added to the cooked raw material. The mixture is fermented in a solid state in a ceramic vat, not contacting the clay, for about 28 days, and then jiupei (fermented grains) with mixed paddy hull is distilled with steam (*31, 33*).

The distilled jiupei is cooled by forced air, mixed with daqu powder, not added to the sorghum, and then refermented in a ceramic vat for 28 days. After fermentation, the jiupei mixed with the rice hull is distilled in zengtong. This distilled jiupei is known as spent grains or jiuzao. This production process is called the qingchafa process (*31, 33*).

**Table 2. Orthonasal Odor Concentrations, Thresholds, and OAV of 24 Odorants in Strong Aroma Type Baijiu (*38*). Reproduced with permission from reference (*38*). Copyright 2010 Nanke Wu.**

| No. | Compounds | Concentration (μg/L) | OAV |
|---|---|---|---|
| 1 | Ethyl hexanoate | 1,850,000 | 33,454 |
| 2 | Ethyl butanoate | 147,900 | 1815 |
| 3 | Ethyl pentanoate | 43,770 | 1633 |
| 4 | Ethyl octanoate | 11,360 | 881 |
| 5 | Butanoic acid | 184,500 | 191 |
| 6 | Hexanoic acid | 368,000 | 146 |
| 7 | Pentanoic acid | 31,840 | 82 |
| 8 | Ethyl acetate | 1,715,000 | 53 |
| 9 | 1-Butanol | 44,280 | 19 |
| 10 | Octanoic acid | 37,640 | 14 |
| 11 | Ethyl lactate | 1,410,000 | 11 |
| 12 | 3-Methylbutanoic acid | 8820 | 8 |
| 13 | 1-Hexanol | 39,230 | 7 |

**Table 2. (Continued). Orthonasal Odor Concentrations, Thresholds, and OAV of 24 Odorants in Strong Aroma Type Baijiu (38)**

| No. | Compounds | Concentration ($\mu g/L$) | OAV |
|-----|-----------|---------------------------|-----|
| 14 | Ethyl 3-phenylpropanoate | 791.2 | 6 |
| 15 | 2-Methylpropanoic acid | 8756 | 6 |
| 16 | Butyl hexanoate | 2923 | 4 |
| 17 | Acetic acid | 582,700 | 4 |
| 18 | 3-Methylbutanol | 601,000 | 3 |
| 19 | Ethyl 2-phenylacetate | 1056 | 3 |
| 20 | Hexyl hexanoate | 3593 | 2 |
| 21 | 3-Methylbutyl hexanoate | 2175 | 2 |
| 22 | 2-Methylpropanol | 43,530 | 2 |
| 23 | 4-Methylphenol | 189.6 | 1 |
| 24 | 2-Phenylethyl hexanoate | 96.45 | 1 |

**Table 3. Omission Experiments from the Complete Recombinate of Strong Aroma Type Baijiu (38). Reproduced with permission from reference (38). Copyright 2010 Nanke Wu.**

| Odorants Omitted from the Complete Recombinate | Similarity [a] |
|-----------------------------------------------|----------------|
| All esters | 0.89 |
| All esters except for ethyl hexanoate | 1.73 |
| Ethyl hexanoate | 1.06 |
| Ethyl acetate | 3.97 |
| Ethyl butanoate | 3.80 |
| Ethyl lactate | 3.29 |
| All acids | 3.85 |
| Hexanoic acid | 3.21 |
| Butanoic acid | 3.33 |
| All alcohols | 2.55 |
| 1-Butanol | 3.45 |
| All aromatic compounds | 4.64 |

[a] 0 = not similar to the strong aroma type baijiu; 5 = similar to strong aroma type baijiu.

The raw distillates are matured in ceramic pottery jars for one to two years. The light aroma type baijiu impacts fruit, flower, and alcoholic aromas (*31*). The light aroma type baijiu are mainly manufactured in the Shanxi, Henan, and Qinghai provinces, and Beijing city, China. The Fenjiu, Baofeng, Qingkejiu, and Erguotou liquors are among the most famous in China.

Before 2008, it was commonly believed that ethyl acetate and ethyl lactate were the key aroma compounds in light aroma type baijiu (*39*). In fact, these esters are the highest level volatile compounds of all light aroma type baijiu. The average concentrations of ethyl acetate were 2.89 g/L for Fenjiu raw liquor, 3.42 g/L for Erguotou raw liquor, and 2.33 g/L for Baofengjiu raw liquor (*40*), and the threshold of ethyl acetate is 32.6 mg/L in 46% by volume in an ethanol–water system (*25*), and their OAVs were approximately 89, 105, and 71, respectively. The average levels of ethyl lactate are 1.79 g/L for Fenjiu raw liquor, 1.61 g/L for Erguotou raw liquor, 0.54 g/L for Baofengjiu raw liquor (*40*), the threshold of ethyl lactate is 128 mg/L (*25*), and the OAVs were approximately 14, 13, and 4, respectively. Both ethyl acetate and ethyl lactate are very important aromas, but ethyl acetate is probably more important due to the higher OAVs in light aroma type baijiu.

From 2006 to 2014, aroma compounds of light aroma type baijiu, including Fenjiu, Baofengjiu, Erguotou, and Qingkejiu, fermented with daqu, were studied by GC–O coupled with GC–MS (*24*, *39*, *41*). Except for ethyl acetate and ethyl lactate, more aroma compounds were identified.

Ethyl octanoate, $\beta$-damascenone, 2-phenylethyl acetate, ethyl acetate, 2-phenylethanol, vanillin, 4-ethyl guaiacol, ethyl 2-phenylacetate, phenylacetaldehyde, ethyl benzoate, ethyl 3-phenylpropanoate, 4-methylphenol, guaiacol, 1-butanol, 2-methylpropanol, 3-methylbutanol, 4-ethylphenol, furfural, acetic acid, 2-methylpropanoic acid, 3-methylbutanoic acid, 2-acetyl-5-methylfuran, and $\gamma$-nonalactone (Osme value of at least 3.00) were very important aromas of Fenjiu liquor discovered by GC–Osme coupled with GC–MS (*24*, *39*).

Ethyl acetate, $\beta$-damascenone, 2-methylpropanol, 3-methylbutanol, ethyl octanoate, ethyl 2-phenylacetate, ethyl 3-phenylpropanoate, 2-phenylethanol, and $\gamma$-nonalactone (Osme value of at least 3.50) were very important aromas of Baofengjiu liquor discovered by GC–Osme and GC–MS (*24*). On the basis of their OAVs, the most important aromas were proposed to be 3-methylbutanal (OAV of 2620), ethyl octanoate (OAV of 472), DMTS (OAV of 237), and 3-methylbutyl acetate (OAV of 127) (*40*).

The very important aromas (Osme of at least 3.50) in the Qingkejiu liquor were ethyl butanoate, $\beta$-damascenone, 1,1-diethoxyethane, 2-methylpropanol, ethyl octanoate, 1-octen-3-ol, 2-methylpropanoic acid, butanoic acid, phenylacetaldehyde, ethyl 3-phenylpropanoate, and 2-phenylethanol (*24*).

On the basis of their OAVs, the most important aromas of Erguotou baijiu were suggested to be 3-methylbutanal (OAV of 3672), ethyl 2-methylpropanoate (OAV of 2291), ethyl octanoate (OAV of 709), $\beta$-damascenone (OAV of 172), DMTS (OAV of 159), ethyl hexanoate (OAV of 158), geosmin (OAV of 132), ethyl 3-methylbutanoate (OAV of 112), and ethyl acetate (OAV of 105) (Table 4) (*40*).

Further research found ethyl octanoate, 3-methylbutanal, ethyl octanoate, 1,1-diethoxyethane, $\beta$-damascenone were very important aromas in Fenjiu liquor, based on their OAVs being greater than 100 (Table 4) (*24*). The results of omission experiments from the complete recombinate showed that $\beta$-damascenone and DMTS were key aroma compounds (Table 5). Ethyl acetate, acetic and propanoic acids, and geosmin were very important aromas (*24*).

**Table 4. Orthonasal Odor Concentrations, Thresholds, and OAVs of 29 Odorants in Light Aroma Type Fenjiu (24). Reproduced with permission from reference (24). Copyright 2014 ACS Publications.**

| No. | Compounds | Concentration (μg/L) | OAV |
|---|---|---|---|
| 1 | 3-Methylbutanal | 29,189 | 1717 |
| 2 | Ethyl octanoate | 5039 | 391 |
| 3 | 1,1-Diethoxyethane | 418,600 | 200 |
| 4 | $\beta$-Damascenone | 19.08 | 159 |
| 5 | Ethyl acetate | 2,124,000 | 65 |
| 6 | 3-Methylbutyl acetate | 5870 | 63 |
| 7 | Ethyl hexanoate | 3089 | 56 |
| 8 | DMTS | 19.52 | 54 |
| 9 | Ethyl lactate | 4,913,000 | 38 |
| 10 | Geosmin | 3.20 | 29 |
| 11 | Ethyl butanoate | 1820 | 22 |
| 12 | Hexanal | 484.2 | 19 |
| 13 | Ethyl 3-methylbutanoate | 118.8 | 17 |
| 14 | Decanal | 798.2 | 11 |
| 15 | Ethyl 2-methylpropanoate | 641.6 | 11 |
| 16 | Ethyl pentanoate | 256 | 10 |
| 17 | Oct-1-en-3-ol | 48.16 | 8 |
| 18 | 2-Methylpropanol | 194,000 | 7 |
| 19 | Phenylacetaldehyde | 1776 | 7 |
| 20 | Ethyl decanoate | 6566 | 6 |
| 21 | 2-Methylpropanoic acid | 7843 | 5 |
| 22 | $\gamma$-Nonalactone | 275.0 | 3 |
| 23 | Guaiacol | 38.53 | 3 |
| 24 | 3-Methylbutanol | 513,600 | 3 |
| 25 | Ethyl dodecanote | 1136 | 3 |
| 26 | 1-Butanol | 6160 | 3 |
| 27 | Acetic acid | 397,400 | 2 |
| 28 | 2-Methylpropyl acetate | 2136 | 2 |
| 29 | 4-Ethylphenol | 202.2 | 2 |

**Table 5. Omission Experiments from the Complete Recombinate of Light Aroma Type Fenjiu (24). Reproduced with permission from reference (24). Copyright 2014 ACS Publications.**

| No. | Odorants Omitted from the Complete Recombinate | $n^{a}$ | Significance[b] |
|---|---|---|---|
| 1 | All esters | 10 | *** |
| 1A | Ethyl octanoate | 4 | |
| 1B | Ethyl acetate | 8 | ** |
| 1C | Ethyl lactate | 7 | * |
| 1D | Ethyl acetate and ethyl lactate | 8 | ** |
| 1E | All esters except ethyl acetate and ethyl lactate | 6 | |
| 2 | 1,1-Diethoxylethane | 5 | |
| 3 | β-Damascenone | 9 | *** |
| 4 | Aromatic compounds | 5 | |
| 5 | Acetic acid and 2-methylpropanoic acid | 8 | ** |
| 6 | All alcohols | 6 | |
| 7 | All aldehydes | 3 | |
| 8 | Geosmin | 8 | ** |
| 9 | DMTS | 9 | *** |
| 10 | γ-Nonalactone | 3 | |
| 11 | 1-Octen-3-ol | 2 | |

[a] number of correct judgments from ten assessors evaluating the aroma difference by means of a triangle test. [b] significance: ***, very highly significant ($\alpha \leq 0.001$); **, highly significant ($\alpha \leq 0.01$); *, significant ($\alpha \leq 0.05$).

### Soy-Sauce Aroma Type Baijiu

In the process of making soy-sauce aroma type baijiu, ground sorghum is added to water and cooked, and high-temperature daqu powder is then added, which naturally ferments from wheat, for 30 days. The cooked sorghum is stacked on the ground for two to three days to initate the fermentation (at a final temperature of 40–50 °C), then moved to a special fermentor for fermenting in a solid state for 30 days. The fermentor is a cuboid vessel made of rectangular stone, and the bottom is coated with a layer of fermentation mud made of clay, spent grain, bean cake powder, and fermentation bacteria (*Clostridium* sp.). After fermentation, the fermented grains are directly distilled in zengtong by steam (*31, 33, 42*).

These base liquors are stored in pottery jars, aged for less than five years, blended, and marketed accordingly. The soy-sauce aroma type baijiu has fruit, soy-sauce-like, and baked aromas (*1, 42, 43*).

The soy-sauce aroma type baijiu are produced in the Chishui River Valley located at the junction of the Guizhou and Sichuan provinces. The famous brands are Moutai (also named Maotai), Langjiu, and Xijiu liquors.

The volatile compounds of soy-sauce aroma type baijiu were the most complex among all aroma type baijiu (*11, 12, 44*). A total of 983 GC peaks were detected using GC×GC-time-of-flight (TOF)-MS in 2007 (*44*). A total of 468 volatiles were identified by GC–MS using LLE and fractionation, including 126 esters, 40 alcohols, 19 aldehydes, 28 ketones, 10 acetals, 27 fatty acids, 56 terpenes, 61 benzoic compounds, 16 volatile phenols, 12 sulfur-containing compounds, 27 furanic compounds, 31 pyrazines, 4 other nitrogen-containing heterocyclic compounds, 6 lactones, 2 alkanes and alkenes, and 3 other compounds (*11, 12*). Although a large number of volatile compounds were identified, the aroma-contributing compounds remained active in the study.

It was postulated in 1960s that 4-ethylguaiacol was the key aroma compound of Moutai baijiu based on paper chromatography identification. Until the 1980s, some researchers found this state was mistaken by addition test (*45*).

We first systematically investigated the aroma-active compounds of soy-sauce aroma type baijiu using fractionation and GC–O technique. The aroma compounds of Moutai and Langjiu liquors were isolated by LLE, and further fractionated into acidic, basic, and neutral fractions. The neutral fraction was separated into seven subfractions by using a normal phase liquid chromatography column based on their polarity. A total of 186 aroma-active compounds were further identified by GC–O and GC–MS. Among these, ethyl hexanoate, hexanoic acid, 3-methylbutanoic acid, 3-methylbutanol, 2,3,5,6-tetramethylpyrazine, ethyl 2-phenylacetate, 2-phenylethyl acetate, ethyl 3-phenylpropanoate, 4-methylguaiacol, and $\gamma$-decalactone had the highest aroma intensities (*1, 14*).

Quantitative results of soy-sauce aroma type baijiu were achieved by Fan and coworkers using SBSE coupled with GC–MS in 2011 (*16*). A total of 76 volatile compounds were quantitated from 14 Chinese liquors, including 25 esters, 10 alcohols, 9 aldehydes and ketones, 8 aromatic compounds, 5 furans, 3 nitrogen-containing compounds, 6 acids, 4 phenols, 3 terpenes, 1 sulfide-containing compound, 1 lactone, and 1 acetal. Principal component analysis could further classify the 14 Chinese liquor samples into three groups. The chemometrics approach revealed that Liang jiu liquor with soy-sauce aroma could exhibit more prominent sauce flavor after extended storage, and its special manufacturing practice was responsible for the soy-sauce flavor. The skeleton compounds in soy-sauce aroma type baijiu were investigated by liquid-liquid microextraction (LLME) (*43*). The total of 48 aroma-active compounds were identified and quantified by GC–O and GC–MS. Of these, 29 compounds were found to be to skeleton compounds. The overall aroma of soy-sauce aroma type liquor, while excluding the typical aromas, could be mimicked by an aroma recombination consisting of 52 compounds. According to omission experiments, esters and alcohols of skeleton compounds contributed greatly to the aroma of Chinese soy-sauce aroma type liquor, and ethyl hexanoate, ethyl acetate, ethyl butanoate, 1-propanol, 2-phenylethanol, and 3-methylbutanol were important skeleton aroma compounds.

In order to elucidate the differences between aroma compounds in liquors with different aroma styles and the reasons, aroma compounds of Xijiu in soy-sauce aroma and strong aroma type were investigated by GC–O and OAV analysis. The research results showed that ethyl hexanoate, butanoic acid, 3-methylbutanoic acid, hexanoic acid, and DMTS were considered to be the most powerful odorants in both liquor samples (aroma intensity of at least 3.5) by GC–O (*26*).

Although many aroma-active compounds have been identified in soy-sauce aroma type baijiu, it is still inconclusive as to which compound(s) are responsible for the characteristic aroma. To find these key aroma compounds, an interesting research study was carried out by Fan in 2012 (*46*). In this experiment, 0.1 N NaOH were dropped wisely and added to 10 mL soy-sauce aroma type baijiu, and the key aromas (i.e., soy sauce–like aroma) slowly disappeared as the pH increased in this liquor sample. When the pH was close to 9, no soy-sauce aroma was smelled. The pH was adjusted with 4 N $H_2SO_4$, and the soy sauce–like aroma was sniffed again. When the pH was 5, the soy sauce–like aroma was strongest, and when the pH dropped to a range of 2.0–2.5, the soy sauce–like aroma would disappear again. The soy sauce–like aroma was detected by GC–O, but key aroma compounds could not be identified by GC–MS because of interference from fatty acids (47). Further studies still need to confirm which compounds give the characteristic aroma of soy-sauce aroma baijiu.

**Sweet-Honey Aroma Type Baijiu**

The strong, light, soy-sauce, and sweet-honey aroma type baijiu are four basic aroma types of Chinese baijiu. The sweet-honey aroma type baijiu great differs from upper three aroma types in three ways:

1) the raw material of sweet-honey aroma type baijiu is rice, not sorghum or other grains;
2) it uses xiaoqu as saccharification and fermentation agents, not daqu; and
3) cooked rice is fermented in a liquid state, while the upper three aroma types are fermented in a solid state.

The xiaoqu, as a starter for xiaoqu baijiu, is a kind of mold culture, which contains *Rhizopus*, *Aspergillus*, yeast, bacteria, and complex enzymes (2). The xiaoqu is typically made from rice. The raw material rice is ground, added to water, And then mixed with certain Chinese herbs. The mixture is pressed in certain shapes such as a round-like shape named a xiaoqu pellet (Figure 1 left) or a cake-like shape named bingqu (Figure 1 middle). The mixture is naturally fermented in a solid state in a special room. The fermenting temperature of xiaoqu is controlled at less than 37 °C. After approximately four days, the xiaoqu fermentation is fisished (31, 48).

*Figure 1. The shape of xiaoqu.*

The selected rice, as a raw material for xiaoqu baijiu fermentation, is added to water, cooked with steam, cooled, and then mixed with about 2% of xiaoqu powder. The mixture is saccharified in a solid state in a pottery jar (approximately 200 L) or in a closed stainless steel vessel. The saccharification temperature is usually at less than 32 °C in summer or less than 36 °C in winter for 16 to 20 hours. After saccharification, the mixture is added to water and fermented in a liquid state in a ceramic pottery vat (approximately 500 L) or a closed stainless steel vessel. The fermenting temperature is less than 30 °C for approximately 96 hours. The distillation is carried out in alambique. The raw distillates are aged in pottery jars (48).

The main producing and sales regions of sweet-honey aroma type baijiu are the Guangdong and Guangxi provinces in southern China. The important brand is Guilin sanhua baijiu.

Historically, 2-phenylethanol and ethyl lactate were believed to be very important aromas of sweet-honey type baijiu (31). Using AEDA coupled with GC–MS, it was determined that ethyl acetate, 3-methylbutyl acetate, ethyl octanoate, acetic acid, propanoic acid, butanoic acid, 2-phenylethanol, 4-ethylguaiacol, and $\gamma$-nonalactone (FD values of at least 81) are important aroma compounds (2).

## Chixiang Aroma Type Baijiu

The manufacturing process of chixiang aroma type baijiu is very similar to that of sweet-honey aroma type baijiu before aging the raw baijiu. The distinctions are as follows (2):

1) The xiaoqu is made from cooked rice and soybean.
2) The semi-solid-state fermentation is typically carried out at 28–32 °C for 20 days under anaerobic conditions in a 50-kL stainless steel vessel.
3) In the fermentation, the starch saccharification and sugar fermentation are carried out simultaneously.
4) The base distillate (Chinese named zhaijiu) mixed with a piece of cooked pork meat (approximately 20% of the distillate by weight) is aged in a sealed pottery jar at room temperature for 30 days. After filtration, the distillate is matured in a stainless steel tank for 20 days.

The chixiang aroma type baijiu is only produced in the Guangdong province. The important brands are Jiujiang shuangzheng and yubingshao baijius.

Before 2015, most studies focused on the volatile compounds of chixiang aroma type baijiu and many papers analyzed 2-phenylethanol, diethyl pimelate (diethyl heptane-1,7-dioate), diethyl suberate (diethyl octane-1,8-dioate), diethyl azelate (diethyl nonane-1,9-dioate), and 3-methylsulfanylpropan-1-ol, and these compounds were thought to be very important aroma compounds (49–52).

The average concentration of 2-phenylethanol of chixiangxing aroma type baijiu was highest among Chinese baijiu with 66.0 mg/L (20.0–127.5 mg/L, n = 59), compared to 22.3 mg/L (n = 1) for soy-sauce baijiu, 2.7 mg/L (1.9–11.5 mg/L, n = 8) for strong aroma baijiu, 6.4 mg/L (4.6–9.2 mg/L, n = 5) for light aroma baijiu, 37.3 mg/L (31.5–43.6 mg/L, n = 5) for sweet-honey aroma baijiu, and 9.9 mg/L (n = 1) for fenxiang aroma baijiu (15, 49). But, the threshold of 2-phenylethanol was 11.06 mg/L in a 46% by volume ethanol–water solution (25). The OAV of 2-phenylethanol is about 6 for chixiang aroma type baijiu.

The average concentration of 3-methylsulfanylpropan-1-ol in chixiang baijiu was the highest among all Chinese baijiu at approximately 0.7 mg/L (0.2–2.0 mg/L, n = 56), compared to 0.4 mg/L for sweet-honey aroma type baijiu and 0.7 mg/L for roasted-sesame-like aroma type baijiu (15, 52). The threshold of 3-methylsulfanylpropan-1-ol was 2.11 mg/L in a 46% by volume ethanol–water solution (2, 25). The OAV of 3-methylsulfanylpropan-1-ol was an average of 0.35 (from 0.09 to 0.95) in chixiang baijiu, so it should be an important aroma contributor.

The average concentrations of diethyl pimelate, diethyl suberate, and diethyl azelate in chixiang aroma type baijiu were 0.9 mg/L (0.3–2.3 mg/L, n = 56), 2.3 mg/L (0.7–6.2 mg/L, n = 56), and 0.8 mg/L (0.2–2.0 mg/L, n = 56), respectively (49). But, these fatty acid diethyl esters have high threshold values of 396 mg/L, 641 mg/L, and 1280 mg/L in a 46% by volume ethanol–water solution (2), respectively. Their OAVs are far less than 1, thus, they do not contribute to the aroma.

The aroma compounds of chixiang aroma type baijiu were studied using both GC–O intensity and AEDA technique. The resulting compounds were (E)-2-octenal, (E)-2-nonenal, (E)-2-decenal, (E)-2-undecenal, hexanal, heptanal, octanal, nonanal, (E,E)-2,4-decadienal, 3-methylsulfanylpropan-1-ol, 3-methylbutanol, 2-phenylethanol, ethyl hexanoate, ethyl 3-phenylpropanoate, γ-nonalactone, acetic acid, butanoic acid, propanoic acid, butanoic acid, pentanoic acid, ethyl 2-methylpropanoate, ethyl acetate, ethyl 3-methylbutanoate, 3-methylbutyl acetate, 3-methylsulfanylpropan-1-ol, and γ-nonalactone (FD of at least 81) (2, 53).

Omission experiments from a complete recombinate showed that (E)-2-octenal and 2-phenylethanol were very important aromas, and (E)-2-nonenal was the key aroma compound for chixiang aroma type baijiu (Table 6) (2).

**Table 6. Omission Experiments from the Complete Recombinate of Chixiang Aroma Type Baijiu (2). Reproduced from reference (2). Copyright 2015 ACS Publications.**

| No. | Odorants Omitted from the Complete Recombinate | $n^a$ | Significance $^b$ |
|---|---|---|---|
| 1 | (E)-2-Octenal, (E)-2-nonenal, (E)-2-decenal, (E,E)-2,4-decadienal, octanal, nonanal | 10 | *** |
| 1–1 | (E)-2-Nonenal | 9 | *** |
| 1–2 | (E)-2-Octenal | 8 | ** |
| 1–3 | (E)-2-Decenal | 4 | |
| 1–4 | (E,E)-2,4-Decadienal | 4 | |
| 1–5 | Octanal | 5 | |
| 1–6 | Nonanal | 3 | |
| 2 | 2-Phenylethanol, ethyl 3-phenylpropanoate, ethyl benzoate | 8 | ** |
| 2–1 | 2-Phenylethanol | 8 | ** |
| 2–2 | Ethyl 3-phenylpropanoate | 6 | |
| 2–3 | Ethyl benzoate | 4 | |
| 3 | Hexanal, heptanal | 7 | * |
| 3–1 | Hexanal | 7 | * |
| 3–2 | Heptanal | 4 | |
| 4 | All esters | 5 | |
| 5 | 2-Methylpropanol, 3-methylbutanol, and 1-propanol | 6 | |
| 6 | Fatty acids | 5 | |
| 7 | $\gamma$-Nonalactone | 3 | |
| 8 | 3-Hydroxy-2-butanone | 5 | |
| 9 | 3-Methylsulfanylpropan-1-ol | 3 | |

$^a$ number of correct judgments from 10 assessors evaluating the aroma difference by means of a triangle test. $^b$ significance: ***, very highly significant ($\alpha \leq 0.001$); **, highly significant ($\alpha \leq 0.01$); *, significant ($\alpha \leq 0.05$).

## Laobaiganxiang Aroma Type Baijiu

The producing process of laobaiganxiang aroma type baijiu is similar to the processs of light aroma type baijiu. The biggest difference are as follows (39):

1) the raw material used for making daqu is wheat, not barley and pea;
2) the fermentation temperature for daqu is 50–55 °C (i.e. using moderate-temperature daqu in the fermentation of sorghum); and

3) the xuchafa process is used instead of the qingchafa process, which produces light aroma type baijiu.

The laobaiganxiang aroma type baijiu are only produced in the Hebai province. The Hengshui laobaiga brand is very famous in China.

The main volatile compounds of laobaiganxiang aroma type baijiu are similar to the light aroma type baijiu. For example, the concentrations of ethyl acetate and ethyl lactate were 1.08 g/L and 1.87 g/L, and their OAVs were approximately 33 and 15 (39). respectively. Recently, most of research has shown that the laobaiganxiang aroma type baijiu belonged to the light aroma type baijiu.

The important aroma compounds in laobaiganxiang liquor were ethyl acetate, ethyl 2-methylpropanoate, 2-methylpropyl acetate, ethyl pentanoate, ethyl decanoate, diethyl butanedioate, phenylacetaldehyde, ethyl benzoate, ethyl 2-phenylacetate, 2-phenylethyl acetate, ethyl 3-phenylpropanoate, 2-phenylethanol, 2-phenylacetic acid, 3-phenylpropanoic acid, 4-ethylguaiacol, 4-methylphenol, vanillin, 3-methylbutanol, 1-octanol, 2-methylpropanoic acid, butanoic acid, 3-methylbutanoic acid, hexanoic acid, heptanoic acid, octanoic acid, 2-ethylfuran, 2-acetyl-5-methylfuran, 2,3,5-trimethylpyrazine, 3-ethyl-2,5-dimethylpyrazine, tetramethylpyrazine, 1,1-diethoxy-3-methylbutane, 1,1,3-triethoxypropane, DMTS, and $\gamma$-nonalactone (Osme value of at least 3.00) based on GC–Osme coupled with GC–MS (39). Further research found that geosmin was the key aroma compound (29, 54).

## Roasted-Sesame-Like Aroma Type Baijiu

The production of roasted-sesame-like aroma type baijiu only began in the 1960s, while other aroma type baijiu have a longer history (55, 56).

This aroma type baijiu uses naturally fermenting daqu, pure-culture moldy brans, or daqu coupled with moldy brans as saccharifying and fermentation agents (56, 57). The moldy brans have three types: bacteria, *Aspergillus hanoi*, and yeast moldy brans (57, 58).

Sorghum (95% by weight) and brans (5% by weight) are the raw materials for baijiu production, which use the xuchafa process. Sorghum is cracked, and mixed with brans and fermented grains (jiupei). The mixture is distilled with steam. The distilled jiupei is added to daqu powder and three moldy brans and then stacked on the ground for one to two days to initate the fermentation (final temperature 44–50 °C). The stacked mixture is moved to a fermentor for fermenting in a solid state for 30 days. The fermentor is a cuboid vessel made of rectangular stone, and the bottom is coated with a layer of fermentation mud made of clay, spent grain, bean cake powder, and fermentation bacteria (*Clostridium* sp.). After fermentation, the jiupei mixed with the raw material is distillated, and the raw liquors are aged in a pottery jar for two to three years (27, 31, 56, 57).

The roasted-sesame-like aroma type baijiu is produced in the Shandong province. The famous brands are Jingzhi and Bandaojing liquors.

In the early 1980s, diethyl butanedioate, ethyl lactate, 2-phenylethanol, and ethyl propanoate were proposed as important aroma compounds for roasted-sesame-like aroma type baijiu due to their high concentrations at 480 mg/L, 1017 mg/L, 244 mg/L, and 949.6 mg/L, respectively (55, 59). The calculated OAVs for these compounds would be 1, 8, 8, and 50, respectively, suggesting that some of them could be important aromas.

3-Methylsulfanylpropan-1-ol of roasted-sesame-like aroma type Jingzhi baijiu was first detected and identified by GC–MS in 1994 (60). Next year, 3-methylsulfanylpropanal and ethyl 3-methylfanylpropanoate were found in roasted-sesame-like aroma type baijiu (60). The concentration of 3-methylsulfanylpropan-1-ol was from 1.7 to 3.8 mg/L (52). The threshold of 3-

methylsulfanylpropan-1-ol was higher at 2.11 mg/L in a 46% by volume aqueous ethanol solution (2), and its OAV was only 0.80–1.80. Thus, 3-methylsulfanylpropan-1-ol was not an important aroma compound for most roasted-sesame-like aroma type baijiu.

GC–O technology was used to study the roasted-sesame-like aroma type baijiu in 2015 (27). Based on the Osme value, the important aroma compounds were ethyl butanoate, ethyl hexanoate, ethyl octanoate, 2-heptanol, benzaldehyde, 2-phenylethanol, furfural, acetic acid, butanoic acid, hexanoic acid, 2,6-dimethylpyrazine, and 2,3,5,6-tetramethylpyrazine (Osme values of at least 2.5) (27). Further studies found that ethyl hexanoate, ethyl octanoate, ethyl butanoate, ethyl 2-methylpropanoate, 3-methylbutanal, DMTS, ethyl 3-methylbutanoate, ethyl pentanoate, ethyl 2-methylbutanoate, 3-methylbutyl acetate, 1,1-diethoxyethane, DMDS, and methanethiol were very important aroma compounds on the basis of their OAVs (OAV of at least 100) (Table 7) (27, 61).

The results of a complete recombinate showed that the sweet, baked, fruity, acidic, flowery, and vegetable aromas were very similar to the real roasted-sesame-like aroma type baijiu, but the roasted-sesame aroma only scored a 3 whereas the real baijiu scored a 5 (27). Apparently, some important aroma-contributing compounds were missing from the recombination list.

**Table 7. Orthonasal Odor Concentrations, Thresholds, and OAVs of 47 Odorants in Roasted-Sesame-Like Type Aroma Baijiu (27). Reproduced with permission from reference (27). Copyright 2015 Qingyun Zhou.**

| No. | Compounds | Concentration (mg/L) | OAV |
|-----|-----------|----------------------|-----|
| 1 | Ethyl hexanoate | 689.1 | 12461 |
| 2 | Ethyl octanoate | 94.58 | 7332 |
| 3 | Ethyl butanoate | 575.7 | 7064 |
| 4 | Ethyl 2-methylpropanoate | 389.6 | 6779 |
| 5 | 3-Methylbutanal | 51.12 | 3007 |
| 6 | DMTS | 0.59 | 1639 |
| 7 | Ethyl 3-methylbutanoate | 9.63 | 1398 |
| 8 | Ethyl pentanoate | 26.35 | 983 |
| 9 | Ethyl 2-methylbutanoate | 5.96 | 331 |
| 10 | 3-Methylbutyl acetate | 21.85 | 233 |
| 11 | 1,1-Diethoxyethane | 382.5 | 183 |
| 12 | DMDS | 1.43 | 157 |
| 13 | Methanethiol | 0.32 | 145 |
| 14 | Ethyl acetate | 1865 | 57 |
| 15 | 3-Methylbutyl butanoate | 50.04 | 55 |
| 16 | Butanoic acid | 43.41 | 45 |
| 17 | Ethyl trans-4-decenoate | 3.34 | 30 |
| 18 | 1-Butanol | 53.09 | 22 |
| 19 | Hexanoic acid | 40.37 | 16 |

| No. | Compounds | Concentration (mg/L) | OAV |
|---|---|---|---|
| 20 | Pentanoic acid | 5.27 | 14 |
| 21 | Ethyl 2-phenylacetate | 5.7 | 14 |
| 22 | Butyl hexanoate | 8.97 | 13 |
| 23 | Ethyl lactate | 1692 | 13 |
| 24 | Ethyl 3-phenylpropanoate | 1.36 | 11 |
| 25 | 2-Phenylethyl acetate | 6.85 | 8 |
| 26 | 3-Methylbutyl hexanoate | 8.80 | 6 |
| 27 | 2-Methylpropanoic acid | 10.25 | 6 |
| 28 | 2-Phenylethyl hexanoate | 0.56 | 6 |
| 29 | Ethyl decanoate | 3.03 | 3 |
| 30 | 2-Methylpropanol | 156.9 | 3 |
| 31 | 3-Methylbutanol | 535.4 | 3 |
| 32 | Octanoic acid | 8.99 | 3 |
| 33 | 3-Methylbutanoic acid | 3.49 | 3 |
| 34 | 2-Phenylethyl butanoate | 2.88 | 3 |
| 35 | Furfural | 126.7 | 3 |
| 36 | Heptanoic acid | 21.97 | 2 |
| 37 | Ethyl benzoate | 3.48 | 2 |
| 38 | Naphthalene | 0.35 | 2 |
| 39 | 1-Hexanol | 12.26 | 2 |
| 40 | 4-Methylphenol | 0.26 | 2 |
| 41 | γ-Nonalactone | 0.20 | 2 |
| 42 | Ethyl nonanoate | 6.81 | 2 |
| 43 | Ethyl dodecanote | 0.62 | 2 |
| 44 | 1-Propanol | 54.61 | 1 |
| 45 | Ethyl propanoate | 25.43 | 1 |
| 46 | Acetic acid | 235.4 | 1 |
| 47 | 2-Methylpropanal | 1.34 | 1 |

The roasted-sesame-like aroma type baijiu was further studied using SPME-AEDA (62). The results showed ethyl hexanoate, ethyl butanoate, ethyl 2-methylbutanoate, ethyl pentanoate, ethyl 4-methylpentanoate, 2-furfurylthiol, 3-methylbutanal, propyl hexanoate, ethyl 2-phenylacetate, 2-phenylethyl acetate, DMTS, terpineol, and 2-phenylethanol (FD of at least 200). The aroma compounds with higher OAVs included ethyl hexanoate, ethyl octanoate, ethyl butanoate, 3-methylbutanal, 2-furfuylthiol, ethyl pentanoate, ethyl 3-methylbutanoate, ethyl 2-methylpropanoate, 3-methylbutyl acetate, 2-methylpropanoic acid, DMTS, ethyl 2-methylbutanoate, and $\beta$-damascenone. The results of omission experiments from complete recombinate showed that 2-furfurylthiol was the typical odorant of roasted-sesame-like aroma type baijiu (62).

**Complex Aroma Type Baijiu**

The complex aroma type baijiu (jianxiang baijiu) has a similar odor to soy-sauce and strong aroma and flavor type baijius. For this reason, you would find that the producing technology is similar to that of soy-sauce type baijiu. The important differences are as follows (31, 33, 58, 63):

1) The ground and cooked sorghum is saccharified and fermented with the mixture of high-temperature and moderate-temperature daqus, which are made from wheat.
2) The distillates are aged in pottery jars for more than three years.

Complex aroma type baijiu are produced in the Hubei and Anhui provinces. The famous brands are Baiyunbian and Kouzijiao liquors.

In 2008, aroma compounds from complex aroma type baijiu samples were extracted using LLE and then separated into four fractionations: acidic, water-solution, neutral, and basic fractions. A total of 90 aromas were detected and identified by GC–Osme coupled with GC–MS in the four fractions, including 13 fatty acids, 11 alcohols, 27 esters, 6 phenols, 10 aromatic compounds, 4 ketones. 3 acetals, 1 sulfur compound, 1 lactones, 7 pyrazines, 5 furans, and 2 unknowns (64). The important aroma compounds contributing to Chinese complex aroma type baijiu were ethyl hexanoate (Osme value of 4.00), 4-vinylguaiacol (Osme value of 3.67), hexanoic acid (Osme value of 3.50), 3-methylbutanol (Osme value of 3.33), ethyl 3-methylbutanoate (Osme value of 3.33), 4-ethylguaiacol (Osme value of 3.17), vanillin (Osme value of 3.17), 2-phenylethyl acetate (Osme value of 3.17), and butanoic acid (Osme value of 3.00).

**Fenxiang Aroma Type Baijiu**

The process of producing fenxing aroma type baijiu is similar to the strong aroma type liquor. The distinctions are:

1) the saccharification and fermentation agents are moderate-temperature daqu made from a mixture of barley and pea (traditional) (33) or moderate-high-temperature daqu from barley, wheat, and peas (current); and
2) the fermentation period is 14 days (for traditional) (33) or 30 days (for current) (65, 66)

The fenxiang aroma type baijiu is mainly produced in the Shaanxi province. The most famous brand is Xifen liquor.

Ethyl acetate, ethyl hexanoate, ethyl octanoate, phenylacetaldehyde, ethyl 3-phenylpropanoate, 2-phenylethanol, 2-phenylacetic acid, 4-ethylguaiacol, vanillin, 3-methylbutanol, 1-octanol, acetic acid, propanoic acid, 2-methylpropanoic acid, butanoic acid, 3-methylbutanoic acid, pentanoic acid, hexanoic acid, and octanoic acid (Osme value of at least 3.50) were important aroma compounds discovered by GC–O and GC–MS (39).

## Herb-Like Aroma Type Baijiu

The production technology for herb-like aroma type baijiu is complex and includes two steps.

The first step is producing the xiaoqu liquor. The xiaoqu (Figure 1 right, 3.5 × 3.5 × 3 cm) is made from ground rice mixed with muqu powder (1% by weight) and 95 Chinese herb powders (5% by weight). The mixture is fermented at less than 40 °C in a solid state for seven days. The sorghum used for liquor-making is soaked with hot water (90 °C) for eight hours, cooked with steam, cooled with forced air, and added to 0.5% (by weight of sorghum) xiaoqu powder. The mixture is kept at 28 °C for saccharification, and then moved to a fermentor for fermentation in a solid state for six to seven days. This fermentor is similar to the one used for strong aroma type liquor. After fermentation, the raw xiaoqu liquor is produced by steam distillation (31, 33, 67, 68).

The second step is the aromatic liquor-making. The wheat xiaoqu (11 ×11 × 3 cm) is made from craked wheat mixed with muqu powder (2% by weight) and 40 Chinese herb powder (5% by weight). The mixture is fermented at a temperature below 40 °C in a solid state during seven days. The spent grains after distillation of jiupei from the first and second steps are mixed with aromatic jiupei from second step. The wheat xiaoqu powder is added to the mixture. After that, the mixture is put into a fermentor for 10–12 months of fermentation. When the fermentation is finished, the aromatic jiupei is distillated with alcoholic steam, which is produced by heating the raw xiaoqu liquor using direct water steam. This distilling process is known as chuanzheng in China (31, 33, 67, 68). The raw liquors are then matured in a pottery jar for approximately two years (33). The main production area for this type of baijiu is the Guizhou province. The most important brand is Dongjiu liquor.

Chinese herb-like aroma type baijiu is a special type of liquor. A total of 40 Chinese herbs were added as adjunct in the wheat xiaoqu-making process, and 95 Chinese herbs were added in the rice xiaoqu-making production process. The finished baijiu product has a herb-like aroma (20).

In an early study, researchers found that the amounts of ethyl butanoate (approximately 316 mg/L), butanoic acid (approximately 1.02 g/L), 1-propanol (approximately 1.27 g/L), and 2-butanol (approximately 676 mg/L) were higher than in other aroma type baijius. The total level of acids was higher than the total level of esters, while this was the opposite in other aroma type baijiu (33).

Fifty-two terpenoids in herb-like aroma type baijiu (Dongjiu liquor) were detected by LLE coupled with GC–MS (69). A total of 41 volatile terpenoids were quantified using HS–SPME in the liquor base and finished samples of Dongjiu baijiu. The total terpenoid content was higher in daqu-based liquor compared to xiaoqu-based liquor. Seventeen terpenoids including D-camphor were found in the daqu-based liquor rather than the xiaoqu-based liquor. The 17 terpenoids were speculated to be mainly derived from daqu-fermented grains, while calamine was found only in xiaoqu-based liquor, suggesting that it is derived from xiaoqu-fermented grains (70).

Dongjiu is a typical herbaceous aroma type liquor, and its aroma compositions were investigated with normal phase chromatography followed by GC–O (20). The dongjiu liquor sample was fractionated into water-acidic and neutral-basic fractions. The neutral-basic fraction was further separated into seven fractionations using silica gel normal phase chromatography (pentane-diethyl ether of 100:0, 95:5, 90:10, 80:20, 70:30, 50:50, and 100% methanol). The results showed that dongjiu liquor aromas were mainly created by fatty acids, esters, terpenoids, alcohols, aromatic compounds, phenols, sulphur compounds, ketones, and aldehydes. According to the Osme values, butanoic acid, ethyl hexanoate, hexanoic acid, DMTS, ethyl butanoate, 2-phenylethyl alcohol, 3-methylbutanoic acid, 4-methylphenol, 4-methylguaiacol, β-damascenone, ethyl pentanoate, (E,Z)-2,6-nonadienal, (–)-borneol, and fenchol had the highest Osme values. Those compounds were the most important aroma compounds in dongjiu liquor of herbaceous aroma type liquor.

**Texiang Aroma Type Baijiu**

The texiang aroma type baijiu is produced by a combination of the soy-sauce and strong aroma type baijiu-making processes. The biggest difference is the raw material as this aroma type baijiu uses flour for daqu-making, not wheat or other types, and rice for liquor-making, not sorghum (3, 71).

The wheat flour (35–40% by weight), wheat brans (40–50%), and spent grains (15–20%) as raw meterials for daqu-making are mixed, added to water, and then pressed into the shape of a brick. The mixture is naturally fermented in a solid state and controlled at a temperature of less than 58–60 °C (3, 58).

The liquor-making process uses the xuchafa process with rice as the raw material. The rice is mixed with the jiupei and distillated by steam. The cooked rice is added to water, cooled, and moved to the fermentor for fermentation for 30 days. The fermentor is very similar to one used for soy-sauce aroma type liquor. After 30 days, the jiupei (fermented grains) is added to the rice and paddy hull and then steamed. The distillates are matured in pottery jars for one to two years (3, 58).

The texiang aroma type baijiu is produced in the Jiangxi province, and its main brand is Sitejiu liquor. However, the production of this aroma type baijiu is rare and has declined in recent years.

In 1994, researchers detected volatile compounds of Sitejiu liquor using GC–MS, GC–FPD, and GC–NPD (72). A total of 121 volatiles were identified, including 15 alcohols, 39 esters, 20 fatty acids, 5 aldehydes, 6 acetals, 4 ketones, 6 aromatic compounds, 2 furans, 18 pyrazines, 2 pyridines, 2 thiazoles, and 2 sulfur-containg compounds. An interesting phenomenon was found: The concentrations of odd-carbon fatty acid ethyl esters (for example, ethyl propanoate, ethyl pentanoate, ethyl heptanoate, and ethyl nonanoate) were higher than in other aroma type baijius.

Up to now, no research on aroma compounds found by GC–O has been published.

## Volatile Compounds of Astringent and Bitter Taste

Lots of volatile and nonvolatile compounds (1-propanol, butanal, 2-methylpropanol, 3-methylbutanol, furfural, acrolein, sulfur-containing compounds, and some amino acids) have astringent and bitter taste (73–75). However, there is no published research reporting on the bitter and astringent thresholds in baijiu. We investigated the bitter and astringent tastes in volatile and nonvolatile fractions of baijiu after vacuum distillation. The volatile fraction had a bitter and astringent taste, but the nonvolatile fraction did not (76).

Some bitter and astringent compounds reported in the literature were investigated using the triangle test (22). Tyrosol (4-hydroxyphenylethanol) is bitter with a threshold of 22 mg/L in water. Its concentration was only 2.72 µg/L in baijiu (77), so it does not contribute to a bitter and astringent taste. Acrolein was also reported to give a bitter taste in baijiu (78). Acrolein is a poisonous compound, and its concentration was 114 µg/L in baijiu (78). No bitterness was tasted at 114 µg/L and 3.08 mg/L in water (76). Thus, it was found that acrolein does not produce a bitter taste in baijiu.

DMDS and 3-(methylthio)-1-popanol had no taste and mouthfeel properties when their concentrations were at 100 mg/L and 113 mg/L in water (76), respectively. The concentrations of these two compounds usually were 1.46–3.71 mg/L and 1.72–3.80 mg/L in baijiu (79).

Some amino acids, including L-histamine, L-valine, L-isoleucine, L-leucine, L-lysine, L-phenylalanine, L-tyrosine, and L-arginine produced bitter tastes, and their bitter thresholds were 6.98, 2.34, 1.31, 1.44, 14.61, 7.43, 0.72, and 1.31 g/L in water (76), respectively. Yet, these amino acids had lower concentrations in Chinese baijiu: 0.30–0.66 mg/L, 0.71–0.99 mg/L, 0.019–1.73 mg/L, 0.17–2.95 mg/L, 0.026–0.39 mg/L, 0.001–0.29 mg/L, 0.017–1.85, and 0.23–1.41 mg/L

(*80*), respectively. Compared with their corresponding thresholds, the dose-over-thresholds (DoTs) of these amino acids were less than 1.

To identify volatile compounds with bitter and astringent taste in baijiu (*81*), the baijiu sample was separated into 4 fractions (fractions A, B, C, and D) using gradient vacuum distillation (Figure 2). The fraction B, with strong astringency and bitterness, was further separated by semipreparative HPLC, and 10 subfractions (B-1–B-10) were obtained. Each subfraction was extracted by pentane (named a "pentane phase" from B-1-I to B-10-I) and then by diethyl ether (named a "ether phase" from B-1-II to B-10-II). The residual was named the "residual water-solution" from B-1-III to B-10-III (Figure 2). Each organic phase was separated into two portions: One was washed with pure water, and each water phase was evaporated by vacuum to remove the solvent for taste dilution analysis (TDA), and the other was concentrated to 250 μL under a gentle stream of nitrogen for GC–MS. The 4 subfractions (B-8-II, B-2-II, B-3-II, and B-4-II) had high taste dilution (TD) factors of 128, 64, 32, and 8 for astringency and 32, 16, 16, and 8 for bitterness, respectively (Figure 3), and 7 taste-active compounds were identified using GC–MS.

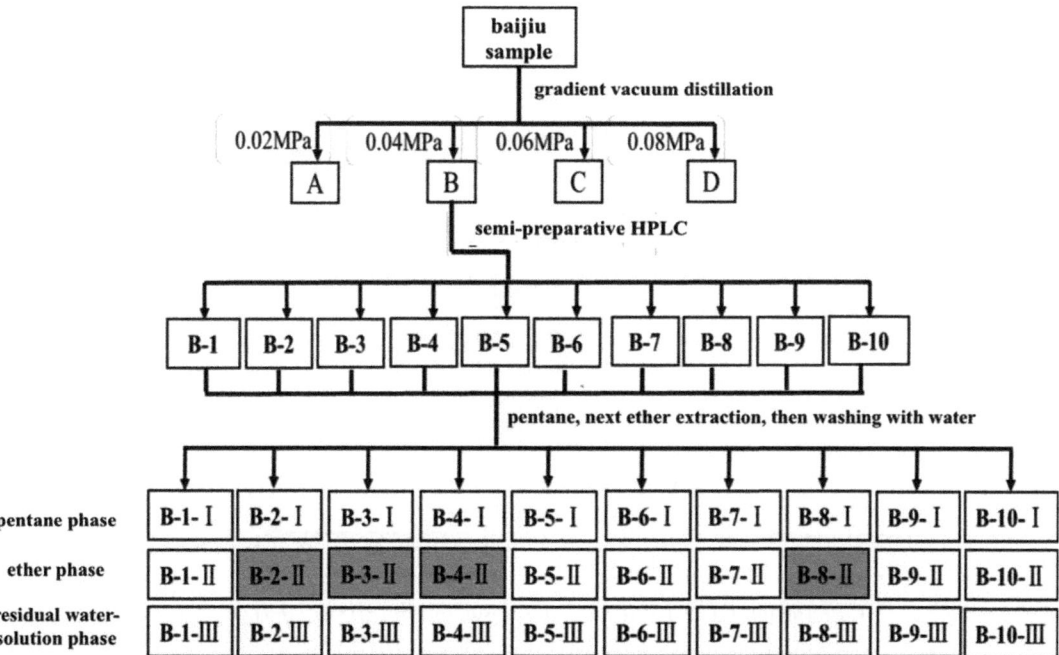

Figure 2. The flow diagram of extraction and separations of bitter and astringency compounds in baijiu. Reproduced with permission from reference (*7*). Copyright 2018 Yinye Wang.

Based on the reference compounds, 2-phenylethanol and ethyl lactate contributed only to astringency, and their astringency thresholds were 2.3 µmol/L (276 µg/L) and 4.71 mmol/L (557 mg/L, see Table 8) in water with a pH of 3.8 (adjusted by $H_3PO_4$) (76), respectively, while furfural, 2-methypropanol, 3-methybutanol, 1-butanol, and 1-propanol contributed to both bitter and astringent tastes, and their bitter thresholds were 0.78 mmol/L (75 mg/L), 2.62 mmol/L (194 mg/L), 5.01 mmol/L (442 mg/L), 2.48 mmol/L (184 mg/L), and 31.46 mmol/L (1891 mg/L) in water with a pH of 3.5 (adjusted by $H_3PO_4$, Table 8) and their astringent thresholds were 54 µmol/L (5.20 mg/L), 310 µmol/L (27.22 mg/L), 310 µmol/L (27.62 mg/L), 300 µmol/L (22.50 mg/L), and 1.97 mmol/L (118 mg/L) in water with a pH of 3.8 (adjusted by $H_3PO_4$, Table 8) (76). Among these compounds, 2-phenylethanol was first reported in the taste property.

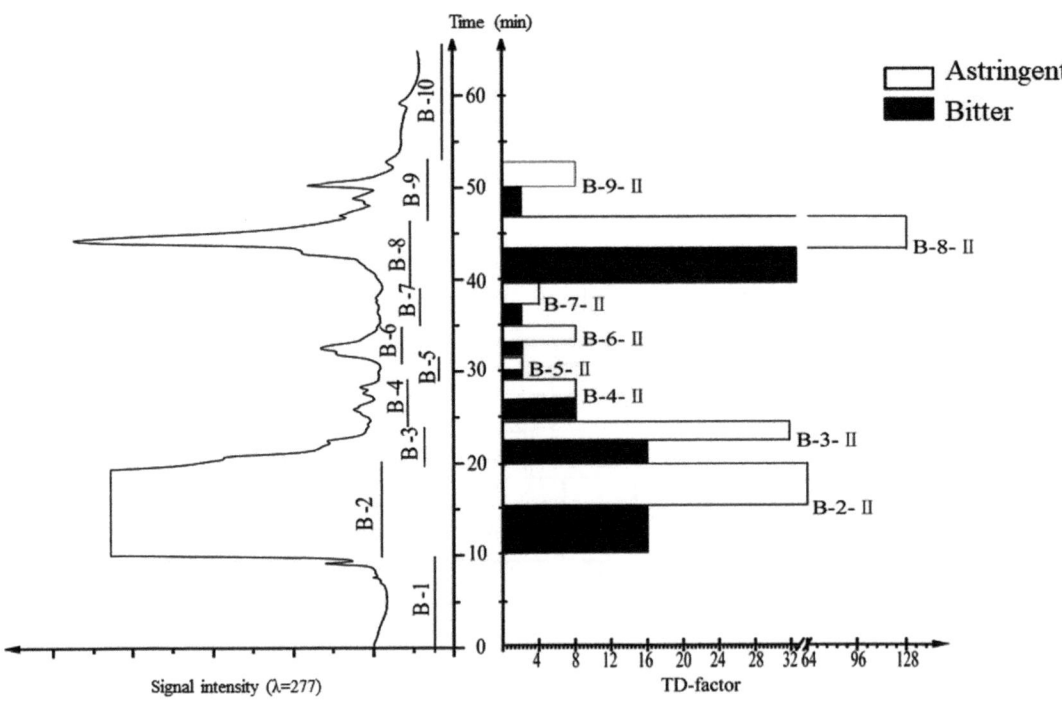

*Figure 3. RP-HPLC of fraction B (left) and TDA of subfraction B-1-II–B-10-II (right). Reproduced with permission from reference (76). Copyright 2018 Yinye Wang.*

**Table 8. The DoTs of Astringent Compounds in Baijiu (76). Reproduced with permission from reference (76). Copyright 2018 Yinye Wang.**

| Compounds | Threshold[a] | | SAB[b] | LAB | SSAB | SHAB | FXAB | CAB | RSAB | LBGAB | HLAB | CXAB | TXAB |
|---|---|---|---|---|---|---|---|---|---|---|---|---|---|
| | mmol/L | mg/L | | | | | | | | | | | |
| 1-Propanol | 1.97 | 118 | 2 | 2 | 4 | 3 | 3 | 2 | 2 | 2 | 3 | 1 | 5 |
| 1-Butanol | 0.30 | 22.50 | 5 | 2 | 4 | 3 | 7 | 5 | 6 | 3 | 10 | 0.7 | 4 |
| Isobutanol | 0.31 | 27.22 | 8 | 9 | 6 | 28 | 18 | 6 | 14 | 7 | 19 | 11 | 6 |
| Isopentanol | 0.31 | 27.62 | 15 | 19 | 9 | 51 | 23 | 8 | 31 | 20 | 29 | 13 | 16 |
| 2-Phenylethanol | 0.0023 | 0.276 | 59 | 108 | 147 | 409 | 124 | 40 | 74 | 110 | 43 | 387 | 58 |
| Furfural | 0.054 | 5.20 | 22 | 15 | 36 | 18 | 25 | 21 | 31 | 23 | 20 | 9 | 21 |
| Ethyl lactate | 4.71 | 557 | 2 | 1 | 2 | 2 | 3 | 2 | 4 | 4 | 2 | 1 | 3 |

[a] determination by the half-tongue test (82).  [b] SAB, strong aroma type baijiu; LAB, light aroma type baijiu; SSAB, soy-sauce aroma type baijiu; SHAB, sweet-honey aroma type baijiu; FXAB, fenxiang aroma type baijiu; CAB, complex aroma type baijiu; RSAB, roasted-sesame-like aroma type baijiu; LBGAB, laobaiganxiang aroma type baijiu; HLAB, herb-like aroma type baijiu; CXAB, chixiang aroma type baijiu; TXAB, texiang aroma type liquor.

**Table 9. The DoTs of Bitter Compounds in Baijiu (76). Adapted with permission from reference (76). Copyright 2018 Yinye Wang.**

| Compounds | Threshold[a] | | SAB[b] | LAB | SSAB | SHAB | FXAB | CAB | RSAB | LBGAB | HLAB | CXAB | TXAB |
|---|---|---|---|---|---|---|---|---|---|---|---|---|---|
| | mmol/L | mg/L | | | | | | | | | | | |
| 1-Propanol | 31.46 | 1891 | <1 | <1 | <1 | <1 | <1 | <1 | <1 | <1 | <1 | <1 | <1 |
| 1-Butanol | 2.48 | 184 | 1 | <1 | 1 | <1 | 1 | 1 | 1 | <1 | 1 | <1 | <1 |
| Isobutanol | 2.62 | 194 | 1 | 1 | 1 | 3 | 2 | 1 | 2 | 1 | 2 | 1 | 1 |
| Isopentanol | 5.01 | 442 | 1 | 1 | 1 | 3 | 1 | 1 | 2 | 1 | 2 | 1 | 1 |
| Furfural | 0.78 | 75 | 2 | 1 | 3 | 1 | 2 | 2 | 2 | 2 | 1 | 1 | 1 |

[a] determination by triangle test (22).  [b] SAB, strong aroma type baijiu; LAB, light aroma type baijiu; SSAB, soy-sauce aroma type baijiu; SHAB, sweet-honey aroma type baijiu; FXAB, fenxiang aroma type baijiu; CAB, complex aroma type baijiu; RSAB, roasted-sesame-like aroma type baijiu; LBGAB, laobaiganxiang aroma type baijiu; HLAB, herb-like aroma type baijiu; CXAB, chixiang aroma type baijiu; TXAB, texiang aroma type liquor.

Based on the DoTs, the important volatile compound for astringency was 2-phenylethanol, and its highest DoT was 409 in sweet-honey aroma type baijiu, followed by chixiang aroma type baijiu (Table 8). The most important volatile compounds for bitterness were 2-methylpropanol, 3-methylbutanol, and furfural (Table 9) (76).

## Conclusions

In conclusion, esters are very important aroma compounds; these include ethyl hexanoate, ethyl butanoate, ethyl acetate, ethyl octanoate, ethyl pentanoate, and ethyl 3-methylbutanoate. Ethyl hexanoate was the key aroma compound in strong aroma and flavor type baijiu, while ethyl acetate was very important in light aroma and flavor type baijiu based on the recombination and omission rest. $\beta$-Damascenone has a rose aroma, and it is a key aroma compound for light aroma and flavor type baijiu, while geosmin contributed to negative characteristic aromas. However, geosmin was a key aroma compounds for laobaiganxiang aroma type baijiu. (E)-2-Enals were very important aroma compounds for chixiang aroma type baijiu, including (E)-2-octenal, (E)-2-nonenal, and (E)-2-decenal. (E)-2-Nonenal was a key aroma compound for chixiang aroma type baijiu. Fatty acids were important aroma compounds for Chinese baijiu, including hexanoic, butanoic, 3-methylbutanoic acids, especially in those baijiu fermented in a special fermentor made with clay and the inside of which is coated with a layer of fermentation mud made of clay, spent grain, bean cake powder, and fermentation bacteria (*Clostridium* sp.). Pyrazines were important aroma compounds in baijiu using high-temperature daqu as saccharification and fermentation agents. These pyrazines included 2,3,5,6-tetramethylpyrazine and 2,3,5-trimethylpyrazine. Many aromatic compounds were important aroma contributors, including 2-phenylethanol, phenylacetaldehyde, ethyl benzoate, ethyl 2-phenylacetate, 2-phenylethyl acetate, and ethyl 3-phenylpropanoate. Phenols were important aromas, including 4-methylguaiacol and 4-ethylguaiacol. Some were also negative aromas, including 4-methylphenol, 4-ethylphenol, and 4-vinyphenol. Some sulfur-containing compounds, acetals, furans, and lactones were important in baijiu, for example, DMTS, 2-furfurylthiol, 1,1-diethoxyethane, 1,1-diethoxy-3-methylbutane, 1,1,3-triethoxypropane, furfural, and $\gamma$-nonalactone. 2-Furfurylthiol was the typical odorant for roasted sesame-like aroma type baijiu. Very important volatile bitter compounds were 2-methylpropanol, 3-methylbutanol, and furfural, and a very important volatile astringent compound was 2-phenylethanol, based on DoTs. Tyrosol, acrolein, DMDS, 3-(methylthio)-1-rpopanol, L-histamine, L-valine, L-isoleucine, L-leucine, L-lysine, L-phenylalanine, L-tyrosine, and L-arginine made no contribution to the bitter and astringent tastes of Chinese baijiu.

## Acknowledgments

Financial support from the Ministry of Science and Technology, P. R. China under Nos. National Key R&D Program of China (2016YFD0400503) are gratefully acknowledged.

## Abbreviations

AEDA, aroma extract dilution analysis
DEDS, diethyl disulfide
DMDS, dimethyl disulfide
DMS, dimethyl sulfide
DMTS, dimethyl trisulfide

DoTs, dose-over-threshold

EtSAc, *S*-ethyl thioacetate

FD, flavor dilution

GC, gas chromatography

GC–FID, gas chromatography–flame ionization detector

GC–FPD, gas chromatography–flame photometric detector

GC×GC-TOF-MS, two-dimensional gas chromatography-time-of-flight-mass spectrometer

GC–MS, gas chromatography–mass spectrometer

GC–NPD, gas chromatograph–nitrogen-phosphorus detector

GC–O, GC–olfactometry

HS–SPME, headspace–solid phase microextraction

LLE, liquid-liquid extraction

LLME, liquid-liquid microextraction

MeSAc, *S*-methyl thioacetate

OAV, odor activity value

PFBHA, *o*-(2,3,4,5,6-pentafluorobenzyl) hydroxylamine

SBSE, stir bar sorptive extraction

SIM, selected ion monitoring mode

TDA, taste dilution analysis

TD, taste dilution

## References

1.   Fan, W.; Xu, Y.; Qian, M. C. Identification of Aroma Compounds in Chinese "Moutai" and "Langjiu" Liquors by Normal Phase Liquid Chromatography Fractionation Followed by Gas Chromatography/Olfactometry. In *Flavor Chemistry of Wine and Other Alcoholic Beverages*; Qian, M. C.; Shellhammer, T. H., Eds.; ACS Symposium Series 1104; American Chemical Society: Washington, DC, 2012; pp 303–338.

2.   Fan, H.; Fan, W.; Xu, Y. Characterization of Key Odorants in Chinese Chixiang Aroma-Type Liquor by Gas Chromatography–Olfactometry, Quantitative Measurements, Aroma Recombination, and Omission Studies. *J. Agri. Food. Chem.* **2015**, *63*, 3660–3668.

3.   Xiong, Z. Development of Te-Type Liquor. *Liquor-Making Sci. Technol.* **2006**, *139*, 102–104.

4.   Fan, W.; Qian, M. C. Characterization of Aroma Compounds of Chinese "Wuliangye" and "Jiannanchun" Liquors by Aroma Extraction Dilution Analysis. *J. Agri. Food. Chem.* **2006**, *54*, 2695–2704.

5.   Fan, W.; Qian, M. C. Identification of Aroma Compounds in Chinese 'Yanghe Daqu' Liquor by Normal Phase Chromatography Fractionation Followed by Gas Chromatography/Olfactometry. *Flav. Fragr. J.* **2006**, *21*, 333–342.

6.   Fan, W.; Xu, Y. The Review of the Research of Aroma Compounds in Chinese Liquors. *Liquor Making* **2007**, *34*, 31–37.

7.   Yu, X.; Yin, J.; Hu, G. Determination of Nitrogen-Containing Compounds from Chinese Liquor. *Niangjiu* **1992**, 71–76.

8.   Lu, J.; Hu, G. Analysis Volatile Sulfur-Containing Compounds of Baijiu by GC-FPD. *Liquor-Making Sci. Technol.* **1994**, *61*, 23–25.

9. Cai, X.; Yin, J.; Hu, G. Research on Volatile Compounds of Chinese Liquor using FFAP Bounded Column by Direct Injection. *Liquor-Making Sci. Technol.* **1994**, *61*, 18–22.

10. Cai, X.; Yin, J.; Hu, G. Determination of Minor Flavor Components in Chinese Spirits by Direct-Injection Technique with Capillary Columns. *Food Ferment. Ind.* **1997**, *15*, 367–371.

11. Fan, W.; Xu, Y. Identification of Volatile Compounds of Fenjiu and Langjiu by Liquid-Liquid Extraction Coupled with Normal Phase Liquid Chromatography (Part One). *Liquor-Making Sci. Technol.* **2013**, *224*, 17–26.

12. Fan, W.; Xu, Y. Identification of Volatile Compounds of Fenjiu and Langjiu by Liquid-Liquid Extraction Coupled with Normal Phase Liquid Chromatography (Part Two). *Liquor-Making Sci. Technol.* **2013**, *225*, 17–27.

13. Fan, W.; Xu, Y.; Zhang, Y. Characterization of Pyrazines in Some Chinese Liquors and Their Approximate Concentrations. *J. Agri. Food. Chem.* **2007**, *55*, 9956–9962.

14. Qian, M. C.; Burbank, H.; Wang, Y. Pre-Separation Technique for Flavor Analysis. In *Sensory-Directed Flavor Analysis*; Marsili, R., Ed.; CRC Press: Boca Raton, FL, 2006; pp 111–154.

15. Fan, H.; Fan, W.; Xu, Y. Liquid-Liquid Micro-Extraction Combine Gas Chromatography-Mass Spectrometrum Analyse Trace Flavor Compounds of Chixiang Aroma Style Liquor. In *2013 International Alcoholic Beverage Culture & Technology Symposium*; Xu, Y., Ed.; China Light Industry Press: Beijing, 2013; pp 97–106.

16. Fan, W.; Shen, H.; Xu, Y. Quantification of Volatile Compounds in Chinese Soy Sauce Aroma Type Liquor by Stir Bar Sorptive Extraction (SBSE) and Gas Chromatography–Mass Spectrometry (GC–MS). *J. Sci. Food Agric.* **2011**, *91*, 1187–1198.

17. Fan, W.; Xu, Y. *A Method of Determination of Sulfide-Containing Compounds in Chinese Liquor.* China Patent, 2007.

18. Chen, S.; Sha, S.; Qian, M.; Xu, Y. Characterization of Volatile Sulfur Compounds in Moutai Liquors by Headspace Solid-Phase Microextraction Gas Chromatography-Pulsed Flame Photometric Detection and Odor Activity Value. *J. Food Sci.* **2017**, *82*, 2816–2822.

19. Cao, C. *Flavor Compounds of Chinese Kongfujia Liquor.* Masters Thesis, Jiangnan University, 2014.

20. Fan, W.; Hu, G.; Xu, Y.; Jia, Q.; Ran, X. Analysis of Aroma Components in Chinese Herbaceous Aroma Type Liquor. *J. Food Sci. Biotechnol.* **2012**, *31*, 810–819.

21. Fan, W.; Qian, M. C. Headspace Solid Phase Microextraction (HS-SPME) and Gas Chromatography-Olfactometry Dilution Analysis of Young and Aged Chinese "Yanghe Daqu" Liquors. *J. Agri. Food. Chem.* **2005**, *53*, 7931–7938.

22. ASTM, *Standard Practice for Determination of Odor and Taste Thresholds by a Forced-Choice Ascending Concentration Series Method of Limits*; ASTM Standard E 679-91 (Reapproved 1997), April 1997.

23. Zhang, C. *Identification and Removal of Off-Flavor Compounds in Chinese Liquor.* Masters Thesis, Jiangnan University, 2012.

24. Gao, W.; Fan, W.; Xu, Y. Characterization of the Key Odorants in Light Aroma Type Chinese Liquor by Gas Chromatography–Olfactometry, Quantitative Measurements, Aroma Recombination, and Omission Studies. *J. Agri. Food. Chem.* **2014**, *62*, 5796–5804.

25. Fan, W.; Xu, Y. Determination of Odor Thresholds of Volatile Aroma Compounds in Baijiu by a Forced-Choice Ascending Concentration Series Method of Limits. *Nanjing* **2011**, *38*, 80–84.

26. Wang, X.; Fan, W.; Xu, Y. Comparison on Aroma Compounds in Chinese Soy Sauce and Strong Aroma Type Liquors by Gas Chromatography-Olfactometry, Chemical Quantitative and Odor Activity Values Analysis. *Eur. Food Res. Technol.* **2014**, *239*, 813–825.

27. Zhou, Q. *Odor Profile of Chinese Roasted-Sesame-Like Aroma Type Liquor*. Masters Thesis, Jiangnan University, 2015.

28. Gao, W. *The Important Odorants of Qingke Liquor and the Aroma Compounds in its Fermented Grains*. Masters Thesis, Jiangnan University, 2014.

29. Du, H.; Fan, W.; Xu, Y. Characterization of Geosmin as Source of Earthy Odor in Different Aroma Type Chinese Liquors. *J. Agri. Food. Chem.* **2011**, *59*, 8331–8337.

30. Fan, W.; Xu, Y. Methodology for Aroma Compounds in Baijiu. *J. Food Sci. Technol.* **2018**, *36*, 1–10.

31. Shen, Y. *Manual of Chinese Liquor Manufacturers Technology*; Light Industry Publishing House of China: Beijing, China, 1996.

32. Fan, W.; Xu, Y. Comparison of Flavor Characteristics Between Chinese Strong Aromatic Liquor (Daqu). *Liquor-Making Sci. Technol.* **2000**, *101*, 92–94.

33. Li, D. Achievements of 50 Years of Production Technology of Liquor (2nd Part). *Liquor Making* **1999**, *131*, 22–28.

34. Xu, Y.; Wang, D.; Fan, W.; Mu, X.; Chen, J. Traditional Chinese Biotechnology. In *Biotechnology in China II: Chemicals, Energy and Enviroment*; Taso, G. T.; Ouyang, P.; Chen, J., Eds.; Vol. 122; Springer: Heidelberg, Germany, 2010; pp 189–233.

35. Fan, W.; Chen, X. Study on Increase of Ratio of Famous Product in Luzhou-Flavor Baijiu by Sandwich Fermentation with Mud. *Niangjiu* **2001**, *28*, 71–73.

36. Fan, W. Improvement the Quality of Luzhou-Flavor Daqu Liquor by the Secondary Fermentation. *Liquor-Making Sci. Technol.* **2001**, *108*, 40–42.

37. AQSIQ; SAC. *Strong Flavour Chinese Spirits*; GB/T 10781.1; China National Standard, China Standard Press: Beijing, China, 2006.

38. Wu, N. *Establishment of the Model Liquor of Strong Aroma Type Liquor and a Preliminary Study on the Interaction of Important Flavor Compounds*. Thesis, Jiangnan University, Wuxi, China, 2010.

39. Ding, Y. *Studies on Characteristic Aroma Compounds in Fen-Liquor*. Masters Thesis, Jiangnan University, Wuxi, Jiangsu, China, 2008.

40. Fan, W.; Xu, Y. Volatile Aroma Compounds from Light Aroma Type Liquors. *Liquor Making* **2012**, *39*, 14–22.

41. Gao, W.; Fan, W.; Xu, Y. Important Volatile Aroma Compounds in the Liquor Made from Highland Barely in Northwest China. *Sci. Technol. Food Ind.* **2013**, *34*, 49–53.

42. Xiong, Z. Research on Three Flavor Type Liquors in China (II): Introduction to Maotai-Flavor Liquor. *Liquor-Making Sci. Technol.* **2005**, *130*, 25–30.

43. Wang, L.; Fan, W.; Xu, Y. Analysis of Capillary Chromatographic Skeleton Compounds in Chinese Soy Sauce Aroma Type Liquor by Liquid-Liquid Microextraction and Aroma Recombination. *Sci. Technol. Food Ind.* **2012**, *33*, 304–308.

44. Ji, K.; Guo, K.; Zhu, S.; Lu, X.; Xu, G. Analysis of Microconstituents in Liquor by Full Two-Dimensional Gas Chromatography/Time of Flight Mass Spectrum. *Liquor-Making Sci. Technol.* **2007**, *153*, 100–102.

45. Zhou, H. 4-Ethylguaiacol. *Niangjiu* **1989**, *16*, 7–9.

46. Fan, W.; Xu, Y. Current Practice and Future Trends of Key Aroma Compounds in Chinese Soy Sauce Aroma Type Liquor. *Liquor Making* **2012**, *39*, 8–15.

47. Shen, H. *Studies on Aroma Compounds of Chinese Soy Sauce Aroma Type Liquor*. Masters Thesis, Jiangnan University, Wuxi, Jiangsu, China, 2010.

48. Huang, M.; Zeng, J.; Guo, Q.; Kou, Q. Production Technology of Traditional Rice-Flavor Wine. *Mod. Food. Sci. Technol.* **2013**, *29*, 845–847.

49. Feng, Z.; Qiu, X. Analysis of Aroma Compounds of Chinese Chixiangxing Aroma Type Liquor. *Niangjiu* **1995**, *109*, 75–82.

50. Jin, P.; Ji, J.; Shen, Y. Identification of Binary Acid Ethyl Esters as the Flavor Constituents of Shaojiu. *Niangjiu* **1990**, *104*, 24–28.

51. Jin, P.; Shen, Y. Technology for Manufacture of Yubingshao, a Traditional Light Baijiu. *Niangjiu* **1989**, *103*, 9–13.

52. Jin, P. 3-Methylthio-1-propanol of Chixiangxing Baijiu. *Liquor Making* **2004**, *31*, 110–111.

53. Fan, H.; Fan, W.; Xu, Y. Characetrization of Volatile Aroma Compounds in Chinese Chixiang Aroma Type Liquor by GC-O and GC-MS. *Food Ferment. Ind.* **2015**, *41*, 147–152.

54. Du, H. *Discovery and Origin of One Kind of Off-Flavors in Chinese Liquors*. Masters Thesis, Jiangnan University, 2009.

55. Wang, H.; Yu, Z. Review and Prospect of Research Work of Distinctive Jingzhibaigang Baijiu. *Niangjiu* **1992**, *19*, 61–70.

56. Shen, Y. The Producing Technology of High Quality Roasted-Sesame-Like Type Baijiu. *Liquor-Making Sci. Technol.* **1993**, *57*, 43–46.

57. Wang, F. Discussion on the Research of Sesame-Flavor Liquor. *China Brewing* **2007**, *175*, 60–61.

58. Li, D. Achievements of 50 Years of Production Technology of Liquor (3rd Part). *Liquor Making* **1999**, *132*, 13–19.

59. Zhou, Q.; Fan, W. Research on Sesame Flavor Liquor Skeleton Compounds. In *2013 International Alcoholic Beverage Culture & Technology Symposium*; Xu, Y., Ed.; China Light Industry Press: Beijing, China, 2013; pp 88–96.

60. Hu, G.; Lu, J. Analysis Research of Sulfur-Containing Compounds of Roasted-Sesame-Like Aroma Type Baijiu. *Liquor-Making Sci. Technol.* **1995**, *72*, 67–68.

61. Zhou, Q.; Fan, W.; Xu, Y. Important Volatile Aroma Compounds in Chinese Roasted-Sesame-Like Aroma Type Jingzhi Liquors. *Sci. Technol. Food Ind.* **2015**, *36*, 62–67.

62. Sha, S.; Chen, S.; Qian, M.; Wang, C.; Xu, Y. Characterization of the Typical Potent Odorants in Chinese Roasted Sesame-Like Flavor Type Liquor by Headspace Solid Phase Microextraction–Aroma Extract Dilution Analysis, with Special Emphasis on Sulfur-Containing Odorants. *J. Agri. Food. Chem.* **2017**, *65*, 123–131.

63. Xiong, X. Summary of the Production Techniques of Maotai-Luzhou-Flavor Baiyunbian Liquor. *Liquor-Making Sci. Technol.* **2007**, *159*, 35–42.

64. Liu, J.; Fan, W.; Xu, Y.; Zhang, G.; Xu, Q.; Ding, Y.; Li, Z. Comparison of Aroma Compounds of Chinese 'Miscellaneous Style' and 'Strong Aroma Style' Liquors by GC-Olfactometry. *Liquor Making* **2008**, *35*, 103–107.

65. Feng, X.; Yan, Z. The Unique Quality of Xifeng Liquor Developed by Traditional Techniques & Special Geographic Environment. *Liquor-Making Sci. Technol.* **2006**, *144*, 102–103.

66. Deng, Q.; Zhang, L. Improve and Create of Tradition of Xifeng Liquor. *Liquor Making* **2005**, *32*, 85–86.

67. Ran, X. Investigation on the Control Factors in the Production of Flavoring Fermented Grains of Dongjiu. *Liquor-Making Sci. Technol.* **2008**, *174*, 62–64.

68. Ran, X.; Qiu, S.; Fan, H.; Wang, Y. Analysis of the Change of Technical Parameters in the Fermentation of Dongjiu in Small Pits. *Liquor-Making Sci. Technol.* **2012**, *217*, 76–78.

69. Hu, G.; Fan, W.; Xu, Y.; Jia, Q.; Ran, X. Research on Terpenoids in Dongjiu. *Liquor-Making Sci. Technol.* **2011**, *205*, 29–33.

70. Fan, W.; Hu, G.; Xu, Y. Quantification of Volatile Terpenoids in Chinese Medicinal Liquor Using Headspace-Solid Phase Microextraction Coupled with Gas Chromatography-Mass Spectrometry. *Food Sci.* **2012**, *33*, 110–116.

71. Xie, X.; Zhu, L. Discussion on Making Jiu Qu and Brewing Technology of Special Type Wine. *Sichuan Food Ferment.* **2005**, 51–52.

72. Hu, G.; Cai, X.; Lu, J.; Yin, J.; Zhu, Y.; Cheng, J. Volatile Compounds of Site Baijiu. *Liquor-Making Sci. Technol.* **1994**, *61*, 9–17.

73. Du, H.; Fan, W.; Xu, Y. In *Review and Prospect of Chinese Liquor Off-Flavor Research*, Proceedings of the 7th International Alcoholic Beverages Culture & Technology Symposium, Beijing, China, 2010; Zhao, G., Ed. China's Textile Press: Beijing, China, 2010; pp 90–94.

74. Wang, H. The Bitter of Baijiu. *Liquor-Making Sci. Technol.* **2007**, *158*, 165–167.

75. Hu, Y. Bitter Taste in Liquor & Its Solutions. *Liquor-Making Sci. Technol.* **2006**, *143*, 67–69.

76. Wang, Y. *A Profile of the Volatile Compounds with Bitter and/or Astringent Taste in Baijiu (Chinese Liquor)*. Masters Thesis, Jiangnan University, 2018.

77. Gong, S. *Comparison of Compounds Between Traditional and Mechanical Raw Baijiu (Chinese Liquor) of Laobaiganand Roasted-Sesame-Like Aroma Type Baijiu*. Masters Thesis, Jiangnan University, 2018.

78. Zhu, M. *Volatile, Toxic and Small Molecular Aldehydes and Their Derivatives in Chinese Liquor*. Masters Thesis, Jiangnan University, 2016.

79. Sha, S. *Volatile Sulfur Compounds in Baijiu*. Masters Thesis, Jiangnan University, 2017.

80. Zhang, Z.; Fan, W.; Xu, Y. Comparative Analysis of Free Amino Acid Content and Composition in Different Aroma Type Chinese Liquors. *Sci. Technol. Food Ind.* **2014**, *35*, 280–284.

81. Wang, Y.; Fan, W.; Xu, Y. Extraction and Isolation Method of Volatile Compounds with Astringent and Bitter Taste in Baijiu (Chinese Liquor). *Food Ferment. Ind.* **2018**, *44*, 240–244.

82. Brock, A.; Hofmann, T. Identification of the Key Astringent Compounds in Spinach (*Spinacia oleracea*) by Means of the Taste Dilution Analysis. *Chemosens. Percept.* **2008**, *1*, 268–281.

# Chapter 13

# Flavor Chemistry in Baijiu with Sesame Flavor: A Review

Juan Wang, Mingquan Huang,* Jinglin Zhang, and Jihong Wu

**Beijing Key Laboratory of Flavor Chemistry, Beijing Advanced Innovation Center for Food Nutrition and Human Health, Beijing Technology and Business University, Beijing, China 100048**

*Phone: 86-10-68984613. Fax: 86-10-68984890. E-mails: huangmq@btbu .edu. cn, hmqsir@163.com.

Sesame-flavor baijiu (SFB), produced since 1960s, is a clear and transparent fermented alcoholic beverage and a well-known distilled spirit in northern China. It can be divided into four flavor styles: pure sesame flavor, sesame flavor mixed with strong aroma, sesame flavor mixed with light aroma, and sesame flavor mixed with Jiang aroma. This chapter summarizes this baijiu's production process, presents recent developments and studies on its flavor chemistry, and discusses the questions in flavor chemistry. More than 463 volatile compounds, including esters, alcohols, ketones, heterocycles, nitrogenous compounds, acids, aldehydes, terpenes, sulfur compounds, acetals, and lactones, were detected in SFB. Among them, 140 aroma-active compounds were identified as important odorants for developing the flavor of SFB, such as ethyl hexanoate, ethyl pentanoate, ethyl butanoate, ethyl acetate, 1-propanol, 1-butanol, 3-methy-1-butanol, acetic acid, butanoic acid, hexanoic acid, 3-methylbutanal, benzeneacetaldehyde, 2-furyl-methanal, diethyl acetal, 4-methylphenol, 3-(methylthio)propanal, dimethyl trisulfide, 2-ethyl-5-methylpyrazine, and 2-ethyl-3,5-dimethylpyrazine. Furthermore, the development trends and future research directions for baijiu with sesame flavor are also discussed and proposed.

## Introduction

Baijiu is a traditional alcoholic beverage in China with a long history, which is regarded as a pearl of ancient Chinese wisdom. Twelve major flavor styles of Chinese baijiu have been recognized since the introduction of classification of baijiu flavor style in 1979. Among them, Jiang (soy sauce), Nong (strong), mild (light), and Mi (rice) flavor styles are considered the four basic flavor styles, while other flavor styles are considered to be derived from these four basic flavor styles. The relationships among the 12 flavor styles of baijius are shown in Figure 1 (*1*, *2*).

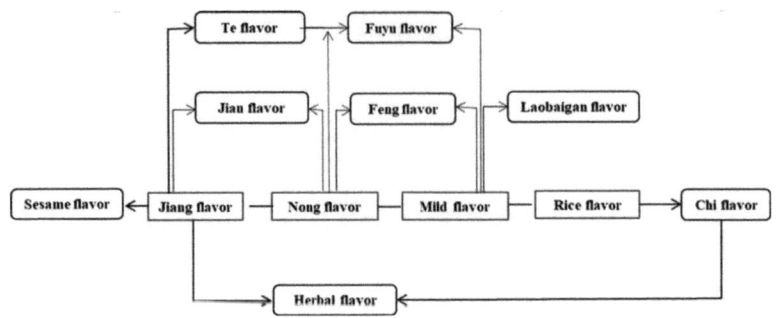

*Figure 1. The relationships among the 12 flavor types of baijius.*

Sesame-flavor baijiu (SFB) has been around since the founding of New China. It is derived from Jiang flavor baijiu (e.g. Maotai liquor) as shown in Figure 1. Currently, SFB is famous particularly in northern China for its unique flavor that combines roasted sesame-like, fruity with sweaty, and floral aromas. There are some representative brands of SFB including Jingzhi (JZ), Guojing (GJ), Meilanchun (MLC), Baotuquan (BTQ), Shengliyuan (SLY), and Yanghu (YH). So far, the development of SFB has gone through the following stages (3):

- First is the discovery stage, which lasted from 1965 to 1975. In the early 1960s, the sesame-flavor began to appear in a batch of Chinese baijiu stored for several years in Shenyang Laolongkou distillery. In 1965, Xiong first explored the sesame flavor of Jingzhi Baigan using paper chromatography (1). Starting in 1966, gas chromatography was used to analyze the volatile components in baijiu. Simultaneously, some mechanical equipment was applied to produce baijiu, such as using an electric grinder instead of manual stone grinding to pulverize raw materials, using boiler pipe steam instead of direct fire distillation, and using a bed machine instead of artificial turning. By using these mechanical processes, Baijiu production had begun to mechanize to replace manual operations to a certain extent (4).

- Then came the preliminary research stage from 1975 to 1985. The Light Industry and Food Bureau of China proposed the combination of "learning and innovation" to develop high-quality baijiu in 1975. At that time, Shen (5) proposed SFB production by using a single bacteria cultivation. Later, he successfully developed a baijiu with a sesame aroma called "Nalsson" at the Wumeng Jining winery. In this period, the fermented grain in some manufacturing processes was conveyed by the bucket of the construction excavator. The electric cart replaced the manual trolley in the transportation of material, and the traditional wooden distillation equipment (Zengtong) was replaced by corresponding stainless steel equipment. The condenser, Tianguo (a condenser full of cold water and placed on the upper level of Zengtong to condensate the alcoholic steam), was replaced by a straight-line condenser. In 1979, gas chromatography–mass spectrometry (GC-MS) was applied in the analysis of volatile compounds from baijiu. In addition, the extensive application of the Daqu preparation machine has promoted large-scale production of Daqu and also greatly reduced the labor intensity of workers (4).

- Next was the development stage from 1985 to 1995. Along with the accumulation of research and production experience, the production technology used for SFB made a great breakthrough during this stage. There were many SFB–producing areas in China, such as Jiangsu, Shandong, Inner Mongolia, and the northeast provinces. In 1994, Hu et al. (6) first identified the 3-(methylthio)-1-propanol in SFB by using a gas

178

chromatography–flame photometric detector (GC-FPD) and later identified it as the key flavor compound for SFB. SFB was formally established as a new flavor type of baijiu, since the enterprise standard QB 2187-1995 of SFB had been drafted in the Shandong Jingzhi Baijiu distillery.

- The wandering stage occurred from 1995 to 2005. During this time, the liquor industry was depressed mainly due to the Asian financial crisis. In 1997, the Asian financial crisis caused the growth rate of Baijiu exports to drop to 0.5%. Meanwhile, the five-year deflation caused the Chinese economy to continually slump. In addition, the government pushed a series of policies to restrict the development of the liquor industry. Therefore, only a limited number of SFB producers survived and most of their production scales were relatively small.

- The final stage, the rapid development stage, has been occurring since 2005. Due to the economic development and the expanded demand for the liquor market, the number and scale of SFB production enterprises have increased sharply along with the wide application of advanced technical equipment and modern technologies. Since 2005, headspace solid phase microextraction (HS-SPME) and gas chromatography–olfactometry (GC–O) have been used to analyze the aroma-active compounds in baijiu (7). Stir bar sorptive extraction (8), thermal desorption (TD) (9), and two-dimensional GC-MS (10) have been used to analyze the volatile compounds in baijiu. GB/T 20824-2007 (11), which was published in 2007, marked the maturity of the flavor style baijiu. In 2011, Jiangnan University verified that the main heterocyclic compound 2,3,5,6-tetramethylpyrazine in Chinese baijiu was mainly derived from the metabolic reaction of *Bacillus subtilis* (12). Ge et al. (13) then separated four strains of the thermophilic bacteria in the high temperature Daqu of SFB; *Thermoactinomyces sp., Bacillus sp., Schlegelella sp.,* and *Streptomyces sp.* were identified. These provided important theoretical guidances for the application of high temperature bacteria in the production of SFB. Meanwhile, technological advancements in production are developing rapidly, such as the automated process control for the production of sesame Jiuqu, the metering control for automatic delivery of baijiu, and the centralized monitoring and energy measurement in the production process (14).

Generally, the brewing technology of SFB has been developed based on the brewing technology of Jiang flavor liquor mixed with parts of the brewing technology for the mild and Nong flavor baijiu at the same time. Therefore, the flavor of SFB is a mixture of three types of Nong, Jiang, and mild flavor liquors (9). The following section introduces the brewing technology for SFB and research progress on its flavor chemistry.

## Brewing Technology for SFB

The definition of SFB (from GB/T 20824-2007) (11) is that it is a kind of baijiu with sesame flavor, produced with sorghum, wheat (bran), and others as raw materials using traditional a solid-state method, including fermentation, distillation, aging, and blending. Furthermore, there should be no added edible alcohol or flavor substances. The brewing process of SFB is shown in Figure 2 below (15, 16). The raw material is grinded and mixed uniformly before adding hot water, thus distilling the grains. The distilled grains are then mixed with Daqu and Fuqu, and after accumulated on the ground for several days, they are put into pits to ferment for several months. After fermentation, the grains are distilled and the distillate (liquor) is obtained. The fresh liquor is then stored and aged in a pot or stainless-steel vessel. When aged for several years, the stored liquor is blended with other liquors to obtain different products.

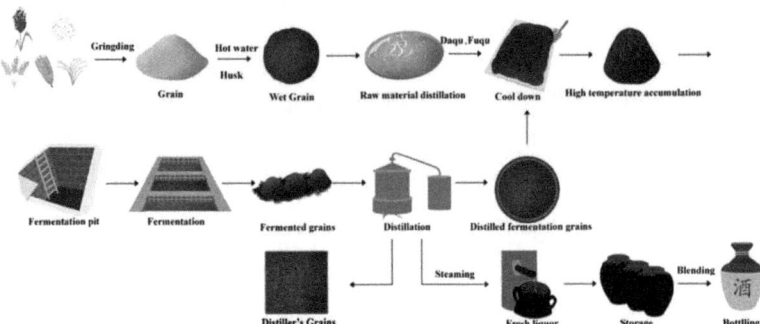

*Figure 2. The brewing process of baijiu with sesame flavor.*

The above process includes (17–23): QingZheng XuCha, a brick cellar with mud in the bottom as a fermentation cellar, the combination of Daqu and Fuqu as a saccharification and fermentation starter, a high nitrogen proportion in raw material, high temperature accumulation, high temperature fermentation, and long storage time.

**Material**

The production materials used for SFB mainly include grains, a fermentation starter, husk, and water, which are the same materials as other baijiu. They all are important for the quality of baijiu. Different grains produce different flavor liquors, and good quality water produces good quality liquor. The fermentation starter is directly related to baijiu flavor and yield. The husk is used as a loosening agent to ferment and distill better and is basically the same for most distilleries. Current water treatment technologies have been applied in many distilleries and ensure the stability of water quality. The water and husk presently have no differences among different baijiu. The material differences of different baijiu are mainly reflected in the starter, as well as the variety and proportion of grains.

*Raw Material: Grain*

Compared to other flavor styles of liquor, the raw materials of SFB can contain higher nitrogen content or richer proteins (24). The richer proteins in fermented grains are hydrolyzed into various amino acids, which provide enough precursors for microbial formation, thermal degradation, and non-enzymatic chemical reaction. These conditions make the sesame flavor of the liquor more typical. However, the nitrogen content is insufficient when sorghum is the only raw material, as its nitrogen content is insufficient. Thus, except for sorghum and a few others, the raw material for SFB usually contains wheat and wheat bran to increase the ratio of nitrogen to carbon in the fermentation material, as wheat bran is rich in protein. Having a higher content of amino acids in the raw materials provides a material basis for the formation of sesame flavor. This is a characteristic of the brewing technology for SFB.

*Fermentation Starter*

The fermentation starter for SBF is high-temperature Daqu mixed with Fuqu. Daqu with a brick shape, made of wheat and peas, is naturally formed in air and contains molds, yeasts, bacteria, and actinomycetes. These microorganism in Daqu come from the air or the surroundings. The three types of Daqu are divided according to their cultivation temperatures into high-temperature (60–70 °C), medium-temperature (50–60 °C), and low-temperature (40–50 °C) Daqu. High-temperature Daqu includes many aroma compounds including tetramethylpyrazine, guaiacol, phenylethanol, propanoic acid, 1,3-butanediol, and acetic acid (22, 23). However, high-temperature Daqu has a lower capacity for saccharification, liquefaction, and fermentation than low-temperature Daqu. A suitable amount of high-temperature Daqu promotes the production of various aroma components in baijiu, especially the sauce aroma. It produces baijiu with a rich and full mouthfeel, but it is difficult to produce liquor with an outstanding sesame flavor (24).

Fuqu uses bran as its main raw material. The raw material is an inoculated single strain aspergillus or other distilled molds. It is then cultivated for multiple days at an artificially controlled temperature and humidity to form Fuqu. The main three microorganisms in Fuqu used for SFB are Hanoi white koji, compound fragrant yeast, and compound bacterium. Hanoi white koji has high saccharification and liquefaction capacity, acid protease activity, and acid resistance. Generally, compound fragrant yeast (containing Hansenula, Candida, spherical yeast, and alcohol yeast) have strong fermenting power and esterification power under high temperature and strong acidic conditions. The screened compound bacterium (including *Hydrogenobacter Thermophilus* and *Phytophthora licheniformis*) not only adapt to higher accumulation fermentation temperatures, but also produce more acid protease to promote Maillard reaction during fermentation. The sesame flavor of SFB is present when pure Fuqu is used, but the mouthfeel is short of the fullness and softness, and is overall thin (1, 24).

Thus, Daqu and Fuqu can complement each other well when combined (3, 25). For example, JZ and MLC SFBs, two typical representative brands, both use the combination of Daqu and Fuqu as a saccharification and fermentation starter. Liu et al. (20) proposed that the quality of the SFB was best when the weight of the added Fuqu was 25% of the total grain weight.

## Raw Material Distillation

The raw materials (grains and husk of rice or sorghum) are distilled separately to remove the miscellaneous flavor without being mixed with fermented grains. The process is called *QingZheng*, which is one characteristic of SFB. It is also the characteristic of mild flavor style baijiu.

## Fermentation

### Xucha

*Xucha* inherits the traditional technology of fermentation, meaning that the freshly distilled grains are mixed with distilled fermentation grains in the last batch before the next batch of fermentation. This is beneficial to the fermentation and accumulation of aroma components, which makes full use of the starch in the raw materials.

*QingZheng XuCha* not only improves the clean mouthfeel of the product, but also highlights the product's elegant flavor characteristic (16).

### Fermentation Cellar

The fermentation cellar is the brick cellar with mud in the bottom, a combination of fermentation cellars of mild, Jiang, and Nong flavor Baijiu (16). A mud cellar with more hexanoic acid bacteria, methane bacteria, and other cellar mud microbes increases the composition of Nong flavor liquor. Shi et al. (26) reported that the microbial groups in the brick cellar of Shuihu SFB also included yeast (*Hansenula, Saccharomyces,* and *Candida*), anaerobic butyric acid bacteria, aerobic *genus Bacillus,* and *Lactobacillus.* Metabolites of these microorganisms (including 4-ethylguaiacol, 4-methylguaiacol, and 3-furanone) contribute to the sesame flavor. Unlike the cement cellar and the stone cellar, the brick cellar does not include a lot of pit mud microbes because it is a setting that is not conducive for microorganisms. The mud cellar is used to produce Nong flavor Baijiu, the pithos is used to produce mild flavor Baijiu, and the stone cellar is used for Jiang flavor Baijiu. Because the used cellar is taken as the combination of the above three cellars, the average content of ethyl hexanoate in SFB was lower than that of Nong flavor Baijiu but higher than that of mild and Jiang flavor Baiiu. Additionally, the content of ethyl acetate was slightly higher than that of Jiang flavor Baijiu. The brick cellar might produce some undiscovered but important microorganisms, which contribute to sesame flavor. More thorough research should be done in the future.

*High Temperature Accumulation*

High temperature accumulation refers to the process where freshly distilled grains, mixed with the fermentation starter and the last batch of distilled fermentation grains, are fermented on the ground before fermentation in cellars with temperatures reaching as high as 40–45 °C. High temperature accumulation is always an important step for soy-sauce flavor liquor, and it is also an important process for producing Nong and Nong-Jiang flavor liquor. The process of high temperature accumulation captures the beneficial microbes in the air and surrounding environment to create conditions for precursor substances in order to produce sesame flavor (27). It is an enriching and growth process for microorganisms in fermented grains, including the enriching of a large amount of yeast and its proliferation, as well as a proliferation process for *Bacillus thermophiles* (28). The appropriate accumulation time is also of importance for sesame flavor. The sesame flavor would be not be as noticeable if the production time is short. For example, two days was deemed an adequate production time for Mengshan Luxiangfang multi-grain SFB (16).

*High Temperature Fermentation*

High temperature fermentation occurs when the fermentation temperature is higher than that of mild flavor or Nong flavor baijiu (close to that of Jiang flavor baijiu) and can rise to 40 °C or higher (17). As previously mentioned, a large amount of high temperature–resistant *Candida cerevisiae,* *bacillus spp,* and other bacteria were cultured using high temperature accumulation. The presence of these microorganisms is convenient for high temperature fermentation, which contributes to a good environment for the formation of flavor components of SFB. The fermentation temperature (40 °C–45 °C) is the optimum temperature for protease and peptidase to decompose proteins. Therefore, high temperature fermentation is also an important procedure for SFB.

**Distillation**

Due to the special production technologies, the SFB is particularly focused on stratified distillation, segment sampling of the liquor, and classified storage. SFB is produced using multi-bacterial microbial flora fermentation. The liquors distilled from the fermentation grains of each layer are different. The bottom layer is affected by the artificial mud, which contains a higher content of ethyl hexanoate, resulting in a liquor flavor that is mixed with strong aroma. The middle layer has a higher content of ethyl acetate that causes the liquor flavor to be mixed with light aroma. The upper layer has a sauce aroma that results in the liquor flavor being mixed with sauce aroma. Therefore, the three layers, each with different flavors, are always distilled separately. At the same time, each layer of the liquor was distilled and collected into 10 fractions. A total of 30 fractions were clustered according to their sensory evaluation results and GC-MS results (29, 30).

## Stored and Aged

The fresh liquors are usually stored for more than three years in order to stabilize the sesame aroma of the liquor (15, 16). After blending different liquors, four flavor styles of SFBs are obtained: pure SFB, SFB mixed with strong aroma, SFB mixed with mild aroma, and SFB mixed with Jiang aroma.

Apart from these factors, the raw material origin, water quality, fermentation time (usually 45 days), and moisture content of fermented grains also have an important influence on the formation of sesame flavor.

SFB is brewed by special technology, a combination of the three technologies of Nong, mild, and Jiang flavor baijiu. The technology uses Fuqu (pure brewed gluten) and Daqu as saccharification and fermentation starters, embodies the combination of traditional and modern technology, and represents the developmental direction of Chinese baijiu. Despite the quality of SFB produced by the mechanization technology being of poorer quality than that of the liquor produced by the traditional technology, SFB brewing of SFB continues to be mechanized and modernized (31, 32).

## Research Progresses on Flavor Chemistry of SFB

### Overall Flavor Characteristics of SFB

According to GB/T 20824-2007 (11), the flavor of SFB is described as "colorless or yellowish, clear and transparent, elegant and prominent sesame, mellow and softness, harmonious in various flavors, long aftertaste." Its typical flavor is an elegant and prominent roasted aroma with ethyl acetate as the main ester (1). The mouthfeel is mellow and refreshing, like the taste of Laobaigan, with a slightly bitter aftertaste.

As mentioned above, there are several SFBs produced by different distilleries, but the flavor profiles of these baijius differ. Using the quantitative descriptive analysis, Liu et al. (33) screened out eight aroma descriptions (sesame, roasted, Chen-aroma, Jiang, strong, light, distilled grain, and Qu-aroma), three taste descriptions (sweet, sour, and bitter), and five mouthfeel descriptions (rich, softness, harmony, clean, and aftertaste) to describe the sensory characteristics of the SFB. These SFB were further divided into four styles: pure SFB, SFB mixed with strong aroma, SFB mixed with light aroma, and SFB mixed with Jiang aroma. The differences in the flavor profiles among these SFB were shown in Figure 3.

According to Figure 3, the sesame and roasted aromas in pure SFB were the most prominent, but the strong aroma was the weakest. Qu-aroma, distilled grain aroma, and light aroma in SFB mixed with light aroma were the most prominent, but Chen-aroma (aging aroma) was the weakest. Strong aroma was the most prominent flavor in SFB mixed with strong aroma, while Qu-aroma, Chen-aroma and Jiang-aroma were the second most prominent and distilled grain aroma was the weakest. Jiang-aroma, Chen-aroma, and sesame-aroma were the most prominent for SFB mixed with Jiang aroma, while Qu-aroma was relatively weaker. In a word, the outstanding sesame and roasted aromas were the most common characteristics among the four SFB, while the other aromas were considered personality characteristics. There were also significant differences in the taste and mouthfeel among different styles of SFB. As a whole, the taste for the SFB is a harmonious combination of sweet, sour, and bitter. The mouthfeels are obvious, such as rich, softness, harmonious, clean, and aftertaste.

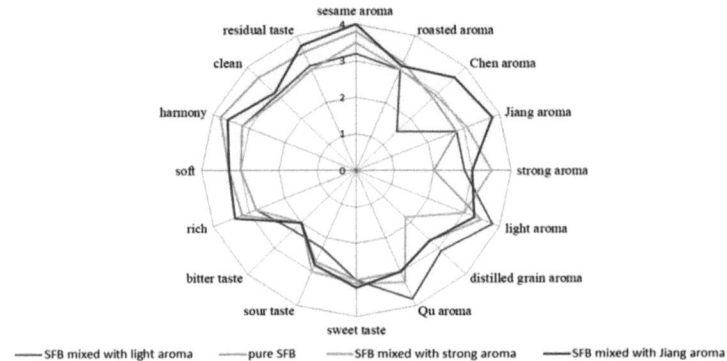

*Figure 3. The aroma, taste, and mouthfeel profile of four different SFB.*

However, there is no better method for evaluating the overall flavor of liquor except the human nose and mouth. Thus, the results obtained by the human nose or mouth have more subjectivity, and different people get different results even when the sample is the same. As mentioned above, there are several SFB from different distilleries (like MLC, JZ, and GJ), and the overall flavor profiles of these baijiu differ as shown in Figure 4.

*Figure 4. The aroma profiles of SFB.*

Figure 4 shows that there were some differences in the two profiles of the same Baijiu (JZ) evaluated by Zheng et al. (34) (JZ-CB) and Sha et al. (35) (JZ-SP). The aroma characteristics described were not completely the same, although five of them were the same: alcoholic, roasted or baked, floral, fruity, and acidic aroma. The other three attributes were different. Additionally, the evaluated aroma intensities were also different among the five aroma characteristics that were the same. The aroma profiles of MLC (36) were also shown in Figure 4, which were different from the overall aroma profiles of JZ. These differences were mainly due to the different sensory evaluators, although the differences in the sample batches were also important. Thus, a better objective evaluation method is needed for future studies.

Though SFB is very welcomed in the market due to its special aroma, taste, and mouthfeel (37), there are still some questions. For example, the entrance taste is very pungent, the aftertaste is bitter, and after drinking the liquor, some people get headaches. Why do these characteristics occur? These questions are possibly related to higher alcohol concentrations and levels of aldehydes, as well as the imbalance between total acids and total esters.

## Volatile Compounds in SFB

### Pretreatment Methods

Because volatile compounds in baijiu have different polarities, solubilities, volatilities, and thermal stabilities, there is no single pretreatment method to effectively extract all the flavor compounds from Baijiu. Currently, direct injection (DI) (38), solid phase extraction (SPE) (39), solid-phase microextraction (SPME) (40, 41), liquid-liquid extraction (LLE) (42), stir bar adsorption extraction (SBE) (43), simultaneous distillation extraction (SDE) (44), headspace (HS) (45), thermal desorption (TD) (9), supercritical carbon dioxide extraction (SCDE) (46), and other pretreatment methods are widely used. These methods all have different advantages and disadvantages. Some methods can be used together to achieve better results.

### Identification

Many methods and instruments have been use to identify volatile compounds in recent years. The methods include: standard comparison method (47), retention index (RI) (48), Gas chromatography (GC), GC-MS (49, 50), GC-O (51–53), two-dimensional gas-mass chromatography (GC×GC), multi-dimensional gas chromatography–mass spectrometry (MDGC-MS) (43, 54), FPD (55), and nitrogen phosphorus detector (NPD) (47). By applying these above extraction and identification methods, many trace, unstable, and low-threshold compounds in liquors were identified in baijiu, such as sulfur compounds and nitrogen compounds. However, a single certain method is also usually limited, but a combination of multiple qualitative means can ensure the accuracy of qualitative results.

### Quantification

Quantifying the aroma compounds is very important for determining their contributions to the flavor of the baijiu. The instruments and pretreatment methods used for quantitation mainly include gas chromatography–flame ionization detector (GC-FID), GC-FPD, GC-NPD, GC-MS, gas chromotography–tandem mass spectrometry (GC-MS/MS), as well as DI, LLE, solid-phase microextraction (SPME), and solvent-assistant flavor evaporation extraction (SAFE). The GC-FID is the most common instrument for quantifying the relatively high content of compounds in baijiu, such as ethyl acetate, ethyl lactate, ethyl hexanoate, 1-propanol, and 1-butanol (56). GC-MS is currently the main instrument for quantitation because of its high sensitivity and a wide range of applications (42). For the quantification of sulfur compounds and nitrogen compounds in baijiu, GC-FPD and GC-NPD methods are the most useful (47, 55). These different methods all have their respective weaknesses and advantages and can be used together to obtain good results.

Quantitative methods mainly include external standard method (ESTD), internal standard method (ISTD), and stable isotope dilution analysis (SIDA), as well as methods such as peak area percentage method (AERA%) peak height percentage method (HEIGHT%), and normalization quantitative method (NORM%). The ESTD corrects the responses of the detector on different compounds, and the result is correct when the peaks of the components are not affected by other compounds. This method has higher requirements for pretreatment, sample injection volume, and instrument status. The ISTD is the main method for quantifying compounds in recent years, which employed several internal standards to eliminate the method errors caused by factors such as injection volumes and the status of instruments. The method has higher requirements for internal standard compounds, such as their volatility, functional group, boiling point, and structure. Furthermore, some internal standard compounds are not readily available. SIDA has both high precision and accuracy, but it is an expensive quantitation method due to the isotope internal standard not being readily available. Many of these standards must also be synthesized manually (57). ISTD will also be the main method for quantifying the components from baijiu in the near future, but SIDA may be the long-term developmental direction.

*Volatile Compounds*

In order to explore the flavor chemicals of SFB, many researchers have studied the volatile compounds in the liquor. Up to now, there have been 463 volatile organic substances reported in SFB (Table 1), including ethyl acetate, ethyl butyrate, ethyl lactate, ethyl hexanoate, propanol, 2-methyl-1-propanol, 3-methyl-1-butanol, acetic acid, butyric acid, pentanoic acid, hexanoic acid, lactic acid, acetaldehyde, acetal, 2,3,5-trimethylpyrazine, 2,3,5,6-tetramethyl pyrazine, dimethyl trisulfide, ethyl 3-(methylthio)propionate, and 3-methyl thiopropionaldehyde among others. Of these compounds, the number of esters (105) is the greatest, followed by hydrocarbon compounds (76), nitrogenous compounds (53), sulfur compounds (47), acids (46), alcohols (45), acetals (18), furans (17), aldehydes (15), ketones (15), terpenes (15), and phenols (11). Most of the hydrocarbon compounds contribute little to the aroma of SFB because of their weak smell and significantly high thresholds, despite being the compound with the second highest total. Other compounds could be important for SFB, especially esters (105), nitrogenous compounds (56), sulfur compounds (46), alcohols (45), and acids (45).

**Table 1. Volatile Compounds in Sesame-Flavor Baijiu (SFB)**

| No. | Compounds | Brands | Pretreatment methods | Identification methods |
|---|---|---|---|---|
| | **Esters** | | | |
| 1 | Ethyl formate | GJ, GZ, MLC, YH | LLE, DI, HS-SPME | MS, S |
| 2 | Propyl formate | - | LLE | MS |
| 3 | Butyl formate | GJ, JZ | DI | MS, RI |
| 4 | Isopentyl formate | MLC | HS-SPME, LLE | MS |
| 5 | Hexyl formate | GJ, JZ | DI | MS, RI |
| 6 | Amyl formate | GJ, JZ | DI | MS, RI |
| 7 | Methyl acetate | GJ, JZ | DI | MS, RI |
| 8 | Ethyl acetate | BTQ, GJ, GZ, MLC, YH | LLE, DI, SPME | MS, RI, S, GC–O |
| 9 | Hexyl acetate | BTQ, GJ, JZ, GZ, MLC | LLE, SPME | MS, RI, S, GC–O, HS-SPME |
| 10 | Propyl acetate | MLC | LLE, DI, SPME | MS, RI, S |
| 11 | Acetic acid, 2-methylpropyl ester | GJ, MLC | LLE, SPME, SAFE | MS, S, GC–O, RI |
| 12 | Butyl acetate | JZ, MLC | SPME, LLE, DI | MS, S, RI, SPME |

**186**

Table 1. (Continued). Volatile Compounds in Sesame-Flavor Baijiu (SFB)

| No. | Compounds | Brands | Pretreatment methods | Identification methods |
|---|---|---|---|---|
| 13 | 3-Methylbutyl acetate | GJ, JZ, JSY, MLC | LLE, SPME, SAFE | MS, S, RI, GC–O |
| 14 | Phenylethyl acetate | BTQ, YH, GJ, JSY, JZ, MLC | LLME, LLE, SPME | MS, S, RI, GC–O |
| 15 | Ethyl propanoate | BTQ, JZ, MLC, YH | LLE, SPME, DI | MS, RI, S, GC–O |
| 16 | Propyl propionate | BTQ | LLE | MS, RI |
| 17 | 2-Methyl acrylate | GJ, JZ | LLE | |
| 18 | Ethyl acrylate | BTQ, JZ, MLC, YH | LLE, SPME, DI | MS, RI, GC–O |
| 19 | Ethyl butanoate | BTQ, GJ, GZ, JSY, JZ, MLC, XJXJ, YH | LLE, SPE, DI, SPME | MS, RI, S, FID, GC–O |
| 20 | Butyl butanoate | GJ, GZ, MLC | LLE, SPME | MS, RI, S |
| 21 | Pentyl butanoate | GJ, GZ | LLE, SPME | MS, S, RI |
| 22 | Isoamyl butyrate | GJ, JZ | LLE | MS, GC–O |
| 23 | Ethyl 2-methylpropanoate | BTQ, GJ, JZ, MLC, YH | LLE, SPME, DI | MS, S, RI, GC–O |
| 24 | Ethyl sec-butyrate | GJ | LLE | MS, RI |
| 25 | 2-Methylpropyl butanoate | GJ | LLE, SPME | MS, S, RI |
| 26 | Methyl valerate | GJ, JZ | DI | MS, RI |
| 27 | Ethyl pentanoate | GJ, JZ, XJXJ | LLE, SPME, DI | MS, RI, S, GC–O |
| 28 | Butyl pentanoate | GJ, GZ | LLE, SPE, SPME | MS, RI, S |
| 29 | Amyl valerate | JZ | LLE, SPME, DI | MS, RI |
| 30 | Ethyl 2-methylbutyrate | BTQ, JZ, MLC, YH | LLE, SPME | MS, S, RI, GC–O |
| 31 | Ethyl 3-methylbutanoate | GJ, JSY, JZ, MLC, XJXJ | LLE, DI, SPME | MS, S, RI, GC–MS, GC–O |
| 32 | Butyl 3-butylbutyrate | GJ, JZ | DI | MS, RI |
| 33 | Isoamyl valerate | GZ | SPME | MS, RI |
| 34 | Methyl caproate | GJ, JZ | DI | MS, RI, GC–O |
| 35 | Ethyl hexanotate | MLC, GJ, JZ, GJ, GZ, BTQ, YH, JSY | LLE, SPME, DI | MS, RI, S, GC–O, HPLC, GC–MS |
| 36 | Propyl hexanoate | GJ, JZ, MLC | LLE, SPME | MS, RI, S, GC–O |
| 37 | Butyl hexanoate | BTQ, GJ, JZ, MLC, YH | LLE, SPME, DI | MS, RI, S, GC–O |
| 38 | Hexyl butanoate | BTQ, GJ, GZ, MLC, YH | LLE, SPME | MS, RI, S |
| 39 | 2-methylpropyl hexanoate | JZ, GJ, MLC, BTQ | LLE, SPME | MS, S, RI |
| 40 | Pentyl hexanoate | GJ, JZ, MLC | LLE, SPME | MS, RI, S, GC–O |

Table 1. (Continued). Volatile Compounds in Sesame-Flavor Baijiu (SFB)

| No. | Compounds | Brands | Pretreatment methods | Identification methods |
|---|---|---|---|---|
| 41 | Isoamyl hexanoate | BTQ, GZ, JZ, JSY, MLC, XJXJ | LLE, DI, SPME | MS, S, RI, GC–O |
| 42 | Hexyl hexanoate | BTQ, GJ, JZ, MLC, XJXJ | LLE, DI, HS-SPME | MS, RI, S |
| 43 | Ethyl heptanoate | GJ, GZ, JSY, JZ, MLC | LLE, DI, HS-SPME | MS, RI, S, GC–O |
| 44 | Ethyl octanoate | BTQ, GJ, GZ, JSY, JZ, MLC, XJXJ, YH | LLE, SPME, DI | MS, RI, S, GC–O |
| 45 | Butyl octanoate | GJ, JZ, MLC | LLE, SPE, SPME | MS, RI, S |
| 46 | Octanoic acid, 3-methylbutyl ester | MLC | LLE, SPME | MS, RI, S, GC–O |
| 47 | Ethyl nonanoate | BTQ, GJ, JSY, JZ, YH | LLE, DI, SPME, SAFE | MS, RI, S |
| 48 | Ethyl decanoate | BTQ, GJ, JZ, MLC, YH | LLE, SPE, DI, SPME | MS, RI, S, GC–O |
| 49 | Ethyl undecanoate | GJ | LLE | MS, RI |
| 50 | Ethyl dodecanoate | BTQ, BTQ, JZ, MLC, YH | LLE, SPE, SPME, DI, | MS, RI, S |
| 51 | Ethyl tetradecanoate | MLC, GJ, JZ, YH | LLE, SPE, SPME, DI, HS-SPME, SAFE | MS, RI, S |
| 52 | Ethyl pentadecanoate | GJ, GZ | LLE, SPE, DI, HS-SPME | MS, RI, S |
| 53 | Ethyl hexadecanoate | BTQ, GJ, GZ, MLC, YH | LLE, SPE, SPME, DI, HS-SPME | MS, RI, S |
| 54 | Ethyl oleate | GJ, GZ, MLC, XJXJ | LLE, SPME, DI, SDE, HS-SPME | MS, RI, S |
| 55 | Isopentyl laurate | GJ, JZ, MLC | HS-SPME, LLE | MS |
| 56 | Ethyl succinate | GJ, JZ, MLC | DI, LLE | MS, RI |
| 57 | Diethyl malonate | JZ | LLE | MS |
| 58 | Diethyl butanedioate | GJ, JSY, JZ, MLC | LLE, SPME, DI | MS, RI, S, GC–O |
| 59 | Diethyl glutarate | JZ | LLE | MS |
| 60 | Diethyl heptanedioate | MLC | LLE, HS-SPME | MS |
| 61 | Diethyl octanedioate | JZ, MLC | LLE, HS-SPME, SAFE | MS, RI, S |
| 62 | Diethyl azelate | MLC | LLE, HS-SPME | MS, HS-SPME |
| 63 | Ethyl benzoate | BTQ, GJ, JZ, MLC, YH | LLE, SPME, HS-SPME | MS, S, RI, GC–O |

Table 1. (Continued). Volatile Compounds in Sesame-Flavor Baijiu (SFB)

| No. | Compounds | Brands | Pretreatment methods | Identification methods |
|-----|-----------|--------|----------------------|------------------------|
| 64 | Ethyl 2-phenylacetate | GJ, GZ, JZ, MLC, YH | LLE, DI, SPME | MS, S, RI, GC–O |
| 65 | Ethyl 3-phenylpropanoate | BTQ, GJ, GZ, JZ, MLC, YH | LLE, HS-SPME | MS, S, RI, GC–O |
| 66 | Butyric acid-2-phenylethyl ester | GJ, JZ | LLE | MS, GC–O |
| 67 | 2-Phenylethyl hexanoate | GZ, MLC | LLE, SPME | MS, S, RI, GC–O |
| 68 | Methyl 2-hydroxypropanoate | GJ, JZ | LLE | MS, RI, S |
| 69 | Ethyl lactate | BTQ, GJ, GZ, JSY, JZ, MLC, XJXJ, YH | LLE, SBSE, DI, HS-SPME | MS, RI, S, GC–O |
| 70 | Propyl lactate | GJ, JSY, JZ | LLE | MS |
| 71 | Butyl lactate | GJ, JZ, MLC | LLE, HS-SPME, DI | MS |
| 72 | 3-Methylbutyl 2-hydroxypropanoate | GJ, JZ, MLC | LLE, HS-SPME, SPME, SAFE | MS, RI, S |
| 73 | Methyl glyoxylate | GJ, JZ | LLE | MS |
| 74 | Methyl hydroxyacetate | - | LLE | MS |
| 75 | Ethyl glycolate | JZ | LLE | MS, RI |
| 76 | Ethyl 2-hydroxy-3-methylbutanoate | GJ, JZ, YH | LLE, SBSE-TDS | MS, RI, GC–O |
| 77 | Ethyl 2-hydroxy-4-methylpentanoate | BTQ, GJ, JZ, MLC, YH | LLE, DI, HS-SPME | MS, GC–O |
| 78 | Butyl 2-methyl pentanoate | GJ | LLE | MS |
| 79 | Ethyl 2-hydroxycaproate | GJ | LLE | MS |
| 80 | 4-oxo- pentanoic acid ethyl ester | GJ | LLE | MS |
| 81 | Trans-4-decenoic acid ethyl ester | GJ, JZ | LLE | MS, GC–O |
| 82 | 2-Ethyl citrate | BTQ, GJ, JZ, MLC, YH | LLE | MS, GC–O |
| 83 | 2-Decyl hexanoate | GJ, JZ | LLE | MS, GC–O |
| 84 | Furfuryl acetate | GJ | LLE | MS, RI |
| 85 | Ethyl citrate | GJ | LLE | MS, RI, GC–O |
| 86 | DL-hydroxy-hexanoic acid ethyl ester | GZ | SPME | MS, RI |
| 87 | 4-Ethyl silicate | GZ | SPME | MS, RI |

## Table 1. (Continued). Volatile Compounds in Sesame-Flavor Baijiu (SFB)

| No. | Compounds | Brands | Pretreatment methods | Identification methods |
|-----|-----------|--------|---------------------|------------------------|
| 88 | Ethyl 2-hydroxy-3-ethylbutanoate | JZ | LLE, SPME, DI | MS, RI |
| 89 | Ethyl 2-hydroxy-butyrate | JZ | LLE, SPME, DI | MS, RI, GC–O |
| 90 | 4-methyl-pentanoic acid ethyl ester | JZ | LLE, SPME, DI | MS, RI |
| 91 | Ethyl linolenate | GJ, JZ | DI | MS, RI |
| 92 | Ethyl 4-ketopentanoate | GJ, JZ, YH | LLE, DI | MS, RI |
| 93 | 2-Hydroxy ethyl hexanoate | GJ, JZ | DI | MS, RI |
| 94 | Phthalate | GJ, JZ | DI | MS, RI |
| 95 | Ethyl 4-methylpentanoate | BTQ, JZ, MLC | LLE | MS, RI, GC–O |
| 96 | Ethyl cyclohexanoate | BTQ | LLE | MS, RI, GC–O |
| 97 | Isopropyl myristate | BTQ | LLE | MS, RI |
| 98 | Ethyl 9-oxodecanoate | BTQ, YH | LLE | MS, RI |
| 99 | Ethyl 2,2-diethoxyacetate | JZ | LLE | MS,RI |
| 100 | Ethyl 9-hexadecenoate | GJ, JZ | LLE, SPME, SPE, HS-SPME | MS, RI, S |
| 101 | 9,12-Octadecadienoic acid ethyl ester | GJ, GZ, MLC, XJXJ | LLE, SPME, DI | MS, RI, S |
| 102 | Γ-butyrolactone | GJ, JZ | DI | MS, RI |
| 103 | Γ-valerolactone | BTQ, JZ, MLC, YH | LLE | MS, RI, GC–O |
| 104 | Γ-caprolactone | JZ, YH | LLE | MS, RI, GC–O |
| 105 | Γ-nonalactone | GJ, JZ | LLE | MS, S, RI, GC–O |
| | hydrocarbons | | | |
| 106 | Pentane | GJ | LLE | MS |
| 107 | 2-Methyl butane | GJ | LLE | MS |
| 108 | 3-Methyl butane | GJ | LLE | MS |
| 109 | 1,1,2-Dimethyl-butane | GZ | LLE | MS |
| 110 | 1,1-Diethoxyisopentane | GZ, YH | LLE | MS |
| 111 | 2-Ethoxy-isopentane | MLC | LLE | MS |
| 112 | 2-Methyl-2-butene | GJ | LLE | MS |
| 113 | 1-Pentene | GJ | LLE | MS |
| 114 | Heptane | GJ | LLE | MS |
| 115 | Cyclohexane | GJ | LLE | MS |
| 116 | Octane | GJ | LLE | MS |

**Table 1. (Continued). Volatile Compounds in Sesame-Flavor Baijiu (SFB)**

| No. | Compounds | Brands | Pretreatment methods | Identification methods |
|---|---|---|---|---|
| 117 | Methylcyclohexane | GJ | LLE | MS |
| 118 | Ethylcyclopentane | GJ | LLE | MS |
| 119 | 3-Methyl-octane | GJ | LLE | MS |
| 120 | 1,3-Dimethylcyclohexane | GJ | LLE | MS |
| 121 | 1-Methoxy-1-Ethoxyethane | GJ | LLE | MS |
| 122 | 1,4-Dimethylcyclohexane | GJ | LLE | MS |
| 123 | 2-Ethoxymethane | GJ | LLE | MS |
| 124 | 1,1-3 Methylcyclohexane | GJ | LLE | MS |
| 125 | 4-Methyloctane | GJ | LLE | MS |
| 126 | Decane | GJ | LLE | MS |
| 127 | 4-Methyl-1-ethyl-cyclohexane | GJ | LLE | MS |
| 128 | 2,3-Dimethyloctane | GJ | LLE | MS |
| 129 | 1,1,-Diethoxypropane | GJ | LLE | MS |
| 130 | Propylcyclohexane | GJ | LLE | MS |
| 131 | 3,6-Dimethyloctane | GJ | LLE | MS |
| 132 | 1,1-Diethoxybutane | GJ | LLE | MS |
| 133 | 4-Methyl decane | GJ | LLE | MS |
| 134 | 1-(Ethoxyethoxy)butane | GJ | LLE | MS |
| 135 | 3-Methyl decane | GJ | LLE | MS |
| 136 | 2-Methyl-1,1-diethoxybutane | GJ | LLE | MS |
| 137 | 1,1-Diethoxypentane | GJ | LLE | MS |
| 138 | 3-Methyl-1,1-diethoxybutane | GJ | LLE | MS |
| 139 | Undecane | GJ | LLE | MS |
| 140 | Tridecane | GJ | LLE | MS |
| 141 | Tetradecane | GJ | LLE | MS |
| 142 | Pentadecane | GJ | LLE | MS |
| 143 | Hexadecane | GJ | LLE | MS |
| 144 | Heptadecane | GJ | LLE | MS |
| 145 | Octadecane | GJ | LLE | MS |
| 146 | Ninecane | GJ | LLE | MS |
| 147 | Eicosane | GJ | LLE | MS |

Table 1. (Continued). Volatile Compounds in Sesame-Flavor Baijiu (SFB)

| No. | Compounds | Brands | Pretreatment methods | Identification methods |
|---|---|---|---|---|
| 148 | 1,1,1-Triethoxypropane | GJ | LLE | MS |
| 149 | Pentylcyclopropane | GJ | LLE | MS |
| 150 | 2,6,10,14-Tetramethylpentadecane | GJ | LLE | MS |
| 151 | 1,1,3,3,-Tetraethoxypropane | GJ | LLE | MS |
| 152 | 2,6,10,14-Tetramethylhexadecane | GJ | LLE | MS |
| 153 | Hexaphenyldodecane | GJ | LLE | MS |
| 154 | Triphenyldodecane | GJ | LLE | MS |
| 155 | Ethylbenzene | GJ | LLE | MS |
| 156 | Paraxylene | GJ | LLE | MS |
| 157 | Meta-xylene | GJ | LLE | MS |
| 158 | Isopropyl = benzene | GJ | LLE | MS |
| 159 | Propylene | GJ | LLE | MS |
| 160 | 2, methyl-1-ethyl-benzene | GJ | LLE | MS |
| 161 | 1,2,4-Trimethylbenzene | GJ | LLE | MS |
| 162 | Styrene | GJ | LLE | MS |
| 163 | 4-Methyl-1-ethyl-benzene | GJ | LLE | MS |
| 164 | Mesitylene | GJ | LLE | MS |
| 165 | 2,2-Diethoxyethylbenzene | GJ | LLE | MS |
| 166 | 2-Methylnaphthalene | GJ | LLE | MS |
| 167 | 1-Pentylheptylbenzene | GJ | LLE | MS |
| 168 | 1-Ethyl-nonylbenzene | GJ | LLE | MS |
| 169 | 1,1-Diethoxy-2-methylbutane | YH | LLE | MS |
| 170 | 2,4,5-Trimethyl-1,3-dioxolane | BTQ, JZ | LLE | MS |
| 171 | ,4,6-Trimethyl-1,3-dioxane | JZ | LLE | MS |
| 172 | P-Dimethylbenzene | GJ | LLE | MS |
| 173 | 4-Ethyltoluene | GJ | LLE | MS |
| 174 | ( 1-Pentyl ) heptylbenzene | GJ | LLE | MS |
| 175 | (1-Ethyl ) decylbenzene | GJ | LLE | MS |
| 176 | Biphenyl | - | LLE | MS |
| 177 | Diphenyl | - | LLE | MS |

## Table 1. (Continued). Volatile Compounds in Sesame-Flavor Baijiu (SFB)

| No. | Compounds | Brands | Pretreatment methods | Identification methods |
|-----|-----------|--------|----------------------|------------------------|
| 178 | Methylnaphthalene | JZ | LLE | MS |
| 179 | Naphthalene | GJ, JZ | LLE | MS |
| 180 | 2,7-Dimethylnaphthalene | - | LLE | MS |
| 181 | 2,6-Dimethylnaphthalene | - | LLE | MS |
| | **Pyrazines** | | | |
| 182 | Pyrazine | GJ, JZ, MLC | LLE, SPE, SPME | MS, RI, S |
| 183 | 2-Methylpyrazine | GJ, JZ, MLC | LLE, SPE,DI, HS-SPME | MS, RI, S, GC–O |
| 184 | 2,3-Dimethylpyrazine | GJ, JZ, MLC | LLE, SPE, HS-SPME, DI | MS, RI, S |
| 185 | 2,6-Dimethylpyrazine | BTQ, GJ, JZ, MLC | LLE, SPE, DI, HS-SPME | MS, RI, S, GC–O |
| 186 | 2,5-Dimethylpyrazine | GJ, JZ, MLC | LLE, SPE, HS-SPME | MS, RI, S |
| 187 | Trimethyl pyrazine | BTQ, GJ, JZ, MLC, YH | LLE, SPE, HS-SPME, DI, S, GC–O | MS, RI, S, GC–O |
| 188 | 2,3,5-Trimethyl-6-ethylpyrazine | GJ | LLE, SPE | MS, RI, S, GC–O |
| 189 | Tetramethylpyrazine | BTQ, GJ, JZ, MLC, XJXJ, YH | LLE, SPE, SPME, HS-SPME, DI | MS, RI, S, GC–O |
| 190 | 3,5-diethyl-2-methyl-Pyrazine | GJ, JZ, MLC | LLE | MS, S |
| 191 | 2-Methyl-5-(1-methylethyl)pyrazine | GJ | LLE | MS, S, RI |
| 192 | 2,5-Dimethyl-3-ethyl pyrazine | JZ | LLE, SPE, DI | MS, RI, S, GC–O |
| 193 | 2,3-Dimethyl-5-ethylpyrazine | GJ, MLC | LLE, SPE, SPME, AEDA, DI, HS-SPME | MS, RI, GC–O, S |
| 194 | 2,5-Dimethyl-3-pentyl pyrazine, | GJ | LLE | MS, S, RI |
| 195 | 2-Ethylpyrazine | GJ | LLE | MS, RI, S |
| 196 | 2,6-Diethylpyrazine | GJ, JZ, MLC | LLE, DI | MS, RI, GC–O, S |
| 197 | 2-Ethyl-3-methylpyrazine | GJ | LLE, SPE | MS, RI, S |
| 198 | 2-Ethyl-5-methylpyrazine | GJ, JZ, MLC | LLE, SPE | MS, RI, S |
| 199 | 2-Ethyl-6-methylpyrazine, 2-methyl-6-ethylpyrazine | GJ, JZ, MLC | LLE, SPE, DI, HS-SPME | MS, RI, S, GC–O |

## Table 1. (Continued). Volatile Compounds in Sesame-Flavor Baijiu (SFB)

| No. | Compounds | Brands | Pretreatment methods | Identification methods |
|---|---|---|---|---|
| 200 | 2-Ethyl-3,5-dimethylpyrazine | GJ, JZ, MLC | LLE, SPE | MS, RI, S |
| 201 | 2-Ethyl-3,6-dimethylpyrazine | GJ | LLE | MS, S, RI |
| 202 | 2,3-diethyl-5-methylpyrazine | GJ, JZ, YH | LLE | MS, S, RI |
| 203 | Pyrazine,2-ethenyl-6-methyl- | GJ | LLE, DI | MS, S, RI |
| 204 | 2-Butyl-3,5-dimethylpyrazine | GJ, JZ | LLE, SPE | MS, S, RI, GC–O |
| 205 | Pyrazine,3-butyl-2,5-dimethyl- | GJ | LLE | MS, S, RI, GC×GC |
| 206 | 2-Ethyl- 2,5 -dimethyl pyrazine | GJ | LLE | MS |
| 207 | 2-Ethyl-3-isobutyl-6-methyl-pyrazine | GJ, MLC | LLE | MS |
| 208 | 3-Ethyl-2,5-dimethylpyrazine | JZ, MLC | LLE | MS |
| 209 | 3-Propyl-5-ethyl-2,6-dimethyl pyrazine | GJ, JZ, MLC | LLE | MS |
| 210 | 3-Isobutyl -2,5-dimethyl pyrazine | MLC | LLE | MS |
| 211 | 3-Isobutyl- 2,5-diethyl-pyrazine | GJ, MLC | LLE | MS |
| 212 | 3-Isobutyl- 2,5-diethyl-pyrazine | GJ | LLE | MS |
| 213 | 3-Isopentyl -2,5-dimethyl pyrazine | MLC | LLE | MS |
| 214 | Azabenzene | GJ, JZ, MLC | LLE, HS-SPME | GC×GC, MS, S, RI |
| 215 | 3-Phenylpyridine | BTQ, GJ, JZ, MLC | LLE | MS, S, RI |
| 216 | 5-Ethyl-2-methylpyridine | MLC | HS-SPME | MS |
| 217 | 2-Acetylpyridine | GJ | LLE, SPE | MS, RI, S, GC–O |
| 218 | 3-Isobutylpyridine | MLC | LLE | MS |
| 219 | Arecoline hydrobromide | | | MS |
| 220 | 1-(1H-pyrrol-2-yl)- ethanone | GJ | LLE, SAFE | MS, RI, S, GC–O |
| 221 | 2,3,4,5-Tetramethylpyrrole | | | MS |
| 222 | 3-Methyl- oxazolidine | JZ | LLE, SPME | MS |
| 223 | 2,4,5-Trimethyloxazole | GJ,MLC | LLE, HS-SPME | MS, GC×GC |
| 224 | 3-Methyl-2-Oxazolidone | GJ | LLE | MS,S,RI |
| 225 | (Z)-13-Docosenamide | JZ | DI | MS, S |

## Table 1. (Continued). Volatile Compounds in Sesame-Flavor Baijiu (SFB)

| No. | Compounds | Brands | Pretreatment methods | Identification methods |
|---|---|---|---|---|
| 226 | Formamide | - | LLE | MS |
| 227 | Ethylenecarboxamide | - | LLE | MS |
| 228 | Benzylhydrazine | - | LLE | MS |
| 229 | 2-butoxyethyl 4-(Dimethylamino)benzoate | - | HS-SPME | MS |
| 230 | Urethane | GJ | LLE | MS, S, RI, HPLC, GC-MS, MS, LC-MS |
| 231 | 2-Hydroxyindole | - | LLE | MS |
| 232 | Diacetone acrylamide | - | LLE | MS |
| 233 | Hexonitrile | BTQ, JZ, YH | LLE | MS |
| 234 | 2-Pyrrolidine | YH | LLE | MS, RI, S, GC–O |
| | **Sulfur compounds** | | | |
| 235 | 3-Methylthiopropanol | BTQ, GJ, JZ, KFJ, LL, MLC, YH | LLE, SPME, FPD, DI | GC-MS/MS |
| 236 | 3-(Methylthio)propionaldehyde | BTQ, GJ, JZ, MLC, YH | LLE, SPME, DI | GC-MS/MS |
| 237 | 3-(Methylthio)butyraldehyde | BTQ, GJ, TS, JSY, JZ, LL | LLE | FPD |
| 238 | S-methyl butanethioate | BTQ, GJ, TS, JSY, JZ, LL | LLE, SPME | FPD |
| 239 | Ethyl (methylthio)acetate | GJ | LLE, SPME | FPD |
| 240 | Ethyl 3-(methylthio)propanoate | BTQ, GJ, JZ, MLC | LLE, SPME | FPD |
| 241 | Dimethyl sulfide | JZ, MLC | LLE, SPME | FPD |
| 242 | Furfuryl methyl sulfide | JZ, MLC | SPME | FPD |
| 243 | Dimethyl disulfide | GJ, JZ | LLE, SPME, FPD | GC-MS/MS |
| 244 | Isopropyl disulfide | GJ | LLE, SPME, FPD | FPD |
| 245 | Difurfuryl disulfide | GJ | LLE, SPME, FPD | FPD |
| 246 | Dimethyl trisulfide | BTQ, GJ, GJ, GZ, JZ, MLC, YH | LLE, SPME, FPD | FPD |
| 247 | Dimethyl tetrasulfide | GJ | LLE, SPME | GC-MS/MS |
| 248 | Thiophene | MLC | LLE | FPD |
| 249 | 4-Methyldibenzothiophene | BTQ, GJ, JSY, JZ, LL, TS | LLE, SPME | FPD |
| 250 | Thiazole | JZ, MLC | LLE, SPME | FPD, GC-MS/MS |

Table 1. (Continued). Volatile Compounds in Sesame-Flavor Baijiu (SFB)

| No. | Compounds | Brands | Pretreatment methods | Identification methods |
|---|---|---|---|---|
| 251 | Benzothiazole | BTQ, GJ, JSY, JZ, LL, TS | LLE | GC-MS/MS |
| 252 | 4-Methyl-5-thiazoleethanol | MLC | LLE, SPME | FPD |
| 253 | Ethyl mercaptan | MLC | LLE, SPME | GC-MS/MS |
| 254 | Furfuryl mercaptan | JZ | LLE, SPME | FPD, GC-MS/MS |
| 255 | S-methyl methanethiolsulfonate | BTQ, GJ, JSY, JZ, LL, TS | LLE, SPME | MS, S, RI |
| 256 | Dimethylthiosulfinate | GJ | LLE, SPME | MS, S, RI |
| 257 | Methyl mercaptan | JZ, GJ | LLE, SPME | LLE |
| 258 | 2-Propanethiol | BTQ, GJ, JSY, JZ, LL, TS | LLE, SPME | FPD |
| 259 | 3-Mercapto-2-butanone | BTQ, GJ, JSY, JZ, LL, TS, | LLE, SPME | FPD |
| 260 | 2,3-Butanedithiol | BTQ, GJ, JSY, JZ, LL, TS | LLE, SPME | FPD |
| 261 | Diethyl trisulfide | BTQ, GJ, JSY, JZ, LL, TS | LLE, SPME | FPD |
| 262 | Benzothiophene | BTQ, GJ, JSY, JZ, LL, TS | SPME | FPD |
| 263 | Methyl thiofurancarboxylate | BTQ, GJ, JSY, JZ, LL, TS | SPME | FPD |
| 264 | Methyl thioacetate | JZ | SPME | GC-MS/MS |
| 265 | 2-Methylthiophene | JZ | SPME | GC-MS/MS |
| 266 | Methyl ethyl disulfide | JZ | SPME | GC-MS/MS |
| 267 | 2-Ethylthiophene | JZ | SPME | GC-MS/MS |
| 268 | Methyl thiobutyrate | JZ | SPME | GC-MS/MS |
| 269 | 2-Isopropylthiophene | JZ | SPME | GC-MS/MS |
| 270 | 2-Propylthiophene | JZ | SPME | GC-MS/MS |
| 271 | Methyl n-butyl disulfide | JZ | SPME | GC-MS/MS |
| 272 | Methyl sec-butyl disulfide | JZ | SPME | GC-MS/MS |
| 273 | Thiophene-3-carbaldehyde | JZ | SPME | GC-MS/MS |
| 274 | 2,3-Dimethylthiophene | JZ | SPME | GC-MS/MS |
| 275 | 2-Isopentylthiophene | JZ | SPME | GC-MS/MS |
| 276 | Methyl thiodecanoate | JZ | SPME | GC-MS/MS |
| 277 | Ethyl 2-thiophenecarboxylate | JZ | SPME | GC-MS/MS |
| 278 | 2-Methylbenzothiophene | JZ | SPME | GC-MS/MS |

Table 1. (Continued). Volatile Compounds in Sesame-Flavor Baijiu (SFB)

| No. | Compounds | Brands | Pretreatment methods | Identification methods |
|---|---|---|---|---|
| 279 | 3-Phenylthiophene | JZ | SPME | GC-MS/MS |
| 280 | 2-Mercaptoacetate | JZ | SPME | MS, RI, S |
| 281 | Benzenemethanethiol | GJ | LLE | MS, S, GC–O |
| | **Alcohols** | | | |
| 282 | Methanol | GJ, GZ, JSY, MLC, XJXJ | LLE, DI | MS, S |
| 283 | 1-Propanol | GJ, GZ, JZ, MLC, XJXJ | LLE, DI, SDE, UHP | MS, RI, S, GC–O, FT-NIR |
| 284 | 2-Methylpropanol | BTQ, GJ, GZ, JZ, MLC, XJXJ, YH | LLE, HS-SPME, SDE, SPME | MS, RI, S, GC-MS |
| 285 | 1-Butanol | GJ, GZ, JSY, MLC, XJXJ | LLE, HS-SPME | MS, RI, S, FT-NIR |
| 286 | 2-Butanol | BTQ, GJ, GZ, JSY, JZ, MLC, XJXJ, YH | LLE, HS-SPME, GC–MS | MS, RI, S |
| 287 | 2-Methylbutanol | GJ, GZ, MLC, XJXJ | LLE, DI, SPME | MS, RI, S |
| 288 | 3-Methyl-1-butanol | BTQ, GJ, GZ, JZ, MLC, XJXJ, YH | LLE, DI, SPME, SPE, SDE, UHP, HS-SPME, SAFE | MS, RI, S, GC–O, FT-NIR, GC-MS |
| 289 | 1-Pentanol | GJ, GZ, BTQ, MLC, XJXJ, YH | LLE/SPE/HS-SPME | MS, RI, S |
| 290 | 2-Pentanol | GZ, MLC, XJXJ | LLE, DI, SPME | MS, RI, S, GC–O |
| 291 | 1-Hexanol | BTQ, GJ, GZ, JZ, MLC, XJXJ, YH | LLE, DI, SPE, SPME, SBSE | MS, RI, S, GC–O |
| 292 | 2-Ethyl-1-hexanol | GJ | LLE, SPME | MS, RI, S, GC–O |
| 293 | 1-Heptanol | BTQ, GJ, JZ, MLC, YH | LLE, HS-SPME, SBSE, DI | MS, RI, S, GC–O |
| 294 | 6-Methyl-2-heptanol | GJ | LLE, DI | MS |
| 295 | 1-Octanol | BTQ, GJ, JZ, MLC, XJXJ, YH | LLE | MS, RI, S, GC–O |
| 296 | 2-Octanol | MLC | LLE, HS-SPME, SBSE | MS, RI, S, GC–O |
| 297 | 3-Octanol | GJ, JZ | LLE | MS, RI |
| 298 | 1-Dodecanol | MLC | SPME, SDE, LLE, HS-SPME | MS, RI, S |
| 299 | 3-Methyl-3-buten-1-ol | GJ, JZ, BTQ, YH | LLE | MS, RI, S |
| 300 | 1-Octen-3-ol | - | LLE | MS, RI, S |

Table 1. (Continued). Volatile Compounds in Sesame-Flavor Baijiu (SFB)

| No. | Compounds | Brands | Pretreatment methods | Identification methods |
|-----|-----------|--------|----------------------|------------------------|
| 301 | Phenylmethanol | BTQ, JSY, JZ, MLC, YH | LLE, DI, SPE, SDE, UHP | MS, S, RI, GC–O |
| 302 | 4-Hydroxybenzyl alcohol | GJ | DI | MS |
| 303 | 2-Phenylethanol | BTQ, GJ, GZ, JSY, JZ, MLC, XJXJ, YH | LLE, DI, SPME | MS, S, RI, GC–O |
| 304 | 1,2-Propanediol | GJ, GZ, MLC, XJXJ | DI, HS-SPME, LLE | MS, S, GC×GC, RI |
| 305 | 2,3-Butanediol | GJ, GZ, MLC, XJXJ | LLE, DI, SAFE, HS-SPME | MS, RI, S |
| 306 | 1,3-Butylene glycol | - | DI, LLE, HS-SPME | MS |
| 307 | Nonamethylene glycol | MLC | LLE | MS |
| 308 | 2,3-Hexamethylene glycol | MLC | LLE | MS |
| 309 | 1,2,3-Propanetriol | - | HS-SPME, LLE, DI | MS |
| 310 | Erythritol | JZ | HS-SPME | MS |
| 311 | Dulcitol | MLC | HS-SPME | MS |
| 312 | Tetrahydrofurfuryl alcohol | JZ | LLE | MS |
| 313 | 2-Heptanol | GJ, JZ | LLE | MS |
| 314 | 3-Phenyl-1-propanol | JZ | LLE | MS |
| 315 | 3-Methoxy-2-butanol | JZ | LLE | MS |
| 316 | 3-Mthoxy-1-propanol | BTQ, MLC | LLE | MS |
| 317 | Geosmin | JZ | LLE | MS |
| 318 | 4-Hydroxyphenylethanol | JZ | LLE | MS |
| 319 | Xylitol | JZ | LLE | MS |
| 320 | Arab arabitol | JZ | LLE | MS |
| 321 | Adonitol | JZ | LLE | MS |
| 322 | Myo-inositol 2 | JZ | LLE | MS |
| 323 | D-glucose | JZ | LLE | MS |
| 324 | Sucrose | JZ | LLE | MS |
| 325 | Trehalose | JZ | LLE | MS |
| 326 | Maltose | JZ | LLE | MS |
| **Acids** | | | | |
| 327 | Formic acid | JZ | LLE | MS |

Table 1. (Continued). Volatile Compounds in Sesame-Flavor Baijiu (SFB)

| No. | Compounds | Brands | Pretreatment methods | Identification methods |
|-----|-----------|--------|----------------------|------------------------|
| 328 | Acetic acid | BTQ, GJ, GZ, JZ, MLC, YH | LLE, SPME, DI | MS, RI, GC–O, S, GC– |
| 329 | Propanoic acid | BTQ, GJ, GZ, JZ, MLC, YH | LLE, SPME, DI | MS, S, RI, GC–O, GC×GC |
| 330 | 2-Methylpropanoic acid/ Isobutyric acid | BTQ, GJ, GZ, JZ, MLC, YH | LLE, SPME, DI | MS, S, RI, AROMA, GC–O |
| 331 | Butanoic acid | BTQ, GJ, GZ, JZ, MLC, YH, JSY | LLE, SPME, DI | MS, S, RI, GC–O, GC×GC |
| 332 | 3-Methylbutanoic acid/ Isovaleric acid | BTQ, GJ, GZ, JZ, MLC, YH | LLE ,SPME ,DI | MS, S, RI, GC–O, GC×GC |
| 333 | Pentanoic acid | BTQ, GJ, GZ, JZ, MLC, YH, JSY | LLE, SPME, DI | MS, S, RI, GC–O |
| 334 | 2-Methylpentanoic acid | - | MS,MS, S, RI, GC–O | |
| 335 | 4-Methylpentanoic acid | BTQ, MLC | MS, RI, A, GC–O, GC–MS | |
| 336 | Hexanoic acid | BTQ, GJ, GZ, JZ, MLC, YH | LLE, SPME, DI | MS, S, RI, GC–O, GC– |
| 337 | Heptanoic acid | GJ, GZ, JZ, MLC, YH | LLE, SPME, DI | MS, S, RI, GC–O |
| 338 | Octanoic acid | BTQ, GJ, GZ, JZ, MLC, YH | LLE, SPME, DI | MS,S,RI ,GC-O |
| 339 | Nonanoic acid | BTQ, MLC | LLE, SPME, DI | MS,S,RI ,GC-O |
| 340 | Decanoic acid | BTQ, GJ, JZ, MLC, YH | LLE, SPME, DI | MS,S,RI ,GC-O |
| 341 | Lauric acid | JZ | LLE ,SPME ,DI | MS, S, RI |
| 342 | Tetradecanoic acid | GJ, JZ | LLE | MS, S, RI |
| 343 | Hexadecanoic acid | JZ | LLE | MS, S, RI |
| 344 | Octadecanoic acid | GJ, JZ | LLE | MS, S, RI |
| 345 | Sorbic acid | MLC | LLE | MS |
| 346 | Linoleic acid | JZ | LLE | MS |
| 347 | Oleic acid | JZ | LLE | MS, RI |
| 348 | Vaccenic acid | GJ | LLE | MS |
| 349 | Lactic acid | GJ, MLC | LLE | MS, RI |
| 350 | Benzoic acid | GJ, JZ | LLE, SPME, DI | MS, S, RI |
| 351 | 3-Phenylpropanoic acid | GJ, JZ | LLE | MS, GC–O, RI, S |
| 352 | Pyruvic acid | JZ | LLE | MS |

Table 1. (Continued). Volatile Compounds in Sesame-Flavor Baijiu (SFB)

| No. | Compounds | Brands | Pretreatment methods | Identification methods |
|---|---|---|---|---|
| 353 | 9,12 Eighteen two acid carbon | GJ | LLE | MS |
| 354 | 2-Butenoic acid | GJ | LLE | MS, RI |
| 355 | Phenylacetic acid | GJ,JZ | LLE | MS, RI |
| 356 | Tannic acid | GJ, JZ | LLE | MS, RI |
| 357 | 2-Methylhexanoic acid | GJ, JZ | LLE | MS, RI |
| 358 | 4-Mydroxybutyric acid | - | LLE | MS, RI |
| 359 | Succinic acid | JZ | LLE | MS, RI |
| 360 | Malic acid | JZ | LLE | MS, RI |
| 361 | 2-Mydroxyisohexanoic acid | JZ | LLE | MS, RI |
| 362 | 2-Mydroxyisovalerate | JZ | LLE | MS, RI |
| 363 | 2,3-Dihydroxypropionic acid | JZ | LLE | MS, RI |
| 364 | 2-Hydroxybutyric acid | JZ | LLE | MS, RI |
| 365 | 3-Hydroxybutyric acid | JZ | LLE | MS, RI |
| 366 | 3-Hydroxypropionic acid | JZ | LLE | MS, RI |
| 367 | DL-3-phenyl lactic acid | JZ | LLE | MS, RI |
| 368 | 4-Hydroxybenzoic acid | JZ | LLE | MS, RI |
| 369 | Glutaric acid | MLC | LLE | MS, RI |
| 370 | Butyric acid-2-pyrene | JZ | LLE | MS |
| 371 | Acetic acid-2-guanidine | JZ | LLE | MS |
| 372 | Hydrocinnamic acid | GJ | LLE | MS, RI, GC–O |
| | **Aldehydes, ketones, acetals** | | | |
| 373 | 2-Propenal | - | | MS |
| 374 | Acetaldehyde | GJ, GZ, JSY, MLC, YH | LLE, DI, SPME | MS, S, RI, RI |
| 375 | Propanal | GJ, MLC, YH | LLE, DI | MS, S, RI |
| 376 | 2-Methylpropanal | GJ, GZ, JSY, MLC, YH | LLE, SPME, DI | MS, S, RI, GC–O |
| 377 | 3-Methylbutanal | BTQ, GJ,GZ, JSYJZ, MLC, YH | LLE, SBSE, DI-SPME, SDE | MS, S, RI, GC–O |
| 378 | Hexanal | JSY, MLC | LLE, DI-SPME | MS, S, RI, GC–O |
| 379 | Nonanal | BTQ, GJ, JZ, MLC, YH | LLE, SBSE, SPE, SPME | MS, S, RI, GC–O |
| 380 | Decanal | MLC | LLE, SBSE, HS-SPME | MS, S, RI |

## Table 1. (Continued). Volatile Compounds in Sesame-Flavor Baijiu (SFB)

| No. | Compounds | Brands | Pretreatment methods | Identification methods |
|-----|-----------|--------|----------------------|------------------------|
| 381 | Benzaldehyde | BTQ, GJ, JZ, MLC, YH | LLE, DI, SPME | MS, S, RI, GC–O |
| 382 | 4-N-pentylbenzaldehyde | - | LLE | MS, RI |
| 383 | Phenylacetaldehyde | BTQ, GJ, JZ, MLC, YHJZ | LLE, SAFE, HS-SPME | MS, S, RI, GC–O |
| 384 | Amylcinnamaldehyde | - | LLE | MS, RI |
| 385 | 2-Phenyl-2-butenal | JZ | LLE | MS, RI |
| 386 | 2-Phenyl crotonaldehyde | BTQ, JZ, YH | LLE | MS, RI |
| 387 | Vanillin | BTQ, JZ, MLC, YH | LLE | MS, RI, GC–O |
| 388 | Acetone | MLC, YH | LLE, DI | MS, S, RI |
| 389 | 2-Pentanone | GJ, GZ, MLC, YH | LLE, SPME, DI | MS, S, RI, GC–O |
| 390 | 2-Heptanone | GJ | LLE | MS, S, RI, GC–O |
| 391 | 2-Nonanone | GZ | LLE, SBSE, HS-SPME | MS, S, RI, GC–O |
| 392 | 2-Undecanone | GZ | LLE, SPME | MS, S, RI |
| 393 | Cyclopentanone | GJ | LLE | MS, S, RI |
| 394 | Cyclohexanone | GJ | LLE | MS, RI |
| 395 | 1-Octen-3-one | | LLE | MS, S, RI |
| 396 | 1-Hydroxy-2-propanone | JZ | LLE, SPE, DI | MS, RI |
| 397 | 3-Hydroxy-2-butanone | BTQ, GJ, GZ, JZ, MLC, YH | LLE, DI, HS-SPME | MS, S, RI, GC–O |
| 398 | 2,3-Butanedione | BTQ, GJ, JZ, MLC, YH | LLE | MS, S, RI, GC–O |
| 399 | 2-Formaldehyde-3-heptanone | GJ | LLE | MS, S, RI, GC–O |
| 400 | 2-Butanone | GJ | LLE | MS, S, RI, GC–O |
| 401 | Ethyl 4-oxopentanoate | GJ, JZ | LLE | MS, S, RI, GC–O |
| 402 | Acetophenone | BTQ, MLC | LLE | MS, S, RI, GC–O |
| 403 | Phenyl-2-propanone | BTQ, YH | LLE | MS, S, RI, GC–O |
| 404 | Diethoxymethane | GJ, MLC | LLE, SPME, DI | MS, S |
| 405 | 1,1-Diethoxyethane | JZ, MLC | LLE, DI, SPME, SI-SPME, HS-SPME | MS, S, RI, GC–O, GC, RI, GC×GC |
| 406 | 1-(1-Ethoxyethoxy)pentane | JZ | LLE, DI, SPME, SDW | MS, RI, GC–O, GC×GC |

Table 1. (Continued). Volatile Compounds in Sesame-Flavor Baijiu (SFB)

| No. | Compounds | Brands | Pretreatment methods | Identification methods |
|---|---|---|---|---|
| 407 | 1,1-Diethoxy-3-methyl-butane | BTQ, GJ, JZ, YH | LLE, DI, SDE, SPME | MS, RI, GC–O, S, GC-GC |
| 408 | (2,2- Dimethoxyphenyl) ethyl benzene | - | LLE | MS, S, RI |
| 409 | Acetal | BT, QGJ, GZ, GJ, JZ, MLC, YH | GC–FID | MS, S, RI, GC–O |
| 410 | Butyraldehyde diethyl acetal | GJ | LLE | MS, S, RI |
| 411 | 1,1,1-Trimethylcyclopentane | GJ | LLE | MS, S, RI |
| 412 | 2-Furfural diethyl acetal | JZ | LLE | MS, S, RI, GC–O |
| 413 | Valeraldehyde diethanol | JZ | LLE | MS, S, RI |
| 414 | (2,2-Diethoxyethyl)-benzene | BTQ, YH | LLE | MS, S, RI |
| 415 | 3-Ethoxypropanal diethyl acetal | BTQ, JZ, ML, CYH | LLE | MS, S, RI |
| 416 | Octanal diethyl acetal | BTQ | LLE | MS, S, RI, GC–O |
| 417 | Isobutyraldehyde diethyl acetal | JZ | LLE | MS, S, RI |
| 418 | 2-Methoxy-1,3-dioxolane | JZ, MLC | LLE | MS, S, RI |
| 419 | 2,4,6-Trimethyl-1,3-dioxane | MLC | LLE | MS, S, RI |
| 420 | 1,1-Diethoxy-2-methylpropane | MLC | LLE | MS, S, RI |
| | **Terpenes** | | | |
| 421 | A-pinene | JZ | SPME | MS |
| 422 | Linalool | JZ | SPME | MS |
| 423 | β-cyclocitral | JZ | SPME | MS |
| 424 | Camphor | JZ | SPME | |
| 425 | (+/-)-Geosmin | JZ | SPME | MS, GC–O |
| 426 | Elemene | JZ | SPME | MS |
| 427 | Anti-geranyl acetone | JZ | SPME | MS |
| 428 | A-curcumene | JZ | SPME | MS |
| 429 | Anti-α-citronene | JZ | SPME | MS |
| 430 | (+)-Flower arborene | JZ | SPM | MS |
| 431 | Orange flower tertiary alcohol | JZ | SPME | MS |
| 432 | Beta-ionone oxide | JZ | SPME | MS |
| 433 | L-α-terpineol | BTQ, JZ, MLC | SPME | MS |

Table 1. (Continued). Volatile Compounds in Sesame-Flavor Baijiu (SFB)

| No. | Compounds | Brands | Pretreatment methods | Identification methods |
|---|---|---|---|---|
| 434 | A-Cedrene | JZ, GJ | LLE | MS |
| 435 | β-Damasone | JZ | LLE | MS, GC–O |
| | Furans | | | |
| 436 | 2-N-butylfuran | GJ | LLE | MS, RI, S |
| 437 | 2-furfural | BTQ, GJ, GZ, JZ, MLC, YH | LLE, SPME, DI | MS, RI, S, GC–O |
| 438 | Sterol | BTQ, GJ, YH, JZ, MLC | LLE | MS, RI, S, GC–O |
| 439 | 2-Ethoxy-5-methylfuran | MLC | LLE | MS, RI, S |
| 440 | 2-Pentylfuran | MLC | LLE | MS,RI,S |
| 441 | 2-Acetyl-furan | BTQ, GJ, YH, JZ, MLC | LLE, SPME, DI | MS, RI, S, GC–O |
| 442 | 5-Methyl-acetaldehyde | MLC | LLE | MS, RI, S, GC–O |
| 443 | 2-Acetyl-5-methyl-furan | BTQ, JZ, MLC, YH | LLE, SPME, DI | MS, RI, S, GC–O |
| 444 | 1,1-Dipropyl hydrazine | GJ | LLE | MS, RI, S |
| 445 | 5-Methyl-2-furaldehyde | GJ, JZ, MLC | LLE | MS, RI, S, GC–O |
| 446 | 2-(Diethoxymethyl)-furan | GJ | LLE | MS, RI, S |
| 447 | 2-Furfural diethyl acetal | GJ | LLE | MS, RI, S, GC–O |
| 448 | 5-Methyl-furfural | BTQ, GJ, JZ, MLC, YH | LLE | MS, RI, S, GC–O |
| 449 | 2-Furan acrolein | JZ | LLE, SPME, DI | MS, RI, S, GC–O |
| 450 | Difluorofuran ether | JZ | LLE, SPME, DI | MS, RI, S, GC–O |
| 451 | 3-(2-Furyl)-2-propenal | BTQ, MLC | LLE | MS, RI, S, GC–O |
| 452 | 2-Propionylfuran | BTQ, JZ, YH | LLE | MS, RI, S |
| | **Phenols** | | | |
| 453 | Phenol | BTQ, JZ, GJ, YH, MLC | LLE | MS, S, RI, GC–O |
| 454 | 3-Methylphenol | JZ | LLE, HS-SPME | MS, S, RI |
| 455 | p-Cresol | BTQ, GJ, JZ, MLC, YH | LLE, SPME, DI | MS, S, RI, GC–O |
| 456 | 4-Ethylphenol | BTQ, GJ, JZ, MLC, YH | DI-SPME, HS-SPME, LLE | MS, S, RI, GC–O |
| 457 | 2,4-Di-tert-butylphenol | BTQ, GJ, JZ, MLC, YH | LLE, HS-SPME, SAFE | MS, RI, S |
| 458 | 2,2'-Methylenebis(6-tert-butyl-4-methylphenol) | - | DI, SHS | S, MS |

## Table 1. (Continued). Volatile Compounds in Sesame-Flavor Baijiu (SFB)

| No. | Compounds | Brands | Pretreatment methods | Identification methods |
|---|---|---|---|---|
| 459 | 4-Ethyl-2-methoxyphenol | BTQ, GJ, JZ, MLC, YH | LLE, HS-SPME, DI-SPME, LLE, SAFE | MS, S, RI |
| 460 | 4-Vinylguaiacol | - | HS-SPME, DI-SPME, LLE, SDE, SPME, DI | MS, RI, S |
| 461 | Methoxyphenol | - | LLE | MS, RI, S |
| 462 | Guaiacol | BTQ, JZ, MLC, YH | LLE, GC–O | MS, RI, S, GC–O |
| 463 | 2,4-Bis(1,1-dimethylethyl)-phenol | GJ | LLE | MS, RI, GC–O |

GJ=Guojing; GZ=Guizhou; MLC=Meilanchun; YH=Yanghu; BTQ=Baotuquan; JSY=Jinshiyuan.

Ester compounds are the most abundant compounds in baijiu, which play an important role in the overall aroma profiles of the liquor. Han et al. (58) analyzed the contents and ratio of important ester compounds in the different flavor styles of baijiu. The results are shown in Figure 5.

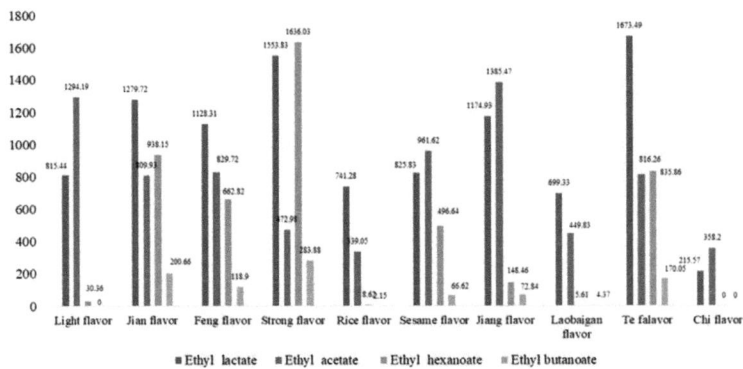

*Figure 5. The content differences of several typical components between SFB and other flavor styles of baijiu (mg/L).*

From Figure 5, it can be seen that the contents of ethyl hexanoate (496.64 mg/L), one of representative flavor components in Nong flavor liquor, was significantly lower in SFB than in Nong flavor liquor (1636.03 mg/L) or Jian flavor liquor (938.15 mg/L) but signficantly higher than in mild flavor (30.36 mg/L) or Jiang flavor baijiu (148.46 mg/L). Ethyl acetate is the main aroma compound of mild flavor baijiu (59) and has a high concentration in SFB (961.62 mg/L), which is lower than that of mild flavor (1294.19 mg/L) and Jiang flavor baijiu (1385.47 mg/L), but higher than Jian flavor (809.93 mg/L), Feng flavor (829.72 mg/L), and Te flavor liquors (816.26 mg/L). Moreover, it was significantly higher than Nong flavor liquor (472.98 mg/L) and others. The ratio of ethyl lactate to ethyl hexanoate in different flavor styles of Baijiu were 0.95 (Nong flavor baijiu), 26.86 (mild flavor baijiu), 7.91 (Jiang flavor baijiu), and 1.66 (SFB). The ratio of ethyl acetate to ethyl hexanoate were 0.29 (Nong flavor baijiu), 42.63 (mild flavor baijiu), 9.33 (Jiang flavor baijiu), and 1.94 (SFB). The ratios of ethyl butanoate to ethyl hexanoate were 0.17 (Nong flavor baijiu), 0 (mild flavor baijiu), 0.49 (Jiang flavor baijiu), and 0.13 (SFB). It is these quantitative relationships among esters, alcohols, acids, and other compounds that make SFB with some flavor characteristics of Nong, Jiang, and mild flavor liquors (60).

When discussing the the different brands of SFB, the content ratios of the traditional four esters are not constant and vary within a certain range. The ratios of ethyl lactate to ethyl hexanoate in different SFBs were 6.98–28.66 (MLC) (61, 62), 1.22–5.65 (GJ) (29, 63, 64), 1.74–4.68 (JZ) (34, 62, 65), 1.61 (BTQ) (66), and 3.24 (YH) (66). The ratios of ethyl acetate to ethyl hexanoate were 6.04–25.21 (MLC) (61, 62), 0.94–5.77(GJ) (29, 63, 64), 2.40–3.30 (JZ ) (34, 62, 65), 1.42 (BTQ) (66), and 2.72 (YH) (66). The ratios of ethyl butyrate to ethyl hexanoate were 0.27–0.68 (MLC) (61, 62). 0.05–0.87 (GJ) (29, 63, 64), 0.57–2.44 (JZ) (34, 62, 65), 0.24 (BTQ) (66), and 0.3 (YH) (66). The average ratios of ethyl lactate to ethyl acetate to ethyl butanoate to ethyl hexanoate were different for each SFB: 14.62:12.92:0.53:1 (MLC), 2.59:2.63:0.31:1 (GJ), 2.86:2.72:1.07:1 (JZ), 1.60:1.43:0.24:1 (BTQ), and 3.25:2.72:0.30:1 (YH). Among them, MLC had the highest ratio.

Based on the reported results by Zhang (30), we conducted a PCA analysis on the volatile components of four kinds of SFB (Figure 6). The cumulative contribution rate of the two principal components was 86.95%, which represented the basis for the samples of the four styles of SFB. As the loading factor analysis results showed, SFB mixed with light aroma and pure SFB were located in the positive semi-axis of the first principal component, which had an important contribution to this first principal component. SFB mixed with strong aroma, pure SFB, and SFB mixed with light aroma were located at the positive principal axis of the second principal component, which had an important contribution to this second principal component. Though the aroma compounds among the four different SFB were almost the same, they were divided into four categories and located in different areas. This was also closely related to the amount of these compounds and their corresponding ratios.

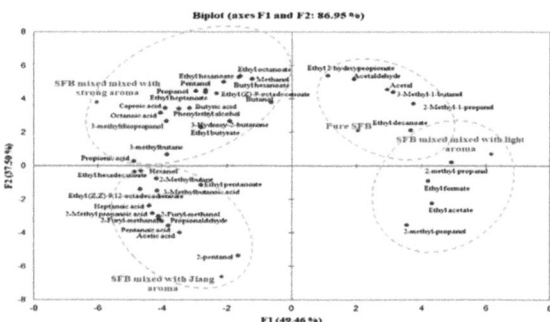

Figure 6. PCA analysis results according to the contents of compounds in four styles of SFB.

As shown in Figure 6, ethyl decanoate, ethyl formate, ethyl acetate, isobutyraldehyde, and 2-methyl-propanol had a positive correlation with SFB mixed with light aroma. Ethyl lactate, acetaldehyde, acetal, isobutanol, and ethyl decanoate had a positive correlation to pure SFB. These 13 aroma compounds had positive correlations to the SFB mixed with strong aroma including ethyl hexanoate, ethyl octanoate, hexanoic acid, butanoic acid, phenylethyl alcohol, propanol, and acetoin. SFB mixed with Jiang aroma was located in the negative semi-axis of the first principal component and the second principal component, and these 13 aroma components were positively correlated with it, including hexanol, ethyl hexadecanoate, 2-methybutane, ethyl pentanoate, 3-methylbutyric acid, heptanoic acid, furanol, furfural, isobutyric acid, pentanoic acid, propionaldehyde, ethyl (Z,Z)-9,12-octadecadienoate, and 2-pentanol. The higher the content of these components, the stronger the Jiang aroma and weaker the other aroma characteristics. These results were consistent with the traditional view that ethyl acetate was important for mild flavor liquor and that ethyl hexanoate was one of the representative compounds in Nong flavor liquor. Furthermore, furfural and furanyl alcohol also had positive contributions to Jiang aroma baijiu.

At the same time, many researchers argue that nitrogenous compounds are important for the "roasted aroma" of SFB, especially pyrazines with the roasted aroma. Zhou et al. (67) reported that the contents of nitrogenous compounds in different flavor baijiu (Jiang aroma, mild flavor, Nong flavor, Jian flavor, and SFB). The number of pyrazines (17) in SFB was basically the same as Jiang flavor baijiu, but higher than Nong flavor baijiu (16) and Jian flavor baijiu (12). The total amount (1520.00 μg/L) of the pyrazines in SFB was much less than the average content of Jiang flavor baijiu (27437.25 μg/L) and Jian flavor baijiu (3389.00 μg/L), but more than Nong flavor baijiu (680.00 μg/L) and mild flavor baijiu (213.00 μg/L). The contents of the nitrogenous compounds in different baijius were analyzed using principal component analysis (PCA), as shown in Figure 7.

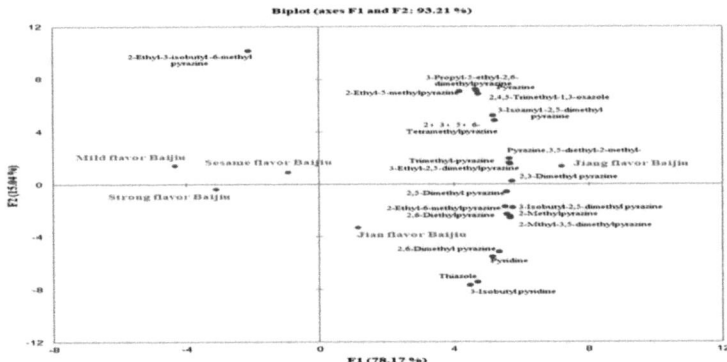

*Figure 7. PCA on the contents of nitrogenous compounds in different flavor baijiu.*

The results in Figure 7 showed that the cumulative contribution rate of the two principal components was 93.21%, which explained the basis for the samples. Jiang flavor and Jian flavor baijiu were located at the positive semi-axis of the first principal component, which was closely associated to high contents of nitrogenous compounds (especially pyrazines). Some nitrogenous compounds in the first quadrant had a positive correlation to Jiang flavor baijiu, while others in the fourth quadrant had a positive correlation to Jian flavor baijiu. However, SFB was located in the middle of other three brands of baijiu, and the nitrogenous compounds had more influence on SFB than on Nong and mild flavor baijiu but less influence on Jiang and Jian flavor baijius as previously described. SFB, Jiang flavor baijiu, and Nong flavor baijiu had very similar kinds of aromatic compounds, but each had different aroma profiles among the four brands of baijiu for their different concentrations and ratios of characteristic components.

Currently, the number of reported compounds (463) in SFB is less than that of Nong flavor baijiu (861), mild flavor baijiu (663), and Jiang flavor baijiu (623) (68). However, the previously mentioned important components are almost the same. The ratio relationships of these important compounds are the critical factor in creating baijiu flavor style (26, 27, 58, 69), including ethyl hexanoate, ethyl butanoate, ethyl lactate, ethyl acetate, hexanoic acid, furfural, benzaldehyde, β-phenylethanol, and nitrogenous compounds among others. Zhu et al. (70) compared the components of two MLC SFB with the other four famous baijiu: BTQ, SLY, Langjiu (LJ), and Maotai (MT). This research found that many compounds were the same in the these baijiu, but their contents were different, and the amounts of 1,1-diethoxyethane, ethyl linoleate, ethyl oleate, ethyl palmitate, lactic acid, and acetic acid in MLC were the highest. Tetramethylpyrazine in MLC was 1.5–4.4 times higher than that of other SFB (SLY, GJ, and JZ), but the content of butanoic acid in MLC was the lowest.

It is worth noting that the contents of some compounds in Baijiu are gradually changing over time. Zhu et al. (70) found that 1,1-diethoxymethane was not detected in all fresh liquors with the sesame flavor, but its content increased over time. Interestingly, a significant decrease in 3-(methylthio)-1-propanol was also observed, thus showing a correlation over time. The regularities of the content changes of these components need to be the subject of future research.

## Aroma-Active Compounds in SFB

It is well known that not all volatile compounds contribute to the flavor of the product. Researchers have studied the aroma-active compounds in SFB by using sensory analysis techniques, such as GC–O, aroma extract dilution analysis (AEDA), odor activity values (OAVs), aroma recombination, aroma addition, and omission experiments. Using GC–O and aroma intensity analysis, Zhou et al. (63) identified that ethyl hexanoate, ethyl octanoate, hexanoic acid, butyric acid, acetic acid, 2-heptanol, 2-phenylethanol, furfural, 2,6-dimethylpyrazine, and 2,3,5,6-tetramethylpyrazine were important odorants in GJ and JZ SFB. Zheng et al. (34) studied the aroma-active compounds in two JZ SFB, and 26 odorants were further confirmed as important odorants due to their OAVs being at least 1. Omission experiments further corroborated that ethyl hexanoate, 3-methylbutanal, ethyl pentanoate, 3-(methylthio)propanal, and dimethyl trisulfide were the key odorants in SFB. Sha et al. (35) also investigated the aroma-active compounds in JZ-CB with sesame flavor and found that the concentrations of 36 odorants were higher than their corresponding odor thresholds. Research shows that 2-furfurylthiol, dimethyl trisulfide, β-damascenone, and 3-(methylthio)propanal could be responsible for the roasted sesame aroma of SFB. In particular, it was proposed that 2-furfurylthiol was the key odorant for SFB per an omission test (35). Sun et al. (62) found 47 odor-active compounds in MLC SFB and proposed that phenols and acetoin were the key odorants, and lactic acid also played a significant role in SFB. Zhang et al. (66) reported that ethyl hexanoate, 3-methylbutanal, ethyl 3-methylbutanoate, dimethyl trisulfide, ethyl pentanoate, ethyl 2-methylpropionate, and 3-(methylthio)propanal may be an important aroma compound in BTQ and YH SFB. Li et at. (64) identified benzenemethanethiol and confirmed that it was an important contributor to the aroma of GJ SFB per GC–O and AEDA.

Currently, a total of 140 aroma-active compounds have been reported in SFB (34–36, 62–64), which is shown in Table 2. These compounds include 46 esters, 15 alcohols, 15 acids, 15 carbonyl compounds, 12 nitrogenous components, 10 sulfur compounds, 9 acetals, 8 phenols, 3 lactones, and 7 others.

**Table 2. Aroma-Active Compounds Identified in Sesame-Flavor Baijiu (SFB)**

| No. | Compounds | Odor | Odor threshold ($\mu$g/L) | Average-FD | Average-OAV | Brands |
|-----|-----------|------|---------------------------|------------|-------------|--------|
| | **Esters** | | | | | |
| 1 | Hexyl acetate | fruity, floral | 670 (45) | - | - | GJ, JZ (70) |
| 2 | Ethyl acetate | nail polish-like, fruity, alcoholic | 32 552 (94) | 32 | 37 | BTQ, GJ, JZ, MLC, YH (75, 36, 62, 64, 35, 66) |

**Table 2. (Continued). Aroma-Active Compounds Identified in Sesame-Flavor Baijiu (SFB)**

| No. | Compounds | Odor | Odor threshold ($\mu g/L$) | Average-FD | Average-OAV | Brands |
|---|---|---|---|---|---|---|
| 3 | Ethyl propanoate | fruity, nail polish-like | 19019(94) | 78 | <1 | BTQ, JZ, MLC, YH (34, 35, 62, 66, 61, 10) |
| 4 | Ethyl acrylate | plastic | 0.2 (95) | 344 | 1134 | JZ, MLC, YH (34, 62) |
| 5 | Ethyl butanoate | fruity, sweet | 82 (94) | 2737 | 1527 | BTQ, GJ, JZ, MLC, YH3436,62,66,64 |
| 6 | Ethyl 2-methylpropionate | fruity, sweet | 58 (94) | 815 | 404 | BTQ, JZ, MLC, YH (35, 62, 66) |
| 7 | Diethyl butanedioate | sweet | 353193 (94) | 329 | <1 | BTQ, GJ, JZ, MLC, YH (34, 35, 62, 66, 64) |
| 8 | 3-Methyl-1-butylbutanoate | fruity | 92075 | - | 4 | GJ, JZ (87, 35) |
| 9 | Ethyl pentanoate | fruity | 27 (94) | 1100 | 443 | BTQ, GJ, JZ, MLC, YH (34, 35, 62, 66, 67) |
| 10 | Ethyl 2-methylbutanoate | fruity | 1874 | 866 | 111 | BTQ, GJ, JZ, MLC, YH (34, 35, 62, 66, 67) |
| 11 | Ethyl 3-methylbutanoate | apple, strawberry-like | 6.99 (94) | 956 | 959 | BTQ, GJ, JZ, MLC, YH (34, 35, 62, 66, 64) |
| 12 | 3-Methylbutyl acetate | fruity | 94 (74) | 81 | 194 | GJ, JZ (35, 64) |
| 13 | Ethyl 4-methyl pentanoate | fruity | 1409 (64) | 160 | <1 | GJ, JZ, MLC (34, 35, 62, 68) |
| 14 | Methyl hexanoate | floral | - | 25 | - | JZ (35) |
| 15 | Ethyl hexanoate | fruity | 55 (94) | 1467 | 6223 | BTQ, GJ, JZ, YH (34, 35, 87, 64–66) |
| 16 | Ethyl nonanoate | wood | | | | |
| 17 | Propyl hexanoate | pineapple, fruity | 13000 (62) | 200 | <1 | GJ, JZ (35) |
| 18 | Butyl hexanoate | fruity | - | 1.5 | - | GJ, JZ (64) |
| 19 | Pentyl hexanoate | fruity | 14000 (62) | - | <1 | JZ (35) |
| 20 | Isoamyl hexanoate | apple | 1,400 (74) | 55 | 2 | BTQ, GJ, JZ (35, 66) |
| 21 | Hexyl hexanoate | fruity, apple, peach | 1900 (75) | 21 | 9.5 | GJ, JZ (35, 64) |

| No. | Compounds | Odor | Odor threshold (μg/L) | Average-FD | Average-OAV | Brands |
|---|---|---|---|---|---|---|
| 22 | Hexanoate-2-phenylethyl ester | floral | 94 (75) | - | - | GJ, JZ (88) |
| 23 | 2-Methylpropyl hexanoate | sweet | 5250.31 (87) | - | <1 | BTQ (87, 66) |
| 24 | Furfuryl hexanoate | - | 24000 (74) | - | <1 | JZ (87, 35) |
| 25 | Ethyl heptanoate | fruity | 13153.17 (95) | - | 1 | BTQ, GJ, JZ (35, 87) |
| 26 | Ethyl octanoate | fruity, fatty | 13 (62) | 392 | 2989 | GJ, JZ (34, 35, 64) |
| 27 | 3-Methylbutyl octanoate | - | 600 (75) | - | <1 | JZ (35) |
| 28 | Ethyl caprate | fruity, floral | 3150.61 (94) | 27 | <1 | BTQ, GJ, JZ, YH (35, 66) |
| 29 | Ethyl decanoate | fruity | 1100 (62) | 9.5 | 21 | GJ, JZ (34, 64) |
| 30 | Ethyl lactate | fruity | 128083.8 (94) | 257 | 8 | GJ, BTQ, MLC, YH (34, 36, 64–66) |
| 31 | Ethyl dodecanoate | medicinal, grassy, sweet, fruity | 640 (74) | 27 | 1 | GJ, JZ (35, 36) |
| 32 | Ethyl tetradecanoate | rancid, sweet, cakey | 46606.731 | 27 | <1 | BTQ, GJ, YH (34, 36, 66) |
| 33 | Ethyl pentadecanoate | rancid | - | 3 | - | GJ (36) |
| 34 | Ethyl hexadecanoate | waxy | - | 81 | - | GJ, JZ (34, 35, 64) |
| 35 | 2-Phenylethyl acetate | floral, rose, honey | 910 (62) | 35 | 4 | GJ, JZ (34, 35, 64) |
| 36 | 2-Phenethyl butanoate | floral | 960 (75) | - | 1 | GJ, JZ (34, 87) |
| 37 | Ethyl 2-hydroxy-3-ethylbutanoate | fruity | - | 24 | - | JZ (34) |
| 38 | Ethyl 2-hydroxy-3-methylbutanoate | fruity | - | | | GJ (67) |
| 39 | Ethyl 2-hydroxybutanoate | fruity | - | 83 | - | JZ (34, 64) |
| 40 | Ethyl 3-phenylpropanoate | floral, fruity, sweet, honey | 125 (94) | 3113 | 67 | BTQ, GJ, JZ, MLC, YH (34, 36, 62, 66, 64) |
| 41 | Ethyl benzoate | fruity | 1400 (62) | 32 | <1 | GJ, JZ (34, 35) |
| 42 | Ethyl phenylacetate | rosy, fruity, sweet | 407 (94) | 45 | 25 | BTQ, GJ, MLC, YH (34–36, 62, 66, 64) |

| No. | Compounds | Odor | Odor threshold ($\mu g/L$) | Average-FD | Average-OAV | Brands |
|---|---|---|---|---|---|---|
| 43 | Ethyl cyclohexanoate | fruity | - | 243 | - | BTQ (66) |
| 44 | Ethyl nicotinate | honey, sweet, | 7781.0543 (69) | 99 | <1 | BTQ,MLC,YH (62, 64) |
| 45 | Ethyl furoate | caramel | 130000 (74) | | <1 | JZ (34, 35) |
| 46 | Ethyl linoleate | rubber | - | | - | GJ (64) |
| | **Lactones** | | | | | |
| 47 | γ-Valerolactone | coconut | 25982 (69) | 3 | <1 | BTQ, GJ (36, 66) |
| 48 | γ-Nonalactone | coconut | 21 (45) | 30 | 21 | GI, JZ (35, 64) |
| 49 | γ-Decalactone | coconut | 21 45 | - | - | GJ,JZ (88, 45) |
| | **Alcohols** | | | | | |
| 50 | 1-Propanol | fruity | 54000 (75) | 100 | 6 | GJ, JZ (35) |
| 51 | 2-Methyl-1-propanol | fruity, malty, roast, nut-like, wine, malty | 28300 (62) | 112 | 5 | BTQ, GJ, JZ, MLC, YH (34, 35, 66, 64) |
| 52 | 1-Butanol | fruity, malty, roast, nut-like | 2, 730 (94) | 328 | 64 | BTQ, GJ, JZ, MLC, YH (62, 66, 64) |
| 53 | 2-Methyl-1-butanol | fruity, alcoholic | - | 27 | - | GJ (36) |
| 54 | 3-Methyl-1-butanol | fruity, alcoholic, malty, roast, nut-like, nail polish | 179191 (62) | 3259 | 2 | BTQ, GJ, JZ, MLC, YH (34–36, 62, 66, 64) |
| 55 | 1-Hexanol | floral, green | 5370 (62) | 22 | 5 | GJ, JZ, MLC, YH (35, 62, 64) |
| 56 | Heptanol | fruity | 2450 (45) | - | -- | GJ, JZ (87) |
| 57 | 2-Heptanol | mushroom | - | - | - | GJ, JZ (87) |
| 58 | 1-Octanol | green, grass | - | - | - | GJ, JZ (87) |
| 59 | 3-Octanol | yeast, truffle | - | - | - | GJ, JZ (87) |
| 60 | Benzyl alcohol | sweet | 40, 900 (62) | 9 | <1 | GJ, JZ (34, 64) |
| 61 | 2-Phenylethanol | rose, honey rose-like floral, rosy | 28923 (94) | 2572 | <1 | BTQ, GJ, JZ, MLC, YH (34–36, 62, 66, 64) |

| No. | Compounds | Odor | Odor threshold (μg/L) | Average-FD | Average-OAV | Brands |
|---|---|---|---|---|---|---|
| 62 | 2-Furanmethanol | caramel, toasted | 2000 (74) | 15 | 6 | GJ,JZ,YH (34, 36, 67, 65) |
| 63 | Alpha-terpineol | floral | 2000 (62) | 25 | <1 | GJ, JZ (35, 25) |
| | **Acids** | | | | | |
| 64 | Acetic acid | vinegar, sour, vinegar-like | 160000 (62) | 86 | 3 | BTQ, GJ, JZ, MLC, YH (34–36, 62, 66, 64) |
| 65 | Propanoic acid | sour | 18100 (74) | 43 | <1 | GJ,JZ (34, 64) |
| 66 | 2-Methylpropanoic acid | sour, rancid | 1600 (74) | 1.5 | 78 | GJ, JZ (35, 64) |
| 67 | Butanoic acid | sweaty, rancid cheesy | 965 (94) | 252 | 57 | BTQ, GJ, JZ, MLC, YH (34, 35, 62, 66, 64) |
| 68 | 3-Methylbutanoic acid | cheese, rancid, sweaty, acid | 1045 (94) | 449 | 16 | BTQ, GJ, JZ, MLC, YH (34, 35, 62, 66, 64) |
| 69 | Pentanoic acid | cheese, rancid, sweaty | 389 (94) | 515 | 45 | BTQ, GJ, JZ, MLC, YH (34, 35, 62, 66, 64) |
| 70 | 4-Methylpentanoic acid | sour | 144 (75) | 3 | 14 | BTQ (66, 64) |
| 71 | Hexanoic acid | sweaty, sour, vinegar-like, acid, rancid | 2517 (75) | 7462 | 41 | BTQ, GJ, JZ, YH, MLC (34, 35, 62, 66, 64) |
| 72 | Heptanoic acid | sweaty | 13821.32 (94) | 170 | <1 | BTQ, GJ, JZ, YH (34, 66, 64) |
| 73 | Octanoic acid | sweaty, rancid | 2701·(94) | 16 | 5 | BTQ, GJ, JZ, MLC, YH (34, 36, 62, 66) |
| 74 | Decanoic acid | sour | 13736 (94) | 27 | <1 | MLC (34, 65) |
| 75 | Benzoic acid | balsam | - | 2 | - | JZ (35) |
| 76 | Phenylacetic acid | sweet, honey | 1430 (62) | 2 | 1.5 | GJ, JZ (34, 64) |
| 77 | Benzenepropanoic acid | floral | - | 32 | - | JZ (34) |
| 78 | Hydrocinnamic acid | floral | 91 (67) | 81 | - | GJ (67) |
| | **Carbonyl odorants** | | | | | |
| 79 | Acetaldehyde | grassy | - | 9 | - | GJ (36) |
| 80 | 3-Methylbutanal | green, toasted, malty, roast, nuts | 16 (94) | 597 | 2159 | BTQ, GJ, JZ, MLC, YH (34, 35, 62, 66) |

| No. | Compounds | Odor | Odor threshold (µg/L) | Average-FD | Average-OAV | Brands |
|---|---|---|---|---|---|---|
| 81 | Nonanal | - | 120 (62) | - | <1 | JZ (35) |
| 82 | Benzaldehyde | floral, fruity | 4203.1 (94) | 7 | <1 | BTQ, GJ, JZ (35, 66) |
| 83 | Benzeneacetaldehyde Phenyl acetaldehyde | rosy, floral, honey | 262 (62) | 43 | 15 | BTQ,GJ,JZ,MLC,YH (36, 62, 66, 64) |
| 84 | Furfural | fruity, floral, sweet, butter, toasted | 44000 (62) | 81 | 4 | GJ, JZ, YH (34–36, 66, 64) |
| 85 | 5-Methylfurfural | grass, baking incense | 466000 (63) | 10 | <1 | GJ, JZ (34, 35, 64) |
| 86 | 3-(2-Furyl)-2-propenal | earthy | - | 4 | <1 | JZ (34) |
| 87 | Vanillin | sweet, creamy, caramel | 438 (94) | 6826 | <1 | GJ, MLC (36, 62, 66, 64) |
| 88 | 2,3-Butandione | butter-like | 5 (95) | 6 | 2 | BTQ, MLC, YH (62, 66) |
| 89 | 1-Octen-3-one | mushroom | - | 5 | - | JZ (35) |
| 90 | 2-Nonanone | - | 480 (75) | - | <1 | JZ (35) |
| 91 | 2-Undecanone | - | 400 (75) | - | <1 | JZ (35) |
| 92 | 3-Hydroxy-2-butanone | butter, yogurt | 259 (74) | 6 | 228 | BTQ, YH (34, 66, 74) |
| 93 | Acetophenone | almond | 9474.9594 (69) | - | <1 | BTQ (66) |
| 94 | Phenyl-2-propanone | mildew, smell, rust | - | 92 | - | BTQ, YH (34, 66) |
| | **Acetals** | | | | | |
| 95 | Diethyl acetal | fruity, green | 2090 (62) | 378 | 59 | BTQ, GJ, JZ, MLC, YH (34, 36, 62, 66) |
| 96 | 1,1,3-Triethoxypropane | fruity | 3700 (75) | 90 | <1 | GJ, BTQ, YH (34, 64) |
| 97 | 1,1-Diethoxy-2-methylpropane | sweet, fruity | - | 27 | - | MLC (62) |
| 98 | 1,1-Diethoxypentane | green | - | 96 | - | JZ (34) |
| 99 | 2-Methoxy-1,3-dioxolane | sweet, cake-like | - | 729 | - | MLC (62) |

| No. | Compounds | Odor | Odor threshold ($\mu g/L$) | Average-FD | Average-OAV | Brands |
|-----|-----------|------|------|------|------|--------|
| 100 | 1,1-Diethoxyoctane | smoked, roasted | - | 81 | - | BTQ (79) |
| 101 | 2,4,6-Trimethyl-1,3-dioxane | sweet, fruity | - | 27 | - | MLC (62) |
| 102 | 2-Furaldehyde diethyl acetal | sweet | - | 4 | - | JZ (34) |
| 103 | (2,2-Diethoxyethyl) benzene | floral | 47 (67) | 122 | <1 | GJ (64) |
| | **Phenols** | | | | | |
| 104 | Phenol | pill medicinal, phenolic | 18900 (94) | 4 | - | GJ, JZ (34, 35) |
| 105 | 4-Methylphenol | animal stinky, fecal, horse, smoky | 167 (94) | 215 | 21 | BTQ, GJ, JZ, MLC, YH (34, 64–66) |
| 106 | M-cresol | leather | - | - | <1 | JZ (66) |
| 107 | 4-Ethylphenol | smoked | 617.68 (94) | 729 | <1 | BTQ, GJ, JZ, YH (35, 66) |
| 108 | Guaiacol | smoky, woody | 13 (94) | 56 | 92 | BTQ, GJ, JZ, MLC, YH (34, 62, 66, 64) |
| 109 | 4-Ethyl-2-methoxyphenol | smoky | 123 (94) | 190 | 10 | GJ, JZ, MLC (34, 62, 64) |
| 110 | 2,4-Di-tert-butylphenol | sweet | 36373.07 (69) | 1 | <1 | YH (34) |
| 111 | 2,4-Bis(1,1-Dimethylethyl)-phenol | phenol | - | - | - | GJ (64) |
| | **Sulfur-containing odorants** | | - | | | |
| 112 | 3-(Methylthio)propanal | roasted potato | 7.12 (74) | 1480 | 64 | BTQ, JZ, MLC, YH |
| 113 | 3-(Methylthio)-1-propanol | vegetable, cooked potato | 2110 (94) | 32 | <1 | GJ, JZ, MLC, YH |
| 114 | Dimethyl sulfide | cooked onion | 17 (3) | 56 | 14 | JZ (35) |
| 115 | Dimethyl disulfide | cooked, onion, vegetable | 0.36 (94) | 3 | 44 | GJ, JZ (36, 62) |

## Table 2. (Continued). Aroma-Active Compounds Identified in Sesame-Flavor Baijiu (SFB)

| No. | Compounds | Odor | Odor threshold ($\mu g/L$) | Average-FD | Average-OAV | Brands |
|-----|-----------|------|----------|------------|-------------|--------|
| 116 | Dimethyl trisulfide | sulfury, meaty, rotten cabbage | 0.36 (75) | 86 | 1952 | BTQ, GJ, JZ, MLC, YH (34–36, 62, 66, 64) |
| 117 | S-methyl thioacetate | rotten cabbage | 21 (65) | 100 | 13 | JZ (35) |
| 118 | Ethyl 2-mercaptoacetate | cooked vegetable | 120 (3) | 100 | <1 | JZ (35) |
| 119 | Ethyl 3-(methylthio)propanoate | sulfur, rotten cabbage | 3100 (62) | 5 | <1 | JZ (35) |
| 120 | 2-Furfurylthiol | roasted, sesame seeds | 0.10 (65) | 400 | 1182 | JZ (35) |
| 121 | Benzenemethanethiol | roasted | 0.0035 | 19683 | 214.5 | GJ (64) |
| **Nitrogenous compounds** | | | | | | |
| 122 | 2,6-Dimethylpyrazine | baked fragrant, woody, roast, nut-like | 790 (94) | 71 | 15 | GJ, JZ, MLC (35, 62, 64) |
| 123 | 2,3,5-trimethylpyrazine | baked fragrant, nut-like, baked, almond-like | 729 (94) | 18 | <1 | BTQ, GJ, JZ, JZ-JZ, MLC, YH (34, 35, 62, 66) |
| 124 | 2,3,5,6-Tetramethylpyrazine | baked fragrant, nut-like | - | 6 | 2 | GJ, JZ (62, 64) |
| 125 | 2-Ethyl-5-methylpyrazine | baking incense | - | - | - | GJ, JZ (87) |
| 126 | 2-Ethyl-6-methyl-pyrazine | nutty | 40 (62) | 23 | 113 | GJ, JZ (34, 64) |
| 127 | 2-Ethyl-3,5-dimethyl pyrazine | roasted potato | 7.5-33 (74) | 64 | 12 | JZ (34) |
| 128 | 2,5-Dimethyl-3-ethylpyrazine | baked | - | 100 | - | JZ (35) |
| 129 | 2,3-Diethyl-5-methylpyrazine | smoky, woody | - | 45 | - | MLC, YH (62, 66) |
| 130 | 2,6-Diethylpyrazine | baked | - | 5 | - | JZ (35) |
| 131 | 2-Acetylpyrrole | nutty | - | 3 | - | GJ (36) |

**Table 2. (Continued). Aroma-Active Compounds Identified in Sesame-Flavor Baijiu (SFB)**

| No. | Compounds | Odor | Odor threshold ($\mu$g/L) | Average-FD | Average-OAV | Brands |
|-----|-----------|------|---------------------------|------------|-------------|--------|
| 132 | Pyrrole-2-carboxaldehyde | nutty | - | 81 | - | YH (66) |
| 133 | 3-Phenylpyridine | sour | 19138 (94) | 27 | <1 | MLC (62) |
| | **Other odorants** | | | | | |
| 134 | Naphthalene | Camphor, musty | 160 (74) | 5 | 3 | GJ, JZ (35) |
| 135 | Geosmin | earthy | 0.11 (69) | 100 | 22 | JZ (35) |
| 136 | Alpha-cedarene | woody | - | - | - | GJ, JZ (35) |
| 137 | B-damascenone | sweet, candy | 0.12 (62) | 100 | 116 | JZ (35) |
| 138 | 2-Acetylfuran | sweet, toasted, nutty | 58504 (95) | 31 | <1 | GJ, JZ (87) |
| 139 | 2-Acetyl-5-methylfuran | sweet, baked | - | 4 | - | JZ (34, 35) |
| 140 | Difurfuryl ether | nutty | - | 4 | - | JZ (34) |

Many of the odorants in Table 2 were also reported as aroma-active compounds in other flavor styles of Chinese baijiu (49, 71–73). Among these odorants, 114 compounds had average flavor dilution factors (FDs) that were greater than 1. In particular, 10 aroma compounds had FDs that were greater than 1000, including benzenemethanethiol (19683), 3-methyl-1-butanol (3259), ethyl 3-phenyl-propionic acid (3113), ethyl butanoate (2737), vanillin (2275), phenylethanol (2572), hexanoic acid (2488), 3-(methylthio)propanal (1480), ethyl hexanoate (1467), and ethyl pentanoate (1100). The remaining 26 odorants were found using other methods such as OAVs. There were 60 compounds with OAVs of at least 1, 40 odorants with OAVs that were less than 1, and 40 compounds without OAVs, which is partially due to no odor thresholds being detected in alcohol solution matrix. These values should be the subject of future research, especially the thresholds in baijiu matrix. The average OAV for ethyl hexanoate (6223) was the highest, followed by ethyl octanoate (2989), 3-methylbutanal (2159), ethyl butanoate (1527), 2-furfurylthiol (1182), ethyl acrylate (1134), ethyl 3-methylbutanoate (959), ethyl pentanoate (443), ethyl 2-methylpropionate (404), 3-hydroxy-2-butanone (228), 2-methylpropanoic acid (228), benzenemethanethiol (215), 3-methylbutyl acetate (194), β-damascenone (116), 2-ethyl-6-methyl-pyrazine (113), and ethyl 2-methylbutanoate (111).

There were 56 odorants with average FDs and OAVs that were greater than 1, including ethyl hexanoate (FD=1467, OAV=6223), ethyl pentanoate (FD=1100, OAV=443), ethyl butanoate (FD=2737, OAV=1527), ethyl octanoate (FD=392, OAV=2989), ethyl acetate (FD=32, OAV=37), 1-propanol (FD=100, OAV=6), 3-methyl-1-butanol (FD=3259, OAV=2), 1-butanol (FD=328, OAV=64), furanmethanol(FD=15, OAV=6), hexanoic acid (FD=2488, OAV=41), 3-methylbutanoic acid (FD=449, OAV=16), butanoic acid (FD=252, OAV=57), acetic acid (FD=86, OAV=3), 3-methylbutanal (FD=597, OAV=2159), furfural (FD=81, OAV=4), benzeneacetaldehyde (FD=43, OAV=5), 3-hydroxy-2-butanone (FD=6, OAV=228), diethyl acetal (FD=378, OAV=59), 4-methylphenol (FD=215, OAV=21), guaiacol (FD=56, OAV=92), 4-ethyl-2-methoxyphenol (FD=190, OAV=10), 3-(methylthio)propanal (FD=1480, OAV=64), dimethyl trisulfide (FD=86, OAV=1952), 2-furfurylthiol (FD=400, OAV=1182), benzenemethanethiol (FD=19683, OAV=215), 2-ethyl-6-methylpyrazine (FD=23, OAV=113), 2-ethyl-3,5-dimethylpyrazine (FD=64, OAV=12), β-damascenone (FD=100, OAV=116), and geosmin (FD=100, OAV=22) among others. These compounds were important to the flavor of SFB. The esters and acetals mainly contributed to the fruity and sweet flavors of SFB, while the alcohols and 3-methylbutanal offered malty and roasted nut-like flavors, the phenethyl alcohol contributed the flowery flavor, the acids provided sour and sweaty flavors, the phenols contributed to the smoky flavor, and 3-(methylthio)propanal offered a cooked potato aroma. The above results were basically consistent with the previously mentioned volatile compounds analysis results. On the contrary, ethyl lactate, lactones, carbonyl compounds, and acetals appear to have little influence on the flavor of SFB so far, despite ethyl lactate being traditionally regarded as the key aroma compound for its particularly high concentrations (74, 75).

A total of 10 sulfur odorants were detected in SFB, including 3-(methylthio)propanal, 3-(methylthio)-1-propanol, dimethyl trisulfide, 2-furfurylthiol, dimethyl sulfide, dimethyl disulfide, s-methyl thioacetate, ethyl 2-mercaptoacetate, ethyl 3-(methylthio)propanoate, and benzenemethanethiol. 3-(methylthio)-1-propanol was considered the key odorant for SFB in the past, but this position was denied by Zheng et al. (34) This was further supported by the results of Sha et al. (35), despite its average FD being 32 and its OAV being 1. In addition, 2-furfurylthiol was only detected in JZ-CB by Sha et al. (35) but was found in neither JZ-CB nor in JZ-BD, or any another SFB, such as MLC, GJ, YH, and BTQ samples. Therefore, the function of 2-furfurylthiol for SFB is questionable at present, but the microbial metabolic mechanism of 2-furfurylthiol has been discovered (76). Its function still needs to be confirmed further by applying yeast to the production of SBF. Additionally, Li et al. (64) reported that the compound benzenemethanethiol, with a maximum FD of 19683, was an important contributor to the flavor of GJ SFB, but the content is too low to be quantitated accurately. Therefore, the research on sulfur odorants will be an important breakthrough for the flavor chemistry of SFB.

There were 12 nitrogenous odorants and 11 compounds had FDs that were greater than 1, but only 4 odorants with OAVs of at least 1 were found in JZ and GJ, including 2,6-dimethylpyrazine (FD=71, OAV=15), 2,3,5,6-tetramethylpyrazine (FD=7, OAV=2), 2-ethyl-6-methylpyrazine (FD=32, OAV=8), and 2-ethyl-3,5-dimethylpyrazine (FD=64, OAV=12). The other six compounds did not have calculated OAVs due to their lack of thresholds. Currently, it seems that these nitrogenous odorants have little contribution to the flavor of SFB. This is inconsistent with the observation that nitrogenous components have an important role on the flavor of SFB because of the high nitrogen proportion in the raw material. The contribution of these nitrogenous odorants, especially pyrazines, need to be researched further in the future. The same is true of the acetals due to their threshold problem.

In consideration of both the FD values and OAVs, it was determined that several compounds with high FD values had relatively smaller OAVs. These compounds include 2-methyl-1-propanol (FD=112, OAV=5), 2-phenylethanol (FD=2572, OAV<1), 3-methyl-1-butanol (FD=3259, OAV=2), 4-methyl pentanoate(FD=160, OAV<1), and vanillin (FD=2276, OAV<1). These values indicate that food matrix possibly has an influence on odorant binding (77). According to the OAVs, 2-phenylethanol contributed little to the aroma of SFB, but it did have an influence on the flavor of SFB (34, 35). This indicated that the contribution of a compound to a liquor's flavor was not only related to its concentration, but also to the interactions among odorants or interactions between odorants and nonvolatile compounds in the food matrix.

## Effects of Nonvolatile Compounds to Aroma Profiles of SFB

Zheng et al. (34) found that the aroma recombination of SFB was weaker in roasted and fermented-like attributes in ethanol than in the commercial baijiu sample and proposed that the nonvolatile compounds in baijiu might impact its overall aroma. Several studies have attempted to investigate the interactions between volatile compounds and nonvolatile compounds. The research team of Huang et al. (78, 79) found several peptides from SFB and reported that Ala-Lys-Arg-Ala (AKRA), Asp-Arg-Ala-Arg (DRAR), Pro-His-Pro (PHP), and Pro-Pro-Asp-Gly (PPDG) changed the volatilities of aroma compounds in GJ SFB. They also investigated the corresponding binding ability between AKRA and aroma compounds in SFB using HS-SPME combined with HS-SPME-GC/MS. In comparison, the aroma molecules in baijiu before and after adding AKRA were the same, only their volatilities were altered, and the volatilities of esters (not all), alcohols, and phenolic compounds decreased. Thus, the AKRA slightly changed the overall flavor of baijiu. The interaction force between AKRA and p-cresol was verified to be the hydrogen bond using HS-SPME-GC/MS, UV absorption spectroscopy, and nuclear magnetic resonance spectrum (1H NMR) analysis. The results showed that the peptides had an effect on the overall flavor of the liquor. This interaction needs to be the focus of further research studies.

## Healthy Flavor Compounds in SFB

SFB contains various flavor compounds, and most of them are beneficial to humans. For example, hexanoic acid, heptanoic acid, octanoic acid, decanoic acid, and oleic acid can inhibit the synthesis of cholesterol (80). Linoleic acid and linolenic acid are currently considered to be essential fatty acids for the human body (81, 82). 2,3,5,6-Tetramethylpyrazine helps to protect the liver from fibrosis, serves as an antioxidant, and inhibits $\alpha$-glucosidase and certain improvement effects on hyperglycemia (83–85). Dimethyltrisulfide is one of the ingredients in the Chinese herb, known as sputum white, which is very effective in inhibiting platelet aggregation (86). 3-Methylthiopropanol, ethyl 3-methylthiopropionate, and 2-furfurylthiol have antioxidant properites (85), and 3-methylthiopropanol also may inhibit $\alpha$-glucosidase activity and has hypoglycemic functions (85). 2-Furfurylthiol can, to some extent, inhibit the oxidation of hexanal to hexanoic acid (87). Terpenes have antibacterial, antiviral, antioxidant, and analgesic properties (88). Ethyl lactate can promote ethanol by stimulating the excitation of cerebral cortex (89). Acid esters with higher fat compositions such as ethyl heptanoate, ethyl linolenate, and ethyl linoleate, are hydrolyzed in the human body to form fatty acids, which inhibit cholesterol synthesis (90). Xu et al. (84) reported that furan compounds had anticancer and antioxidant properties. In addition, 4-methylguaiacol and 4-ethylguaiacol were also reported that they can effectively prevent diseases, delay aging, and, in the case of 4-ethylguaiacol, reduce blood glucose activity (91). The whole health function of baijiu and its functional compounds will be major research points in the future, although there have recently been some disadvantageous reports on baijiu.

## 4. Conclusions

The flavor chemistry of SFB has been discovered by modern instruments and sensory analysis technologies. More than 463 volatile substances and 140 active aroma compounds have been reported in SFB, and the esters, alcohols, acids, carbonyl compounds, sulfur compounds, and phenols have been deemed important for the flavor of the liquor. The contributions of nitrogenous compounds (especially pyrazines), acetals (except for diethyl acetal), and sulfur compounds

217

(furfurylthiol and benzenemethanethiol) should be further researched in the future. The microbial metabolism processes of important aroma components and the thresholds of some odorants in the alcoholic solution matrix should also be studied. A more accurate quantitative method ofodorants should be explored, and the isotope IS method will be a good choice for this future research. The functional components and the influence of non-volatile compounds on flavor are also a focus of research on sesame-flavor liquor (92, 93). In the future, SFB will develop along with research on its flavor and health benefits.

## Funding

The financial support from National Key Research & Development Program of China (2017YFC1600401-3) and National Natural Science Foundation of China (31471665 and 31871749) are gratefully acknowledged.

## Notes

The authors declare no competing financial interest.

## Abbreviations and Nomenclature

SFB, namely Zhima-aroma type baijiu, is a liquor with a special baked sesame flavor; Jiang-flavor baijiu, namely soy sauce–aroma type baijiu; Nong-flavor baijiu, namely strong-aroma type baijiu; mild-flavor baijiu, namely light-aroma type baijiu; Mi-flavor baijiu, namely rice-aroma type or sweet- and honey-aroma type baijiu. Jian-flavor baijiu, namely the combination of Nong- and Jiang-aroma type baijiu; Feng-flavor baijiu, a kind of baijiu with light- and strong-aroma that is popular in the Shannxi Province of China; Laobaigan-flavor baijiu, a kind of baijiu similar to mild-flavor baijiu, but it has its special style; Chi-flavor baijiu, a kind of baijiu that is popular in the Guangdong Province of China and South East Asia. Te-flavor baijiu, a kind of baijiu fermented from the whole rice; Dong-flavor baijiu, namely an herblike-aroma type baijiu. (These kinds of baijius are the represenrive baijiu in China and had national standards. Also, there are other kinds of Jiu in China.) Chen-aroma, is a kind of aroma formed during baijiu aging. Daqu (natural inoculation) and Fuqu (artificial inoculation), two kinds of mold culture, which contains *Rhizopus*, *Aspergillus*, yeast, bacteria, and complex enzymes spontaneously proliferated on wheat. They are usually used as a saccharifying agent and a fermentation starter for initiating the fermentation process. AEDA, aroma extract dilution analysis; OAV, odor activity value; GC-MS/O, gas chromatography-mass spectrometry/olfactometry; GC-MS, gas chromatography-mass spectrometry; FD factor, the flavor dilution factor.

## References

1. Yu, Q. W. *Traditional Baijiu Brewing Technology*, 1st ed.; China Light Industry Press: Beijing, 2016; pp 61.

2. Zheng, X. W.; Han, B. Z. Baijiu (白酒), Chinese liquor: History, Classification and Manufacture. *J. Ethnic Foods* **2016**, *3*, 19–25.

3. Gao, C. Discussion on the Production Techniques of Sesame-Flavor Liquor and Its Style Positioning. *Liquor-Making Sci. Technol.* **2014**, *4*, 60–64.

4. Qian, C.; Liao, Y. H.; Zhang, X.; Ma, H.; Xu, B.; Fan, J.; Zhang, Y. X. Application Progress of Mechanical Automation Liquor-Production Technologies. *China Brewing* **2013**, *32*, 5–8.

5.  Shen, Y. F. Production Technology of Sesame-Flavored High-Quality Liquor. *Liquor-Making Sci. Technol.* **1993**, *3*, 43–46.

6.  Hu, G.; Lu, J.; Cai, X. Characterization of Zhima Aroma-Type Baijiu. *Liquor-Making Sci. Technol.* **1994**, *4*, 75–77.

7.  Fan, W. L.; Qian, M. C. Headspace Solid Phase Microextraction and Gas Chromatography–Olfactometry Dilution Analysis of Young and Aged Chinese "Yanghe Daqu" liquors. *J. Agric. Food Chem.* **2005**, *53*, 7931–7938.

8.  Fan, W. L.; Shen, H.; Xu, Y. Quantification of Volatile Compounds in Chinese Soy Sauce Aroma Type Liquor by Stirbar Sorptive Extraction and Gas Chromatography–Mass Spectrometry. *J. Sci. Food and Agric.* **2011**, *91*, 1187–1198.

9.  Wang, B. X.; Hou, Y.; Yang, L.; Xu, J. C.; Liu, J.; Zou, Y. Study on the Determination of Ester Components in Liquor by SBSE-TDs-GC-MS. *Sci. Tech. Food Industry* **2008**, *7*, 250–253.

10. Chen, S.; Xu, Y. Characterization of Volatile Compounds in Chinese Roasted Sesame-Like Flavor Type Liquor by Comprehensive Two-Dimensional Gas Chromatography/Time-Offlight Mass Spectrometry. *Food. Ferm Industries* **2017**, *43*, 207–213.

11. AQSIQ. *Zhima-Flavor Chinese Spirits*; Standardization Publishing House of China: Beijing, China, 2007; pp 1−6.

12. Xu, Y.; Wu, Q.; Fan, X. L.; Zhu, B. F. The Discovery & Verification of the Production Pathway of Tetramethylpyrazine (TTMP) in Chinese Liquor. *Liquor-Making Sci. Technol.* **2011**, *7*, 37–40.

13. Ge, Y. Y.; Yao, S.; Liu, Y.; Cao, Y. H.; Zhang, F. G.; Xin, C. H.; Xu, L.; Cheng, C. Analysis on Thermophilic Bacterial Communities in High Temperature Daqu of Sesame Flavor Liquor. *Food. Ferm Industries* **2012**, *38*, 16–19.

14. Zhao, D. Y; Han, B.; Li, J. G.; Wu, F. X.; Que, K. Y. The Application of Automation Systems in Jingzhi Distillery. *Liquor-Making Sci. Technol.* **2014**, *6*, 91–97.

15. Xu, X. Viewpoints of the Present Status and the Development of Sesame-Flavor Liquor in Shandong. *Liquor-Making Sci. Technol.* **2012**, *3*, 111–117.

16. Zhang, P. F.; Wang, Z. E.; Fu, W. Q.; Dong, W.; Wang, B. B.; Zhao, J. K. Discussion of Production Technology and Style Characteristics of Luxiangfang Sesame Flavor Liquor. *Liquor Making* **2013**, *40*, 47–52.

17. Wang, J.; Wang, Z.; Wang, S. The Process of Sesame Flavor Liquor. *Liquor Making* **2013**, *40*, 81–85.

18. Ma, X. L.; Liang, Z. W.; Zhang, X. L.; Ma, J.; Li, J. R.; Yan, H. W. Production of Sesame Flavor Liquor. *Liquor Making* **2017**, *44*, 54–55.

19. Wang, J. L. Discussion on the Causes of FuYun Sesame Fragrant. *Liquor Making* **2013**, *40*, 28–30.

20. Liu, M. M.; Wang, J. G.; Sun, P. P.; Wang, K. B. Application of Bran Starter in the Production of Sesame-Flavor Liquor. *Liquor-Making Sci. Technol.* **2013**, *3*, 69–74.

21. Yue, T. F.; Cheng, W.; Zhang, J.; Sun, L. L.; Pan, T. Q.; Guan, Y. Q.; Ding, P. F.; Zhang, Z. Y.; Wang, X. S. Research Progress on Brewing and Technological Innovation of Sesame Flavor Liquor. *Liquor Making* **2018**, *45*, 6–10.

22. Jin, G.; Zhu, Y.; Xu, Y. Mystery Behind Chinese Liquor Fermentation. *Trends Food Sci. Technol.* **2017**, *63*, 18–28.

23. Wang, C. L.; Shi, D. J.; Gong, G. L. Microorganisms in Daqu: A Starter Culture of Chinese Maotai-Flavor Liquor. *World J. Microbiol. Biotechnol.* **2008**, *24*, 2183–2190.

24. Qi, Y. M. New Understanding of Sesame Flavor Liquor. *Liquor Making* **2015**, *42*, 96–98.

25. Lai, A. G.; Zhao, D. Y.; Cao, J. Q. History, Status and Development Trend of Zhima-Flavor Chinese Spirits. *Liquor Making* **2009**, *36*, 91–93.

26. Shi, A. H.; Lu, G. G.; Xiao, H. J.; Liu, S. J. Distribution and Preliminary Identification of Microorganisms in Daqu and Mud of Shuizhi Sesame-Flavored Baijiu. *Liquor Making* **1997**, *3*, 19–21.

27. Wan, Q. H.; Xie, S. K.; Gao, D. W.; Zhang, G. S.; Han, L.; Xia, H. F.; Chen, J. X. Effects of Two Kinds of Accumulated Grains on Fermentation Characteristics and Aroma Quality of Sesame-Flavor Liquor. *Food. Ferm. Industries* **2017**, *43*, 9–15.

28. Li, X. D.; Gao, D. Y.; Tian, Q. Z.; Cai, C. J.; Xia, H. F.; Chen, J. X. Effects of Sesame-Flavor Liquor Accumulation on Cellar Fermentation Process and Liquor Quality. *Food. Ferm. Industries* **2018**, *44*, 63–69.

29. Zhang, F. G. Blending and Flavoring of Multiple-Grains Sesame-Flavor Liquor. *Liquor-Making Sci. Technol.* **2008**, *10*, 62–64.

30. Zhang, F. G. Production of Sesame Flavor Style Liquor by Complex Grains. *Liquor Making* **2009**, *36*, 11–14.

31. Sun, B. G.; Wu, J. H.; Huang, M. Q.; Sun, J. Y.; Zheng, F. P. Recent Advances of Flavor Chemistry in Chinese Liquor Spirits (Baijiu). *J. Chinese Institute Food Sci. Technol.* **2015**, *15*, 1–8.

32. Gong, S. B.; Fan, W. L.; Xu, Y. Comparison of Volatile and Non-Volatile Compounds Between Traditional and Mechanical Raw Baijiu of Roasted-Sesame-Like Aroma Type Baijiu (Chinese Liquor). *Food. Ferm. Industries* **2018**, *44*, 239–245.

33. Liu, C. H.; Liu, M.; Zhong, Q. D.; Xiong, Z. H.; Meng, Z.; Liu, L.; Lv, Z. Y.; Li, X. H. Study on Typical Sensory Characteristics of Zhimaxiang Baijiu (Sesame-Flavor Liquor) by Quantitative Description Analysis. *Liquor-Making Sci. Technol.* **2014**, *6*, 10–15.

34. Zheng, Y.; Sun, B. G.; Zhao, M. M.; Zheng, F. P.; Huang, M. Q.; Sun, J. Y.; Sun, X. T.; Li, H. H. Characterization of the Key Odorants in Chinese Zhima Aroma-Type Baijiu by Gas Chromatography–Olfactometry, Quantitative Measurements, Aroma Recombination, and Omission Studies. *J. Agric. Food Chem.* **2016**, *64*, 5367–5374.

35. Sha, S.; Chen, S.; Qian, M. C.; Wang, C. C.; Xu, Y. Characterization of the Typical Potent Odorants in Chinese Roasted Sesame-Like Flavor Type Liquor by Headspace Solid Phase Microextraction–Aroma Extract Dilution Analysis, with Special Emphasis on Sulfur-Containing Odorants. *J. Agri. Food Chem.* **2017**, *65*, 123–131.

36. Sun, J. Y.; Zhao, D. R.; Zhang, F. G.; Sun, B. G.; Zheng, F. P.; Huang, M. Q.; Sun, X. T.; Li, H. H. Joint Direct Injection and GC–MS Chemometric Approach for Chemical Profile and Sulfur Compounds of Sesame-Flavor Chinese Baijiu (Chinese Liquor). *Eur Food Res. Technol.* **2017**, *9*, 1–16.

37. Zhang, F. G. Some Thoughts on the Production of Sesame-Flavor Baijiu. *Liquor Brewing* **2018**, *45*, 8–9.

38. Wu, Z. Z.; Fan, Z. Y.; Zuo, G. Y.; Wang, H. H. GC-MS Direct Analysis of Qualitative and Quantitative Sampling of Liquor. *Liquor Making* **2009**, *36*, 88–90.

39. Li, T. C.; Hui, R. H.; Hou, D. Y. Analysis of the Volatile Compounds from Chinese Wine by Solid Phase Extraction and GC/MS. *J. Chinese Mass Spectrometry Society* **2007**, *28*, 96–100.

40. Fan, W. L.; Xu, Y. Determination of Aroma Compounds of Chinese Liquors by Direct Immersion-Solid Phase Microextraction (DI-SPME). *Liquor Making* **2007**, *34*, 18–21.

41. Yang, C. X.; Liao, Y. H.; Hu, J. H.; Hu, J. Y.; Xie, J. C. Comparison of Aroma Compounds in Erguotou Liquor by Liquid-Liquid Extraction and Solid Phase Microextraction. *Sci. Technol. Food Industry* **2012**, *8*, 68–74.

42. Wu, J. H.; Huang, M. Q.; Sun, B. G.; Zheng, F. P.; Sun, J. Y. Analysis of Volatile Compounds in Jingzhi Baigan Liquor by Liquid-liquid Extraction (LLE) and Gas Chromatography-Mass Spectrometry (GC-MS). *Food Sci.* **2014**, *35*, 41–45.

43. Xu, Z. C.; Chen, Y.; Wang, S. Analysis of Flavoring Compositions in Liquor by SBSE Absorption Technology Coupled with GC×GC/TOFMS Technology. *Liquor-Making. Sci. Technol.* **2012**, *7*, 50–55.

44. Wang, L. H.; Li, J. F. Research on the Extraction of Concentrated Trace Flavoring Substances in Liquor by Simultaneous Distillation Extraction. *Liquor Making* **2010**, *9*, 25–27.

45. Liang, D.; Huang, M. Analysis of Volatile Components in Maotai and XO Brandy by Purge and Trap-GC/MS. *Liquor Making* **2014**, *8*, 96–104.

46. Liu, Y. P.; Huang, M Q.; Zheng, F P.; Chen, H. T.; Sun, B. G. Recent Advances in Extraction and Analysis of Volatile Flavor Compounds in Chinese Liquor. *Food Sci.* **2010**, *31*, 437–441.

47. Wang, B. W.; Li, H. H.; Zhang, F. G.; Xin, C. H.; Sun, J. Y.; Huang, M. Q.; Sun, B. G. Analysis of Nitrogen-Containing Compounds of Guojing Sesame-Favour Liquor by Liquid-Liquid Extraction Coupled with GC-MS and GC-NPD. *Food Sci.* **2014**, *35*, 126–131.

48. Liao, Y. H.; Zhao, S.; Zhang, Y. B.; Zhang, X.; Tong, R. N.; Xu, M. Analysis of Flavor Substances in Erguotou Wine by LLE, SDE, SPME and GC-MS Combined with Kovats Retention Indices. *J. Chinese Institute of Food Sci. Technol.* **2014**, *14*, 220–228.

49. Hu, F. Y.; Zhang, Q. Y.; Zheng, M. M; Shao, C. Z.; Han, Q. Q. Analysis of Aroma Components of White Wine by DI-SPME and GC-MS (I). *Liquor Making* **2012**, *39*, 39–43.

50. Wang, X. X.; Fan, W. L.; Xu, Y. Characterization of Volatile Aroma Components in Chinese Soy Sauce Aroma Type Xijiu Liquor by GC-O and GC-MS. *Food Ferme. Industries* **2013**, *39*, 154–160.

51. Yuan, H.; Yi, B.; Shen, C. P. Analysis of Volatile Compounds by GC/MS in Nong-Xiang Type Baijiu (Liquor) Produced by Different Grains. *Liquor-Making Sci. Technol.* **2014**, *3*, 44–49.

52. Liu, J.; Fan, W. L.; Xu, Y.; Zhang, G. Q.; Xu, Q. Q.; Ding, Y. L.; LI, Z. Q. Comparison of Aroma Compounds of Chinese "Miscellaneous Style" and "Strong Aroma Style" Liquors by GC-Olfactometry. *Liquor Making* **2008**, *35*, 103–107.

53. Wang, Y.; Xu, Y.; Fan, W. L.; Wei, J. W. Determination of Aroma Compounds in Niulanshan Erguotou Liquor by GC-O. *Liquor-Making Sci. Technol.* **2011**, *2*, 74–79.

54. Zhu, S. K.; Lu, X.; Ji, K. L.; Guo, K. L.; Li, Y.; Wu, C. Y.; Xu, G. W. Characterization of Flavor Compounds in Chinese Liquor Moutai by Comprehensive Two-Dimensional Gas Chromatography/Time-of-Flight Mass Spectrometry. *Analytica Chimica Acta* **2007**, *2*, 340–348.

55. Sun, X. T.; Zhang, F. G.; Dong, W.; Sun, B. G.; Sun, J. Y. GC-FPD Analysis of 3-Methylthiopropanol in Sesame-Flavor Liquor. *J. Food Sci. Technol.* **2014**, *32*, 27–34.

56. Fan, W. L.; Xu, Y. Comparison of Sichuan Base Liquor and Jianghuai Drainage Area Base Liquor by the Use of GC-FID And Cluster Analysis. *Liquor-Making Sci. Technol.* **2007**, *11*, 75–78.

57. Preininger, M. Quantitation of Potent Food Aroma Compounds by Using Stable Isotope Labeled and Unlabeled Internal Standard Methods. *Developments Food Sci.* **1998**, *40*, 87–97.

58. Han, S. N.; Niu, W.; Hou, J. G. Investigation on the Difference Between Taoxiang Baijiu and Baijiu of Other Flavor Types. *J. Brewing Sci. Technol.* **2017**, *1*, 62–64.

59. Gao, W J.; Fan, W L.; Xu, Y. Characterization of the Key Odorants in Light Aroma Type Chinese Liquor by Gas Chromatography–Olfactometry, Quantitative Measurements, Aroma Recombination, and Omission Studies. *J. Agric. Food Chem.* **2014**, *62*, 5796–5804.

60. Li, D. H.; Liu, N.; Nai, D. H. *Training Course on Baijiu Brewing (Baijiu Worker, Brewer and Taster)*, 1st ed.; China Light Industry Press: Beijing, 2013; pp 544.

61. Zhu, S. L.; Gao, C. Q.; Cui, G. Y. Analysis of Trace Compositions of Meilanchun Sesame-Flavor Liquor. *Liquor-Making Sci. Technol.* **2012**, *6*, 106–110.

62. Sun, J. Y.; Li, Q. Y.; Luo, S. Q.; Zhang, J. L.; Huang, M. Q.; Chen, F.; Zheng, F. P.; Sun, X. T.; Li, H. H. Characterization of Key Aroma Compounds in Meilanchun Sesame Flavor Style Baijiu by Application of Aroma Extract Dilution Analysis, Quantitative Measurements, Aroma Recombination, and Omission/Addition Experiments. *RSC Adv.* **2018**, *8*, 23757–23767.

63. Zhou, Q. Y. *Study on Flavor Substances of Sesame-Flavor Liquor.* Master Thesis, Jiangnan University, Jingsu, China, 2015.

64. Li, H. H.; Qin, D.; Wu, Z. Y.; Sun, B. G.; Sun, X. T.; Huang, M. Q.; Sun, J Y.; Zheng, F. P. Characterization of Key Aroma Compounds in Chinese Guojing Sesame-Flavor. *Food Chem.* **2019**, *284*, 100–107.

65. Wang, Q.; Wang, J. Q.; Guo, W. W.; Wang, R. M. Determination of the Content of Ethyl Lactate, Ethyl Acetate, Ethyl Caproate and Ethyl Butyrate in Liquor by HPLC. *Food Industry.* **2016**, *37*, 273–276.

66. Zhang, Q.; Li, Q. Y.; Huang, M. Q.; Wu, J. H.; Li, H. H.; Sun, J. Y.; Sun, X. T. Zheng, F. P.; Sun, B. G. Analysis of Odor-Active Compounds in 2 Sesame-Flavor Chinese Baijius. *Food Sci.* Published Online Nov 9, 2018. http://kns.cnki.net/kcms/detail/11.2206.TS.20181108.1325.042.html (accessed Apr 11, 2019).

67. Zhou, L.; Zhao, D.; Lai, A.; Cao, J.; Liu, J. The Technics and Style Characteristic of Jingzhishenniang Liquor. *Liquor Making.* **2008**, *35*, 27–29.

68. Liu, H. L; Sun, B. G. Effect of Fermentation Processing on the Flavor of Baijiu. *J. Agric. Food Chem.* **2018**, *66*, 5425–5432.

69. Qian, C.; Liao, Y. H.; Liu, M. Y.; Xu, W.; Liu, L.; Yu, L. Cluster Analysis and Principal Components Analysis of Different Flavor Types of Liquor. *J. Chinese Institute Food Sci. Technol.* **2017**, *17*, 243–255.

70. Zhu, M. X.; Fan, W. L.; Xu, Y.; Zhou, Q. Y. 1,1-Diethoxymethane and Methanethiol as Age Markers in Chinese Roasted-Sesame-Like Aroma and Flavor Type Liquor. *Eur. Food Res. Technol.* **2016**, *242*, 1985–1992.

71. Wang, P. P.; Li, Z.; Qi, T. T.; Li, X. J.; Pan, S. Y. Development of a Method for Identification and Accurate Quantitation of Aroma Compounds in Chinese Daohuaxiang Liquors Based on SPME Using a Sol-Gel Fibre. *Food Chem.* **2015**, *169*, 230–240.

72. Fan, H. Y.; Fan, W. L.; Xu, Y. Characterization of Key Odorants in Chinese Chixiang Aroma-Type Liquor by Gas Chromatography—Olfactometry, Quantitative Measurements, Aroma Recombination, and Omission Studies. *J. Agric. Food Chem.* **2015**, *63*, 3660–3668.

73. Wang, X. X; Fan, W. L; Xu, Y. Comparison on Aroma Compounds in Chinese Soy Sauce and Strong Aroma Type Liquors by Gas Chromatography–Olfactometry, Chemical Quantitative and Odor Activity Values Analysis. *Eur. Food Res. Technol.* **2014**, *239*, 813–825.

74. Wang, Y. T. Effects of the Main Trace Components and Its Quantity Relative Ratio Relationship of Fen-Flavor Liquor on the Sensory Quality. *Liquor-Making Sci. Technol.* **2004**, *3*, 27–29.

75. Shen, Y. *Manual of Chinese Liquor Manufactures Technology*; Light Industry Publishing House of China: Beijing, 1996.

76. Zha, M. S.; Sun, B. G.; Yin, S.; Mehmood, A.; Cheng, L.; Wang, C. T. Generation of 2-Furfurylthiol by Carbon-Sulfur Lyase from the Baijiu Yeast *Saccharomyces cerevisiae* G20. *J. Agric. Food Chem.* **2018**, *66*, 2114–2120.

77. Zhang, R.; Wu, Q.; Xu, Y. Lichenysin, a Cyclooctapeptide Occurring in Chinese Liquor Jiannanchun Reduced the Headspace Concentration of Phenolic Off-Flavors Via Hydrogen-Bond Interactions. *J. Agric. Food Chem.* **2014**, *62*, 8302–8307.

78. Huang, M. Q.; Huo, J. Y.; Wu, J. H.; Zhao, M. M.; Zheng, F. P.; Sun, J. Y.; Sun, X. T.; Li, H. H. Interactions Between p-Cresol and Ala-Lys-Arg-Ala (AKRA) from Sesame Flavor-Type Baijiu. *Langmuir* **2018**, *34*, 12549–12559.

79. Wu, J. H. *Antioxidant Activity of Peptides from Sesame Flavor-type Baijiu and Their Interactions with Aroma Compounds*. Ph.D. Thesis, South China University of Technology, Guangzhou, China, 2018.

80. Zhuang, M. Y. Analysis on the Physiological Activity of Trace Elements in Chinese Liquor. *Liquor Making.* **2000**, *5*, 23–25.

81. Xu, Z. C.; Chen, Y.; Zhou, Z. H.; Tang, Q. L. Study on Healthy and Functional Compositions in Jian Nan Chun Liquor. *Liquor-Making Sci. Technol.* **2008**, *5*, 41–44.

82. Li, J. X.; Wu, Y.; Chen, X.; Tu, Y.; Zhou, Y. H. Vegetable Oils Rich in Polyunsaturated Fatty Acids and Their Health-Beneficial Effects: A Review. *Food Sci.* **2014**, *35*, 350–354.

83. Poojab, D. Green Tea Catechins: Defensive Role in Cardiovascular Disorders. *Chinese J. Natural Medicines* **2013**, *11*, 345–353.

84. Gao, C. Q.; Tan, T. T.; Xin, Y. W. The Activities of the Extracts of Zhimaxiang Baijiu (Sesame-Flavor Liquor) and the 4 Kinds of Characteristic Compounds. *Liquor-Making Sci. Technol.* **2015**, *4*, 61–64.

85. Sun, X. S.; Li, W. H.; Li, R.; Gao, C. Q.; Cui, G Y. The Antioxidative Activity of Sulfide and Pyrazine Compositions in Sesame-Flavor Liquor. *Liquor Making* **2013**, *4*, 57–60.

86. Wu, T. T.; Zhu, S. L.; Sun, X. S.; Zhao, W.; Cui, G. Y. Analysis of Health Factors of Meilanchun Sesame-Flavor Liquor. *Liquor-Making Sci. Technol.* **2013**, *8*, 125–130.

87. Eiserich, J. P.; Wong, J. W.; Shibamoto, T. Antioxidative Activities of Furan and Thiophenethiols Measured in Lipid Peroxidation Systems and by Tyrosyl Radical Scavenging Assay. *J. Agri. Food Chem.* **1995**, *43*, 647–650.

88. Fan, W. L.; Xu, Y. Review of Important Functional Compounds Terpenes in Baijiu (Chinese Liquor). *Liquor Making* **2013**, *6*, 11–16.

89. Zhang, J. X; Li, J.; Guo, F. W. The Effect of Trace Components in Liquor on the Human Body. *Liquor-Making Sci. Technol.* **2014**, *10*, 143–148.

90. Xu, Y.; Fan, W. L.; Ge, X. Y.; Huang, Y. G. Scientific Recognition of Biofunctional Components in Chinese Liquors. *Liquor-Making Sci. Technol.* **2013**, *9*, 1–6.

91. Yang, T.; Li, G. Y.; Wu, L. Y.; Zhuang, M. Y. Research on Health Factor in Liquor and Breeding of ealth Factor-Producing Bacteria and Its Application in Liquor Production ( I ). *Liquor-Making Sci. Technol.* **2010**, *12*, 65–69.

92. Wu, J. H.; Huo, J. Y.; Huang, M. Q.; Zhao, M. M.; Luo, X. L.; Sun, B. G. Structural Characterization of a Tetrapeptide from Sesame Flavor-Type Baijiu and Its Preventive Effects Against AAPH-Induced Oxidative Stress in HepG2 Cells. *J. Agric. Food Chem.* **2017**, *65*, 10495–10504.

93. Wu, J. H.; Sun, B. G.; Luo, X. L.; Zhao, M. M; Zheng, F. P.; Sun, J. Y.; Li, H. H; Sun, X. T.; Huang, M. Q. Cytoprotective Effects of a Tripeptide from Chinese Baijiu Against AAPH-Induced Oxidative Stress in HepG2 Cells Via Nrf2 Signaling. *RSC Adv.* **2018**, *8*, 10898–10906.

94. Fan, W. L.; Xu, Y. Determination of Odor Thresholds of Volatile Aroma Compounds in Baijiu by a Forced-Choice Ascending Concentration Series Method of Limits. *Liquor Making* **2011**, *38*, 80–84.

95. Burdock, G. A. *Fenaroli's Handbook of Flavor Ingredients*, 5th ed.; CRC Press: Washington, DC, 2005; Vol. 411, pp 543.

## Chapter 14

# Aroma Comparison of Tibetan "Qingke" Liquor with Other Chinese Baijiu

Yueqi An,[1,2] Yanping Qian,[*,1,3] Shuang Chen,[1,4] and Michael C. Qian[1]

[1]Departmet of Food Science and Technology, Oregon State University, Corvallis, Oregon 97331, United States

[2]College of Food Science and Technology, Huazhong Agricultural University, Wuhan, Hubei Province 430070, P. R. China

[3]Department of Crop and Soil Science, Oregon State University, Corvallis, Oregon 97331, United States

[4]College of Bioengineering, Jiangnan University, Wuxi, Jiangsu 214122, P. R. China

[*]E-mail: yan.ping.qian@oregonstate.edu.

The aroma composition of a Tibetan Qingke (QK) liquor, which is made from highland barley exclusively grown at an altitude greater than 4000 m, was compared with 16 other Chinese baijiu of varying aroma styles. A total of 63 aroma compounds were quantified. Results showed that strong-aroma baijiu has much higher concentrations of ethyl hexanoate, ethyl heptanoate, ethyl octanoate, and hexanoic acid than those of QK liquor and the remaining baijiu. In contrast, QK liquor, along with other light-aroma baijiu and Mao Tai (MT), Si Te Jiu (STJ), and Heng Shui Lao Bai Gan (HS), have higher levels of ethyl acetate and ethyl lactate. QK liquor has the highest concentrations of linalool, β-damascenone, and β-ionone. Dimethyl trisulfide (DMTS) is the most important sulfur-containing compound in Chinese baijiu. Strong-aroma baijiu, along with MT, STJ, and Bai Yun Bian (BYB), have a higher amount of DMTS than the other baijiu. Odor activity values (OAVs) were calculated to illustrate the potency of aroma compounds. Strong-aroma baijiu, along with BYB, STJ, and Gu Jing G6 (GJG6), have OAVs of ethyl butanoate, ethyl hexanoate, ethyl pentanoate, ethyl octanoate, and hexanoic acid that are greater than 100, indicating that they were the key aroma compounds. For light-aroma baijiu, alog with MT, Hong Xing Er Guo Tou (HX), Gui Lin San Hua (GL), and Guang Dong Jiu Jiang (GDJJ), isoamyl acetate was the most important aroma compound, contributing fruity and floral notes. A principal component analysis (OAV of at least 1) illustrated that QK liquor has a similar aroma composition to other light-aroma baijiu (CYW, C20) and HS, but each individual baijiu has its own aroma profile. Strong-aroma baijiu were well separated from light-aroma due to their high concentrations of ethyl hexanoate, ethyl heptanoate, hexanoic acid, and octanoic acid. MT, as the only soy-sauce aroma baijiu, stands alone with its unique aroma profile.

## Introduction

Chinese liquor, also known as baijiu, is a traditional distillate that is mainly solid-state fermented from sorghum, wheat, sticky rice, and corn. Qingke (QK) liquor, a famous Tibetan alcoholic beverage, is made of "Qingke," hull-less highland barley exclusively grown in the Qinghai-Tibetan Plateau at an average altitude of 4000 m (1). Highland barley is rich in essential nutrients, including protein, β-glucan, flavonoids, various amino acids, abundant vitamins (such as vitamin B, B6, and E), trace elements, and dietary fiber (2, 3).

QK liquor is made from three unique raw materials: hull-less highland barley, "Qingke Daqu," and glacial water from the Himalayan mountains (elevation higher than 3000 m). The Daqu used for QK liquor fermentation is prepared by stirring a mixture of highland barley and peas (7:3) into water (40%), cultivating the mixture for 21 to 28 days at 10–16 °C to allow for the growth of crude microorganisms (i.e., Rhizopus, Aspergillus, yeast, and bacteria), aging the mixture for six months for maturation, and then finishing with a complex drying process. Daqu serves as an important saccharifying and fermenting agent with a mixed microflora of fungi, bacteria, and microbial enzymes, which has a significant impact on the flavor of QK liquor (4, 5).

The traditional manufacturing of QK is a "Four Times Steaming-Clearing" process which includes steaming raw materials (highland barley) and excipients, fermenting in a cubic pit underground, and distilling. Due to the combination of its special raw materials, unique fermentation techniques, and geographical environment, the flavor of QK liquor is full-bodied and refreshing with a pure highland barley aroma. Also, QK has a lingering aftertaste, providing a distinctly different flavor from other light-aroma Chinese baijiu.

Chinese baijiu has very distinctive aromas depending on the raw materials, fermentation starters, brewing technology, region, climatic conditions, and natural resources (6). In general, Chinese baijiu is classified into soy sauce–aroma style, strong-aroma style, light-aroma style, rice-aroma, and other miscellaneous styles, such as Lao Bai Gan–aroma style and Te-aroma style, based on its aroma characteristics (6, 7). Each style has its own unique flavor and taste due to its distinguished aroma composition.

Sauce-aroma style baijiu, also known as Maotai-flavor baijiu, is one of the finest Chinese baijiu made from sorghum fermented in pits lined with stone bricks. Of all the distilled liquors, soy sauce–flavor baijiu is the most exclusive due to its eight cycles of fermentation and distillation that occur over the course of a year with a unique Daqu as a starter culture, but only a few of cycles get fresh grains added (8). The traditional production method of Daqu involves three stages: material mixing and shaping, ripening, and drying. Flavor compounds are mainly produced during the ripening period (9). After mixing sorghum and Daqu, the "High-Temperature Gathering and Fermentation" process is another original and core technology used for soy sauce–flavor baijiu (10, 11). The flavor of soy sauce–aroma baijiu is rich, full-bodied, and sauce-like with notes of mushroom and caramel. Also, soy sauce–aroma style baijiu contains large amounts of ester compounds, which impart a layered umami flavor (10, 12).

Strong-aroma Chinese baijiu has a sweet-tasting, soft mouthfeel and a gentle, lasting fragrance which is created by the high levels of ethyl esters (primarily ethyl hexanoate) in balance with ethyl acetate, ethyl lactate, and ethyl butanoate, giving the liquor a strong taste of fruity and anise notes

(*13*). The most common raw material for this style of liquor is wheat, but some pea and barley can be mixed with wheat. The differences between strong-flavor baijiu and soy sauce–flavor baijiu are the result of the temperature during the Daqu preparation and the method of fermentation. In general, the temperature for making Daqu for strong-aroma baijiu is 55–60 °C, which is 5–10 °C lower than that of soy sauce–aroma baijiu. During fermentation, previously-cooked grains are mixed with Daqu powder, and fermentation is typically carried out around 30 °C for 60 days under anaerobic conditions in a solid state. After fermentation, the liquor is distilled out and the distillate is collected and aged in sealed pottery jars for one to three years to develop the balanced aroma (*14*).

Light-aroma baijiu is made from sorghum and qu (made from barley and peas) with short production cycles and minimal aging periods. As a result, light-aroma baijiu is usually described as having a light and mild flavor, mellow sweetness, and a clean mouthfeel. The major aroma compounds of the light-aroma style baijiu are ethyl acetate and ethyl lactate, which contribute to the taste of dried fruit with floral notes (*15*). Two famous light-aroma baijiu are Fen Jiu from Shanxi and Hong Xing Er Guo Tou from Beijing (*12*, *16*).

Rice-aroma style baijiu (such as San Hua Jiu from Guilin) is distilled from glutinous rice fermented with rice-based xiaoqu (*17*, *18*). This baijiu has a light body with notes of sweet, floral, and honey.

Miscellaneous-aroma baijiu (such as Te-aroma or Lao Bai Gan–aroma) is a blend of two or more varieties of baijiu, mainly strong-aroma and sauce-aroma baijiu. This class of baijiu can vary widely in flavor and mouthfeel.

Aroma compounds reported in Chinese baijiu include esters, alcohols, fatty acids, aldehydes, ketones, lactones, acetals, terpenoids, pyrazines, phenolics, sulfur-containing compounds, and heterocyclic compounds (*9*, *16*, *19–21*). However, each individual baijiu has its own aroma profile. Although the aroma characterizations of many famous brands of Chinese baijiu have been studied extensively (*7*), there is limited research on QK liquor. The objectives of this study were to quantify aroma active compounds in QK liquor and other Chinese baijiu in order to better understand the aroma chemistry of QK liquor and evaluate the feasibility of characterizing the aromas of Chinese baijiu using a principal component analysis (PCA).

## Materials and Methods

### Chemicals

All the volatile compounds used were analytical reagent grade except the following: acetaldehyde (>99%), ethyl acetate (anhydrous grade, 99.8%), ethyl lactate (>98%), ethyl propanoate (99.7%), methyl propanoate (>98%), ethyl 2-methylpropanoate (>98%), isoamyl acetate (>95%), propyl acetate (>98%), hexyl acetate (>98%), isobutyl acetate (>98%), octyl acetate (>98%), ethyl butanoate (>98%), ethyl 2-ethylbutanoate (>98%), ethyl 3-methylbutanoate (>98%), ethyl pentanoate (>98%), ethyl hexanoate (>98%), propyl hexanoate (>98%), isobutyl hexanoate (>98%), isopentyl hexanoate (>98%), ethyl 3-hydroxyhexanoate (>98%), ethyl heptanoate (>98%), ethyl octanoate (>98%), ethyl decanoate (>98%), ethyl dodecanoate (>98%), diethyl succinate (>98%), ethyl phenylacetate (>98%), phenethyl acetate (>98%), ethyl 3-phenylpropanoate (>98%), propanol (>99.5%), 2-methyl-1-butanol (>98%), 1-hexanol (>98%), 2-ethyl-1-hexanol (>98%), 2-heptanol (>98%), 2-methyl-1-propanol (>99%), 3-methyl-1-butanol (>99%), (Z)-3-hexen-1-ol (>90%), 1-octanol (>98%), 1-octen-3-ol (>97%), benzyl alcohol (>98%), phenylethyl alcohol (>99%), nerol (>98%), linalool (>97%), geraniol (>97%), α-terpineol (>97%), β-citronellol (>97%), β-damascenone (>98%), β-ionone (>97%), 2-heptanone

(>99%), 2-nonanone (>99%), 2-methylbutanoic acid (99%), 3-methylbutanoic acid (99%), hexanoic acid (≥98.0%), octanoic acid (≥98.0%), nonanoic acid (99%), decanoic acid (>99%), were purchased from Sigma-Aldrich (St. Louis, MO, USA). Butanoic acid (>99%) was purchased from Mallinckrodt Pharmaceuticals (St. Louis, MO, USA). Methyl thioacetate (>97%) and ethyl thioacetate (>97%) were purchased from Alfa Aesar (Ward Hill, MA, US.A). Ethyl methyl sulfide (>99%), disopropyl disulfide (>99%), dimethyl sulfide (>99%), dimethyl trisulfide (>99%) and dimethyl disulfide (>99%) were purchased by Tokyo Chemical Industry Co., Ltd. (Tokyo, Japan). Milli-Q quality water was obtained from a Milli-Q purification system (Millipore Corporation, Boston, MA, USA). Methanol (HPLC grade) was obtained from EMD Chemicals, Inc. (Gibbstown, NJ, USA). Dichloromethane (HPLC grade) from Burdick & Jackson (Muskegon, MI, USA) was distilled before use. Ethanol was purchased from AAPER Alcohol and Chemical Co. (Shelbyville, KY, USA). Anhydrous sodium sulfate and sodium chloride (99.9%, ACS certified) were supplied by Mallinckrodt Baker, Inc. (Phillipsburg, NJ, USA).

Isotope internal standards (IS), ethyl butanoate-4,4,4-$d_3$ (99.8%), ethyl 3-methylbutanoate-$d_9$ (99%), ethyl hexnaoate-$d_{11}$ (98%), ethyl octanoate-$d_{15}$ (98.5%), linalool-$d_3$ (99%), alpha-terpineol-$d_3$ (98%), 2-phenyl-$d_5$-ethan-1,1,2,2-$d_4$-ol (99.2%), butanoic acid-$d_7$ (98%), octanoic acid-$d_{15}$ (98.2%), decanoic acid-$d_{19}$ (98.6%), and hexanoic acid-$d_{11}$ (98.5%) were purchased from CDN Isotopes Inc. (Quebec, Canada).

## Materials

The Chinese baijiu in Table 1 were commercialized products from China and stored at 4 °C before analysis.

**Table 1. Chinese Baijiu with Their Defined Flavor and Alcohol Content**

| Brand | Flavor Style | Alc. Vol. |
|---|---|---|
| Guo Jiao1573 (GJ) | strong aroma | 52% |
| Yang He (YH) | strong aroma | 52% |
| Gu Jing Gong Jiu 16-years aging (GJGJ) | strong aroma | 50% |
| Lu Zhou Lao Jiao (LZLJ) | strong aroma | 52% |
| Jian Nan Chun (JNC) | strong aroma | 54% |
| Cao Yuan Wang (CYW) | light aroma | 53% |
| Cao Yuan Wang 20-years aging (C20) | light aroma | 65% |
| Tian You De Qingke liquor (QK) | light aroma | 52% |
| Fen Jiu (FJ) | bran starter light aroma | 52% |
| Hong Xing Er Guo Tou (HX) | bran starter light aroma | 52% |
| Mao Tai (MT) | soy sauce aroma | 53% |
| Bai Yun Bian (BYB) | miscellaneous aroma | 45% |
| Si Te Jiu (STJ) | Te-aroma | 55% |
| Heng Shui Lao Bai Gan (HS) | Lao Bai Gan aroma | 67% |
| Gui Lin San Hua (GL) | rice aroma | 55% |

**Table 1. (Continued). Chinese Baijiu with Their Defined Flavor and Alcohol Content**

| Brand | Flavor Style | Alc. Vol. |
|-------|-------------|-----------|
| Guo Jing G6 (GJG6) | Guo Jing aroma | 46% |
| Guang Dong Jiu Jiang (GDJJ) | Chi aroma | 29% |

## Synthesis of Esters

Butyl butanoate, octyl butanoate, methyl hexanoate, and ethyl decanoate-$d_{19}$ were individually synthesized by reacting to their respective acids and alcohols. To synthesize the compounds, alcohols and acids were added in a 50 mL glass vial at a 1:1 (v/v) ratio, then 2–3 drops of hydrochloric acid (4 mol/L) were added to the vial. The vial was oscillated at 150 rpm overnight in a water bath (at 50 °C). Esters were then extracted with freshly distilled dichloromethane, and sodium hydrogen carbonate (0.1 mol/L) and distilled water were used to get rid of the extra alcohol and acids using a separating funnel. After being dried with anhydrous sodium sulfate, dichloromethane was evaporated via small steam of nitrogen.

The concentration of the synthesized compounds was calculated by dissolving the compounds in 1 mL of methanol. Gas chromotography–mass spectrometry (GC–MS) (with a split ratio of 100:1) was used to analyze 1 µL of the extract for confirmation of ester identity.

## Quantitation by Static Headspace-GC-Flame Ionization Detection (HS-GC-FID)

Quantification of highly volatile compounds, including acetaldehyde, ethyl acetate, propanol, isobutyl alcohol, isoamyl acetate, and isoamyl alcohol, was conducted using a static headspace-GC-flame ionization detector (HS-GC-FID). The baijiu sample was diluted five times with Milli-Q water due to its high alcohol content. Of the sample, 1 mL was pipetted into a 20 mL auto-sampler vial, and 20 µL of the IS (2.5 mg/mL of methyl propionate) was added into the vial as well. The vial was then tightly capped with a Teflon-faced silicone septum. The sample mixture was then incubated at 50 °C for 15 min and agitated at 250 rpm to achieve the equilibrium between the sample mixture and the headspace. A Varian CP-3800 gas chromatography (GC) equipped with a flame ionization detector (FID, Varian, Inc., Palo Alto, CA, USA) was used for sample analysis. The GC system was used in combination with a CombiPAL autosampler (CTC Analytics, Zwingen, Switzerland) coupled with a 1 mL syringe (Hamilton Company, Reno, Nevada, USA) for sample injection. The syringe was also kept at 50 °C, the same as the incubation temperature. After equilibration, 0.5 mL of sample headspace was directly injected using the syringe into the GC system with a 1:10 split ratio. Separation of analytes was achieved by a DB-WAX capillary column (30 m length, 0.25 mm i.d., 0.5 µm film thickness; Agilent Technologies, Inc., Palo Alto, CA, USA). A constant nitrogen flow of 2 mL/min was applied. The initial oven temperature was 35 °C, which was held for 4 min, then ramped to 150 °C at the rate of 10 °C/min, and held for 5 min. The inlet temperature was 200 °C and the FID temperature was 250 °C. Peak identification of the volatile components was conducted by comparing the retention time with authentic pure standards, and concentration ranges of each standard and linearly dependent coefficient ($R^2$) of standard curves are shown in Table 2.

**Table 2. Calibration Curves Used for Quantification of Highly Volatile Compounds and Sulfur-Containing Compounds**

| Compounds | Equations | Range | $R^2$ |
|---|---|---|---|
| methyl propionate (IS) | | | |
| acetaldehyde | y=0.0311x+0.0042 | 5–1000 mg/L | 0.9993 |
| ethyl acetate | y=0.0689x-0.0022 | 5–1000 mg/L | 0.9987 |
| propanol | y=0.0074x+0.0004 | 5–1000 mg/L | 0.9999 |
| isobutyl alcohol | y=0.0139x-0.0076 | 5–1000 mg/L | 0.9967 |
| isoamyl acetate | y=0.0971x-0.0022 | 0.2–40 mg/L | 0.9963 |
| isoamyl alcohol | y=0.0101x-0.0023 | 10–2000 mg/L | 0.9983 |
| ethyl methyl sulfide (IS) | | | |
| dimethyl sulfide | y=0.0101x-0.1504 | 20–200 µg/L | 0.9917 |
| methyl thioacetate | y=0.2997x-7.3754 | 20–200 µg/L | 0.9919 |
| ethyl thioacetate | y=0.1296x+0.0667 | 3–30 µg/L | 0.9978 |
| diisopropyl disulfide (IS) | | | |
| dimethyl trisulfide | y=0.0089x+0.0164 | 0.2–4 µg/L | 0.9980 |
| dimethyl disulfide | y=0.6216x+0.0154 | 0.2–4 µg/L | 0.9935 |

**Quantification by HS-SPME-GC-Pulsed Flame Photometric Detection**

Volatile sulfur-containing compounds were quantified by headspace-solid phase microextraction and gas chromatography–pulsed flame photometric detector (HS-SPME-GC-PFPD). The analyses were made using a Varian CP-3800 gas chromatograph equipped with a PFPD detector (Varian, Walnut Creek, CA, USA) operating in sulfur mode. An original baijiu sample was diluted five times with Milli-Q water. A 1-mL diluted sample was added into 9 mL of saturated sodium chloride solution with 2% ethanol in an autosampler glass vial. Twenty µL of an IS of 0.5 mg/L of ethyl methyl sulfide and 0.004 mg/L of diisopropyl disulfide were then added. The vial was tightly capped with a Teflon-faced silicone septum. The samples were first equilibrated at 30°C for 15 min in the incubator. A Carboxen-PDMS fiber (1 cm x 85 µm, Supelco, Inc., Bellefonte, PA, USA) was then inserted into the headspace using the autosampler (Gerstel, Inc., Linthicum, MD, USA) and extracted for 20 min. After extraction, the SPME fiber was directly injected into the GC injection port in splitless mode at 300 °C and held for 7 min. The separation was performed using a DB-FFAP capillary column (30 m length, 0.32 mm i.d., 1 µm film thickness; Agilent, Palo Alto, CA, USA). The oven temperature was programmed as follows: 35 °C (initial hold for 3 min), ramp at 10 °C/min to 150 °C (hold for 5 min), and then ramp at 20 °C/min to 220 °C (final hold for 3 min). The carrier gas was nitrogen with a constant flow rate of 2 mL/min. The temperature of the detector was 300 °C, and the detector was supplied with 14 mL/min of hydrogen, 17 mL/min of air 1, and 10 mL/min of air 2. The detector voltage was 500 V, the gate delay for sulfur compounds was 6 ms, and the gate width is 20 ms. All sulfur compounds were identified by comparing their retention times with those of the pure standards. Concentration ranges of each standard and the linearly dependent coefficient ($R^2$) of standard curves are shown in Table 2. The sulfur responses of specific compounds were calculated by the square root of the peak area.

## Quantification by Solid-Phase Microextraction-GC-MS

The rest of the major volatile compounds were quantified using the headspace solid phase microextraction-gas chromatography-mass spectrometry (SPME-GC-MS) method. Chinese baijiu (0.5 mL) was diluted with 9.5 mL of a saturated sodium citrate buffer (pH 3.2) in a 20 mL glass vial. An isotope IS mixture (Table 3) of 10 μL was then added into the diluted sample and the vial was tightly capped with a Teflon-faced silicone septum. Samples were first equilibrated at 45 °C in a thermostatic bath for 5 min. A preconditioned 2 cm–50/30 μm DVB/CARTM/PDMS–coated SPME fiber (Supelco, Bellefonte, PA, USA) was then inserted into the headspace by the autosampler (Gerstel, Linthicum, MD, USA) at the same temperature and extracted for 30 min. The sample was stirred with a magnetic stir bar at 500 rpm during the extraction. The GC and MS conditions were the same as the GC-O. The unique quantification of mass ion and the qualifying of mass ions were carefully selected to give the highest response and lowest interference for each compound.

## Statistical Analysis

The quantification tests were performed in triplicate. Analysis of variance (ANOVA) was done using Statistical Analysis System (SAS Institute Inc., Cary, NC, USA). Differences among mean values were established using the Duncan multiple range test (DMRT) at $P<0.05$. Principal component analysis (PCA) was carried out by SPSS Statistics software and figures were made by Origin 9.0.

**Table 3. Chemical Standards, Target Ions, and Calibration Curves Used for Quantification of Volatiles in Chinese Baijiu**

| Compounds | m/z | Equations | Range (μg/L) | $R^2$ |
|---|---|---|---|---|
| ethyl butanoate-d3 (IS) | 74 | | | |
| ethyl propanoate | 57 | y=0.0912x+0.046 | 0.1–1000 | 0.9980 |
| propyl acetate | 73 | y=0.161x+0.108 | 0.1–1000 | 0.9957 |
| ethyl butanoate | 71 | y=0.0755x+0.0419 | 0.1–1000 | 0.9994 |
| ethyl 2-methylpropanoate | 71 | y=0.035x+0.0018 | 0.1–1000 | 0.9964 |
| isobutyl acetate | 56 | y=0.0324x+0.0009 | 0.1–1000 | 0.9987 |
| ethyl pentanoate | 101 | y=0.503x+0.26 | 0.1–1000 | 0.9992 |
| butyl butanoate | 71 | y=1.2x+0.595 | 0.1–1000 | 0.9993 |
| ethyl 3-methylbutanoate-$d_9$ (IS) | 94 | | | |
| ethyl 2-methylbutanoate | 102 | y=0.486x+0.382 | 0.1–1000 | 0.9997 |
| ethyl 3-methylbutanoate | 88 | y=0.331x+0.202 | 0.1–1000 | 0.9969 |
| ethyl hexnaoate-$d_{11}$ (IS) | 91 | | | |
| ethyl hexanoate | 88 | y=0.0959x-0.0337 | 0.1–1000 | 0.9963 |
| hexyl acetate | 69 | y=0.0277x-0.0156 | 0.1–1000 | 0.9926 |
| methyl hexanoate | 74 | y=0.84x+2.27 | 0.1–1000 | 0.9978 |
| ethyl 3-hydroxyhexanoate | 117 | y=0.051x-0.105 | 0.1–1000 | 0.9878 |

| Compounds | m/z | Equations | Range ($\mu g/L$) | $R^2$ |
|---|---|---|---|---|
| propyl hexanoate | 117 | y=0.151x+0.306 | 0.1–1000 | 0.9997 |
| isobutyl hexanoate | 117 | y=0.224x+0.301 | 0.1–1000 | 0.9983 |
| ethyl heptanoate | 113 | y=0.0991x+0.157 | 0.1–1000 | 0.9969 |
| isopentyl hexanoate | 117 | y=0.199x+0.985 | 0.1–1000 | 0.9990 |
| ethyl lactate | 75 | y=0.0035x+0.015 | 1–100 | 0.9998 |
| ethyl octanoate-d$_{15}$ (IS) | 91 | | | |
| ethyl octanoate | 88 | y=0.0716x+0.013 | 0.1–1000 | 0.9994 |
| octyl acetate | 70 | y=0.0207x-0.0015 | 0.1–1000 | 0.9998 |
| octyl butanoate | 89 | y=0.0205x+0.0281 | 0.1–1000 | 0.9937 |
| ethyl decanoate-d$_{19}$ (IS) | 91 | | | |
| ethyl decanoate | 88 | y=0.0605x+0.0334 | 0.1–1000 | 0.9985 |
| methyl decanoate | 74 | y=0.0022x+0.0001 | 0.1–1000 | 0.9985 |
| diethyl succinate | 101 | y=0.0129x+0.0047 | 0.1–1000 | 0.9974 |
| ethyl phenylacetate | 91 | y=0.0443x-0.0039 | 0.1–1000 | 0.9993 |
| phenethyl acetate | 91 | y=0.0147x-0.0019 | 0.1–1000 | 0.9990 |
| ethyl dodecanoate | 183 | y=0.0836x-0.0844 | 0.1–1000 | 0.9991 |
| ethyl 3-phenylpropanoate | 104 | y=0.258x-0.828 | 0.1–1000 | 0.9872 |
| linalool-d$_3$ (IS) | 124 | | | |
| linalool | 121 | y=0.0843x+0.0745 | 0.1–500 | 0.9988 |
| nerol | 69 | y=0.233x-0.0768 | 0.1–500 | 0.9946 |
| geraniol | 69 | y=0.257-0.0675 | 0.1–500 | 0.9978 |
| alpha-terpineol-d$_3$ (IS) | 124 | | | |
| alpha-terpineol | 121 | y=0.0679+0.0445 | 0.1–500 | 0.9932 |
| beta-citronellol | 95 | y=0.153x-0.0017 | 0.1–500 | 0.9993 |
| beta-damascenone | 121 | y=0.655x+0.0426 | 0.1–500 | 0.9987 |
| beta-ionone | 177 | y=1.21x+0.0361 | 0.1-500 | 0.9995 |
| 2-phenyl-d$_5$-ethan-d$_4$-ol (IS) | 98 | | | |
| 2-heptanone | 58 | y=1.04x-1.44 | 1–500 | 0.9891 |
| 2-methyl-1-butanol | 57 | y=0.501x+0.0062 | 1–500 | 0.9928 |
| 2-heptanol | 45 | y=4.48x+0.0272 | 1–500 | 0.9947 |
| 2-phenyl-d$_5$-ethan-d$_4$-ol (IS) | 98 | | | |
| 1-hexanol | 56 | y=0.13x+0.0093 | 1–500 | 0.9975 |

**Table 3. (Continued). Chemical Standards, Target Ions, and Calibration Curves Used for Quantification of Volatiles in Chinese Baijiu**

| Compounds | m/z | Equations | Range (μg/L) | $R^2$ |
|---|---|---|---|---|
| 2-nonanone | 58 | y=2.65x-1.3 | 1–500 | 0.9979 |
| 2-ethyl-1-hexanol | 57 | y=1.33+0.0761 | 1–500 | 0.9985 |
| 1-octen-3-ol | 57 | y=1.17x+0.0724 | 1–500 | 0.9977 |
| 1-octanol | 56 | y=0.0681x+0.168 | 1–500 | 0.9991 |
| benzyl alcohol | 79 | y=0.0228x+0.0001 | 1–500 | 0.9972 |
| phenethyl alcohol | 91 | y=0.0653x+0.0447 | 1–500 | 0.9964 |
| butanoic acid-$d_7$ (IS) | 63 | | | |
| butanoic acid | 60 | y=0.691x+0.117 | 1–1000 | 0.9964 |
| 2-methyl-butanoic acid | 74 | y=1.29x-0.0191 | 1–1000 | 0.9986 |
| 3-methyl-butanoic acid | 60 | y=0.0621x+0.0989 | 1–1000 | 0.9928 |
| hexanoic acid-$d_{11}$ (IS) | 77 | | | |
| hexanoic acid | 60 | y=0.213x+0.0165 | 1–1000 | 0.9992 |
| octanoic acid-$d_{15}$ (IS) | 77 | | | |
| octanoic acid | 73 | y=0.846x-0.012 | 1–1000 | 0.9913 |
| nonanoic acid | 73 | y=1.85x-0.0253 | 1–1000 | 0.9988 |
| decanoic acid-$d_{19}$ (IS) | 77 | | | |
| decanoic acid | 73 | y=1.78x-0.0244 | 1–1000 | 0.9915 |

## Results and Discussion

### Aroma Compounds in Chinese Baijiu

A total of 63 aroma compounds were quantified including 29 esters, 12 alcohols, 7 terpenoids, 7 acids, 5 sulfur-containing compounds, 2 ketones, and 1 aldehyde by Headspace-GC-FID, Headspace-SPME-GC-MS, and Headspace-SPME-GC-PFPD (Tables 4–12).

Esters are responsible for fruity, floral, and sweet aromas, which are mainly synthesized through esterification of alcohols and acids during the fermentation, distillation, and aging processes (14). Esters, especially ethyl esters, are reported as key aroma compounds in Chinese baijiu (7, 8, 12, 14). Results in Tables 4–6 showed that strong-aroma baijiu has a much higher concentration of ethyl hexanoate, ethyl heptanoate, and ethyl octanoate than those of QK liquor and the other baijiu. In contrast, light-aroma baijiu, MT, STJ, and HS have higher concentrations of ethyl acetate and ethyl lactate than strong-aroma baijiu. Ethyl acetate is reported as a very potent aroma compound in Chinese liquor (8, 12). Its concentration in light-aroma baijiu was about twice as high as that of strong-aroma baijiu. QK liquor has the second highest concentration (374.1 mg/L) of ethyl acetate next to Maotai (531.9 mg/L). Ethyl lactate is another key aroma compound in Chinese baijiu, especially in light-aroma baijiu (12). FJ, a bran starter light-aroma baijiu, showed the highest ethyl lactate concentration (1626.6 mg/L), followed by STJ, Te-aroma (1560.4 mg/L). Light-aroma baijiu (CYW and C20), sauce-aroma baijiu (MT), and Lao Bai Gan–aroma baijiu (HS) also have high concentrations of ethyl lactate (at least 1000 mg/L). Other potential aroma-contributing esters were ethyl decanoate, propyl acetate, hexyl acetate, octyl acetate, isobutyl acetate, ethyl 3-methylbutanoate, phenyl acetate, and phenethyl acetate. Butyl butanoate was detected in QK liquor, CYW, and C20, but methyl hexanoate was only found in QK liquor. Quantified Baijiu samples showed much lower concentrations of branched-chain esters than straight-chain esters.

Fatty acids in Chinese baijiu were mainly produced by the microbial fermentation process and are major precursors for esterification (20). It is well known that short-chain fatty acids contribute to rancid, sweaty, and fatty odors (12). Seven fatty acids including butanoic acid, 2-methylbutanoic acid, 3-methylbutanoic acid, hexanoic acid, octanoic acid, nonanoic acid, and decanoic acid were quantified in this study (see Tables 4–6). All strong-aroma baijiu, BYB, STJ, and GJG6 had extremely high concentrations of hexanoic acid that ranged from 569.7 mg/L to 1142.3 mg/L, corresponding with their very high concentrations of ethyl hexanoate. Strong-aroma baijiu, MT, BYB, and STJ contained relatively higher concentrations of octanoic acid and 3-methylbutanoic acid compared to the other baijiu. Light-aroma baijiu (QK, CYW, and C20) with HS have a slightly higher concentration of 2-methylbutanoic acid than other baijiu. QK and HS have high concentrations of butanoic acid (12.4 mg/L and 30.7 mg/L, respectively).

Alcohols were the second dominant group of volatile compounds (Tables 7–9). They formed mainly during the fermentation stage from sugar or amino acids under aerobic or anaerobic conditions (14), which produces a fruity, floral, and alcoholic-like flavor. In this study, the concentration of alcohols varies among individual baijiu. The concentrations of alcohols in QK were in the middle range compared to the rest of the baijiu, but with a relatively high amount of phenethyl alcohol (60.3 mg/L), which produces a very rosy note. CYW and C20 also had high concentrations of phenethyl alcohol with 96.8 mg/L and 99.6 mg/L, respectively. The highest concentrations of propanol were found in GJ (810.6 mg/L), followed by JNC (564.1 mg/L), and MT (560.9 mg/L), imparting a fermented and musty odor. HS contained a particularly high concentration of isobutyl alcohol (157.8 mg/L), which contributes a sweet note. CYW, C20, HS, and GL contained high concentrations of 2-methyl-1-butanol ranged from 47.8 mg/L to 62.7 mg/L, thus giving a roasted flavor while GJGJ, LZLJ, and GJG6 had high concentrations of 1-hexanol, which produce a typical green odor. 1-octen-3-ol was detected in higher concentrations in all strong-aroma baijiu, and GJG6 compared with the rest of the baijiu (imparting mushroom note) while GJGJ and LZLJ contained high benzyl alcohol with a mild aromatic odor.

Terpenoids, such as linalool, nerol, geraniol, α-terpineol, β-citronellol, β-damascenone, and β-ionone, are important aroma compounds in Chinese baijiu due to their low sensory threshold (21), which contribute to its unique floral notes. QK liquor had the highest concentrations of linalool, β-damascenone, and β-ionone, which contributed to floral and honey flavors (Table 11). In addition, QK, MT, STJ, and HS had a higher amount of β-citronellol than the other Chinese baijiu. However, strong-aroma baijiu with BYB, HS, and GJG6 had extremely high concentrations of geraniol which ranged from 684 µg/L in HS to 13559 µg/L in LZLJ, while GJGJ and LZLJ had relatively higher amounts of nerol.

Acetaldehyde, which contributies to floral and green odors, was an important aroma compound in Chinese baijiu and was probably formed by yeast (20) (see Tables 10–12). There were few differences in the ketone levels among samples except the amount of 2-nonanone in GIG6 (417 µg/L) is 5 to 10 times higher than the other baijiu.

**Table 4. Concentration (mg/L ± SD) of Esters and Acids in Strong-Aroma Chinese Baijiu**

| Compounds | Threshold (mg/L) | GJ | YH | GJGJ | LZLJ | JNC |
|---|---|---|---|---|---|---|
| *Esters* | | | | | | |
| ethyl acetate* | 32.6[a] | 140.2±5 | 121.1±5.9 | 132.2±0.3 | 149.4±8.7 | 97.3±1.4 |
| ethyl propanoate | 19[a] | 3±0.1 | 4±0.2 | 6.4±1.1 | 8.4±0.2 | 3.4±0.1 |
| ethyl butanoate | 0.0815[a] | 158.5±2.8 | 79.9±4 | 287.9±12 | 456.6±7.9 | 118.1±2.2 |
| ethyl pentanoate | 0.0268[a] | 5.6±0.6 | 5.5±0.4 | 9.4±2.1 | 17.7±2.7 | 7.8±0.7 |
| ethyl hexanoate | 0.0553[a] | 10130±875 | 8993±907 | 7853±584 | 10080±502 | 14270±1662 |
| ethyl heptanoate | 13.2[a] | 122.6±13.7 | 248.1±34.7 | 116±7.4 | 131.6±10.3 | 42.6±2.5 |
| ethyl octanoate | 0.0129[a] | 113.5±12.4 | 339.5±36.4 | 185.5±9 | 126±5.9 | 64.6±9 |
| ethyl decanoate | 1.12[a] | 0.4±0 | 0.8±0.1 | 3.2±0.1 | 3.2±0.1 | 0.8±0 |
| ethyl dodecanoate | 0.64[a] | 0.3±0 | 0.4±0 | 0.8±0.1 | 1.2±0.3 | 1.3±0.2 |
| propyl acetate | 11[b] | 0.6±0.1 | 1.2±0.1 | 3.9±0.2 | 1.3±0.2 | 8.6±0.3 |
| hexyl acetate | 5.56[a] | 1.6±0.1 | 4.2±0 | 1.9±0.6 | 1.5±0 | 5.2±1.4 |
| octyl acetate | 0.047[b] | 0.6±0.1 | 17.7±2.8 | 3.1±0.6 | 2.8±0 | 10.8±0.3 |
| methyl hexanoate | 0.087[b] | ND | ND | ND | ND | ND |
| butyl butanoate | 0.1[a] | ND | ND | ND | ND | ND |
| propyl hexanoate | 13[a] | ND | ND | 146.4±25.4 | ND | 36.4±2.2 |
| octyl butanoate | nf | 5.2±0.6 | 4.9±0.4 | 4.9±0.4 | 5.6±0.4 | 5.9±0 |
| ethyl lactate | 128[a] | 665±20 | 778±87 | 973±19 | 773±28 | 699±29 |
| diethyl succinate | 353[a] | 0.2±0.1 | 0.2±0 | 1.9±0.6 | 1.8±0.6 | 0.2±0 |
| ethyl 3-hydroxyhexanoate | 51.4[a] | ND | 0.02±0 | ND | ND | 0.02±0 |
| isoamyl acetate* | 0.043[b] | 0.1±0 | 0.4±0 | 0.4±0 | 0.2±0 | 0.2±0 |

**Table 4. (Continued). Concentration (mg/L ± SD) of Esters and Acids in Strong-Aroma Chinese Baijiu**

| Compounds | Threshold[a] (mg/L) | GJ | YH | GJGJ | LZLJ | JNC |
|---|---|---|---|---|---|---|
| isobutyl acetate | 0.922[a] | ND | 0.2±0 | 0.9±0.1 | 0.3±0.1 | 0.2±0 |
| ethyl isobutanoate | 0.0575[a] | 8.4±0.3 | 3.4±0.2 | 22±1.3 | 16.6±0.6 | 7.6±0.2 |
| ethyl 2-methylbutanoate | 0.018[a] | 0.4±0.1 | ND | 0.8±0.1 | 0.7±0 | 0.1±0 |
| ethyl 3-methylbutanoate | 0.0069[a] | 1.7±0.1 | 0.5±0 | 2.1±0.2 | 2.6±0.1 | 1.1±0.1 |
| isobutyl hexanoate | NF | ND | ND | ND | ND | 29.5±5.8 |
| isopentyl hexanoate | 1.4 | ND | ND | ND | ND | 131.6±25.4 |
| phenyl acetate | 0.407[a] | 1.8±0.5 | 1.9±0.3 | 2.6±0.8 | 11.8±3.7 | 3.4±0.3 |
| phenethyl acetate | 0.909[a] | 0.02±0 | 0.06±0 | 0.1±0 | 0.2±0.1 | 0.04±0 |
| ethyl 3-phenylpropanoate | 0.13[a] | 0.9±0 | 0.9±0 | 1.9±0.1 | 1.9±0.2 | 1.1±0 |
| *Acids* | | | | | | |
| butanoic acid | 0.964[a] | 1.5±0.1 | 0.3±0.1 | 1.2±0.3 | 5.7±0.5 | 0.4±0 |
| 2-methylbutanoic acid | 1.58[a] | 0.1±0 | 0.1±0 | 0.3±0.1 | 0.2±0.1 | 0.1±0 |
| 3-methylbutanoic acid | 1.05[a] | 30.8±1.9 | 30.4±1.3 | 17.4±0.5 | 32±1.3 | 22.9±1.6 |
| hexanoic acid | 2.52[a] | 1112±42 | 842±15 | 858±16 | 776±18 | 1142±34 |
| octanoic acid | 2.7[a] | 9.1±1 | 15.6±0.5 | 16.8±0.3 | 7.8±0.5 | 9.7±0.2 |
| nonanoic acid | 3.56[a] | 0.18±0 | 0.21±0 | 0.19±0 | 0.16±0 | 0.16±0 |
| decanoic acid | 13.7[a] | 0.07±0 | 0.10±0 | 0.18±0 | 0.09±0 | 0.13±0 |

[a] Odor threshold was determined in a hydroalcoholic solution at 48% ethanol by volume (25). [b] Odor threshold was obtained from the literature (26). *Quantified by HS-GC-FID; NF: Threshold was not found in references; ND: Compound was not detected; SD: Standard deviation.

## Table 5. Concentration (mg/L ± SD) of Esters and Acids in Light-Aroma Chinese Baijiu

| Compounds | Threshold (mg/L) | CYW | C20 | QK | FJ | HX |
|---|---|---|---|---|---|---|
| **Esters** | | | | | | |
| ethyl acetate* | 32.6[a] | 211.6±12.1 | 231.2±11.1 | 374.1±19.2 | 361.7±4.4 | 274.7±2 |
| ethyl propanoate | 19[a] | 15.3±0.8 | 16±2.1 | 5.1±0.3 | 0.6±0.1 | 1.7±0.1 |
| ethyl butanoate | 0.0815[a] | 180.9±10.9 | 202.2±22.4 | 230±6.2 | 2.7±0.1 | 3.6±0.1 |
| ethyl pentanoate | 0.0268[a] | 6.1±0.6 | 7.3±0.4 | 1±0.2 | ND | ND |
| ethyl hexanoate | 0.0553[a] | 70.3±3.9 | 80.8±3.4 | 42.8±1.3 | 4±0.3 | 2.6±0.2 |
| ethyl heptanoate | 13.2[a] | 3±0.1 | 4.9±0.9 | 3.5±0 | 0.1±0 | 0.1±0 |
| ethyl octanoate | 0.0129[a] | 26.6±2.7 | 16.1±1.6 | 30.4±1.6 | 2.8±0.3 | 2.5±0.2 |
| ethyl decanoate | 1.12[a] | 15.6±0.1 | 12.1±0.9 | 7.5±0.1 | 1.3±0 | 1.6±0 |
| ethyl dodecanoate | 0.64[a] | 0.8±0.1 | 0.4±0 | 1.2±0 | 0.2±0 | 0.2±0 |
| propyl acetate | 11[b] | 0.9±0.1 | 1.5±0.2 | 0.3±0.1 | ND | ND |
| hexyl acetate | 5.56[a] | 0.4±0 | 0.6±0 | 0.3±0.1 | 0.04±0 | 0.02±0 |
| octyl acetate | 0.047[b] | 0.04±0 | 0.1±0 | 0.1±0 | <0.01 | <0.01 |
| methyl hexanoate | 0.087[b] | ND | ND | 0.1±0 | ND | ND |
| butyl butanoate | 0.1[a] | 0.7±0 | 1.6±0.2 | 1.4±0 | ND | ND |
| propyl hexanoate | 13[a] | 0.1±0 | 0.3±0.1 | ND | ND | ND |
| octyl butanoate | NF | ND | ND | 1.1±0 | ND | ND |
| ethyl lactate | 128[a] | 1088±36 | 1294±98 | 858±64 | 1626±112 | 540±7 |
| diethyl succinate | 353[a] | 0.04±0 | 0.03±0 | 0.04±0 | <0.01 | <0.01 |
| ethyl 3-hydroxyhexanoate | 51.4[a] | <0.01 | 0.02±0 | 0.02±0 | 0.02±0 | 0.02±0 |
| isoamyl acetate* | 0.043[b] | 0.8±0 | 1.1±0.1 | 0.9±0 | 0.9±0 | 0.7±0 |
| isobutyl acetate | 0.922[a] | 4.4±0.3 | 3.9±0.4 | 7.3±0.3 | 0.8±0 | 0.8±0 |
| ethyl isobutanoate | 0.0575[a] | 4.8±0.3 | 4.8±0.2 | 6.8±0.2 | 1.4±0.1 | 1.2±0 |

237

**Table 5. (Continued). Concentration (mg/L ± SD) of Esters and Acids in Light-Aroma Chinese Baijiu**

| Compounds | Threshold (mg/L) | CYW | C20 | QK | FJ | HX |
|---|---|---|---|---|---|---|
| ethyl 2-methylbutanoate | 0.018[a] | ND | ND | 0.1±0 | ND | ND |
| ethyl 3-methylbutanoate | 0.0069[a] | 0.4±0 | 0.3±0 | 0.6±0 | ND | ND |
| isobutyl hexanoate | NF | ND | ND | ND | ND | ND |
| isopentyl hexanoate | 1.4[a] | ND | 0.2±0 | ND | ND | ND |
| phenyl acetate | 0.407[a] | 3.9±0.9 | 1.4±0.2 | 4.5±0.6 | 0.2±0.1 | 0.1±0 |
| phenethyl acetate | 0.909[a] | 2.6±0.2 | 0.9±0 | 2.4±0.3 | 0.2±0 | 0.2±0 |
| ethyl 3-phenylpropanoate | 0.13[a] | 0.9±0 | 0.9±0 | 0.9±0 | 0.8±0 | 0.8±0 |
| *Acids* | | | | | | |
| butanoic acid | 0.964[a] | 1.6±0.4 | 1.6±0 | 12.4±1.9 | ND | ND |
| 2-methylbutanoic acid | 1.58[a] | 4.2±0.7 | 3.7±0.3 | 5.1±0.6 | 0.6±0.1 | 0.3±0 |
| 3-methylbutanoic acid | 1.05[a] | 3.3±0.4 | 3.3±0.1 | 4.2±0 | 4.6±0.4 | 1.9±0.2 |
| hexanoic acid | 2.52[a] | 10.8±1.5 | 9.4±0.7 | 4.1±0.1 | 2.1±0.3 | 4±0.7 |
| octanoic acid | 2.7[a] | 0.7±0.1 | 0.3±0.1 | 0.9±0.1 | 1.2±0 | 1.3±0.1 |
| nonanoic acid | 3.56[a] | 0.07±0 | 0.10±0 | 0.10±0 | 0.14±0 | 0.09±0 |
| decanoic acid | 13.7[a] | 0.16±0 | 0.04±0 | 0.09±0 | 0.21±0 | 0.29±0 |

[a] Odor threshold was determined in a hydroalcoholic solution at 48% ethanol by volume (25). [b] Odor threshold was obtained from the literature (26). * Quantified by HS-GC-FID; NF: Threshold was not found in references; ND: Compound was not detected; SD: Standard deviation.

**Table 6. Concentration (mg/L ± SD) of Esters and Acids in Other Aroma Styles Chinese Baijiu**

| Compounds | Threshold (mg/L) | MT | BYB | STJ | HS | GL | GJG6 | GDJJ |
|---|---|---|---|---|---|---|---|---|
| *Esters* | | | | | | | | |
| ethyl acetate* | 32.6[a] | 531.9±6.7 | 179±3.9 | 203.2±6.2 | 209.3±7.4 | 11.6±0.5 | 105.3±2.5 | 62.4±0.2 |
| ethyl propanoate | 19[a] | 24.5±1.1 | 16.5±1.4 | 16.2±0.7 | 9.7±1.3 | ND | 9.3±0.3 | 0.1±0 |
| ethyl butanoate | 0.0815[a] | 52.1±1.5 | 101±7.6 | 87.4±4.2 | 54.1±3.2 | 0.4±0.1 | 275.6±6 | 1.2±0 |
| ethyl pentanoate | 0.0268[a] | 3.1±0.1 | 3.3±1.8 | 7.1±1.7 | 3.1±0.2 | ND | 19.9±3 | ND |
| ethyl hexanoate | 0.0553[a] | 18.5±0.7 | 7715±19 | 4609±46 | 46.4±4.7 | 1.1±0 | 5852±105 | 3.5±0.1 |
| ethyl heptanoate | 13.2[a] | 2.5±0.7 | 59.5±5.1 | 96.6±5.1 | 2.3±0.5 | ND | 41±5.3 | 0.1±0 |
| ethyl octanoate | 0.0129[a] | 1.6±0.1 | 58.4±1.3 | 73.2±1.1 | 29.8±3 | 0.9±0.1 | 47.2±4.3 | 0.4±0 |
| ethyl decanoate | 1.12[a] | 0.6±0 | 0.07±0 | 0.9±0.1 | 20.3±0.2 | 0.8±0 | 0.6±0 | 0.04±0 |
| ethyl dodecanoate | 0.64[a] | 0.3±0 | 0.2±0 | 0.4±0 | 3.2±0 | 0.3±0 | 0.4±0 | 0.1±0 |
| propyl acetate | 11[b] | 2.9±0.1 | 1.1±0.1 | 0.4±0 | 0.8±0.1 | ND | 1.6±0.1 | 0.09±0 |
| hexyl acetate | 5.56[a] | 0.1±0 | 2.8±0.1 | 6.8±0.2 | 0.2±0 | ND | 1.5±0 | 0.01±0 |
| octyl acetate | 0.047[b] | 0.02±0 | 1.82±0.5 | 0.80±0 | 0.06±0 | 0.01±0 | ND | 0.01±0 |
| methyl hexanoate | 0.087[b] | ND | ND | ND | ND | ND | ND | ND |
| butyl butanoate | 0.1[a] | ND | ND | ND | ND | ND | ND | 0.1±0 |
| propylhexanoate | 13[a] | ND | 135.6±5.2 | 21.1±1.6 | ND | ND | ND | ND |
| octyl butanoate | NF | 0.05±0 | 4.4±0.2 | 4.2±0.2 | ND | 0.03±0 | 4.0±0.1 | 0.03±0 |
| ethyl lactate | 128[a] | 1115±40 | 477±76 | 1560±28 | 1182±33 | 987±18 | 475±25 | 376±29 |
| diethyl succinate | 353[a] | 0.72±0 | 0.11±0 | 0.28±0 | 0.08±0 | <0.01 | <0.01 | <0.01 |
| ethyl 3-hydroxyhexanoate | 51.4[a] | 0.02±0 | 0.02±0 | 0.02±0 | 0.02±0 | 0.02±0 | ND | 0.02±0 |
| isoamyl acetate* | 0.043[b] | 1.3±0.1 | 0.4±0 | 0.7±0 | 0.8±0 | 0.1±0 | 0.1±0 | 0.1±0 |
| isobutyl acetate | 0.922[a] | 1.4±0.1 | 0.2±0 | 0.3±0 | 5.4±0.3 | 0.2±0 | 0.4±0 | 0.2±0 |
| ethyl isobutanoate | 0.0575[a] | 25.5±0.4 | 7.4±0.5 | 10.1±0.4 | 16.6±1.1 | 0.9±0 | 12.4±0.3 | 1.2±0 |

## Table 6. (Continued). Concentration (mg/L ± SD) of Esters and Acids in Other Aroma Styles Chinese Baijiu

| Compounds | Threshold (mg/L) | MT | BYB | STJ | HS | GL | GJG6 | GDJJ |
|---|---|---|---|---|---|---|---|---|
| ethyl 2-methylbutanoate | 0.018[a] | 0.9±0 | 0.1±0 | 0.3±0 | 0.6±0 | ND | 0.6±0.1 | ND |
| ethyl 3-methylbutanoate | 0.0069[a] | 4.7±0.1 | 1.2±0.1 | 1.9±0.1 | 1.6±0.1 | ND | 2.4±0.1 | ND |
| isobutyl hexanoate | NF | ND | 22.8±1.1 | 2±0.2 | ND | ND | ND | ND |
| isopentyl hexanoate | 1.4[a] | ND | 74.5±7.9 | 18±2.2 | ND | ND | ND | ND |
| phenyl acetate | 0.407[a] | 5.9±1.5 | 1.1±0.2 | 1.6±0.1 | 2.7±0.2 | 0.2±0 | 3.4±0.9 | 0.1±0 |
| phenethyl acetate | 0.909[a] | 0.1±0 | 0.04±0 | 0.1±0 | 1.4±0.1 | 0.08±0 | 0.09±0 | 0.04±0 |
| ethyl 3-phenylpropanoate | 0.13[a] | 0.8±0 | 0.8±0 | 1.0±0 | 0.9±0 | 0.8±0 | 1.1±0.1 | 0.8±0 |
| *Acids* | | | | | | | | |
| butanoic acid | 0.964[a] | 0.1±0 | 0.1±0 | 0.2±0 | 30.7±0.6 | ND | 1±0.2 | ND |
| 2-methylbutanoic acid | 1.58[a] | 0.1±0 | 0.1±0 | 0.1±0 | 9.5±0.2 | 0.4±0 | 0.1±0 | 0.1±0 |
| 3-methylbutanoic acid | 1.05[a] | 26.4±3.1 | 11.6±3.8 | 17.1±1.2 | 7±0.4 | 3.2±0.3 | ND | ND |
| hexanoic acid | 2.52[a] | 33.6±4.4 | 989.5±57.1 | 569.7±5.8 | 2.7±0.5 | 1.9±0 | 661.7±34 | 7.9±1.2 |
| octanoic acid | 2.7[a] | 1.1±0 | 4.4±0.2 | 6.2±0.2 | 1.5±0 | 1.1±0.1 | 4.6±0.3 | 0.8±0.1 |
| nonanoic acid | 3.56[a] | 0.14±0 | 0.18±0 | 0.33±0 | 0.09±0 | 0.08±15 | 0.12±0 | 0.19±0 |
| decanoic acid | 13.7[a] | 0.59±0 | 0.06±0 | 0.12±0 | 0.38±0 | 0.68±41 | 0.02±0 | ND |

[a] Odor threshold was determined in a hydroalcoholic solution at 48% ethanol by volume (25).  [b] Odor threshold was obtained from the literature (26).  * Quantified by HS-GC-FID; NF: Threshold was not found in references; ND: Compound was not detected; SD: Standard deviation.

**Table 7. Concentration (mg/L ± SD) of Alcohols in Strong-Aroma Chinese Baijiu**

| Compounds | Threshold (mg/L) | GJ | YH | GJGJ | LZLJ | JNC |
|---|---|---|---|---|---|---|
| propanol* | 54[a] | 810.6±30 | 146.5±8 | 52.1±3.4 | 0.6±0.1 | 564.1±13 |
| isobutyl alcohol* | 28.3[a] | 5.7±0.2 | 20.4±1.2 | 21.3±0.3 | 7.2±0.8 | 42.5±4.8 |
| isoamyl alcohol* | 179[a] | 19.9±0.4 | 60.1±2.1 | 61.3±1.9 | 26.2±2 | 87±8.7 |
| 2-methyl-1-butanol | 1.4[b] | 1±0 | 3.2±0.1 | 21.3±1.3 | 11.9±0.1 | 4±0.2 |
| 2-heptanol | 1.43[a] | 0.02±0 | 0.21±0 | 1.24±0.1 | 0.83±0.1 | 0.16±0 |
| 1-hexanol | 5.37[a] | 24.4±0.7 | 26.6±1.3 | 108.5±6 | 244.7±12 | 19.7±0.3 |
| (Z)-3-hexen-1-ol | 0.07[a] | 0.05±0 | 0.05±0 | 0.21±0 | 0.24±0 | 0.07±0 |
| 1-octen-3-ol | 0.00612[a] | 7±0.3 | 26.2±0.9 | 120.3±1 | 69.1±1.9 | 8.9±0 |
| 2-ethyl-1-hexanol | 0.13[a] | 0.04±0 | 0.04±0 | 0.35±0 | 0.38±0 | 0.02±0 |
| 1-octanol | 1.1[a] | 0.9±0 | 1.4±0.1 | 26.2±0.3 | 16.4±0.3 | 1.2±0 |
| phenethyl alcohol | 28.9[a] | 1.5±0 | 3.8±0.1 | 11.5±1.6 | 15.3±0.5 | 3.4±0.1 |
| benzyl alcohol | 40.9[a] | 11.5±0.1 | 23.3±0.5 | 262.1±26 | 176.6±4 | 46.3±0.5 |

[a] Odor threshold was determined in a hydroalcoholic solution at 48% ethanol by volume (25).   [b] Odor threshold was obtained from the literature (26).   * Quantified by HS-GC-FID; ND: Compound was not detected; SD: Standard deviation.

**Table 8. Concentration (mg/L ± SD) of Alcohols in Light-Aroma Chinese Baijiu**

| Compounds | Threshold (mg/L) | CYW | C20 | QK | FJ | HX |
|---|---|---|---|---|---|---|
| propanol* | 54[a] | 80.2±4.6 | 92.5±2.4 | 29.7±0.4 | 43.6±0.7 | 100.3±2 |
| isobutyl alcohol* | 28.3[a] | 36.4±1.5 | 31.8±0.6 | 54.6±2 | 27.1±0.4 | 51.7±0.8 |
| isoamyl alcohol* | 179[a] | 118.7±7 | 142.7±5 | 95.9±0.3 | 84.1±1.5 | 134.5±2 |
| 2-methyl-1-butanol | 1.4[b] | 60.9±0.3 | 47.8±5.5 | 14.2±0.1 | 4.2±0.9 | 6.7±0.3 |
| 2-heptanol | 1.43[a] | 0.27±0 | 0.49±0.1 | 0.34±0 | 0.02±0 | 0.02±0 |
| 1-hexanol | 5.37[a] | 62.4±0.4 | 77.2±4.7 | 36.5±0.2 | 3.8±0.1 | 2.3±0.1 |
| (Z)-3-hexen-1-ol | 0.07[a] | 0.15±0 | 0.14±0 | 0.05±0 | 0.06±0 | 0.05±0 |
| 1-octen-3-ol | 0.00612[a] | 0.7±0 | 1.2±0 | 3.6±0.1 | 0.1±0 | 0.2±0 |
| 2-ethyl-1-hexanol | 0.13[a] | 0.12±0 | 0.57±0 | 0.20±0 | 0.01±0 | 17±1 |
| 1-octanol | 1.1[a] | 12.9±0.8 | 16.1±1.2 | 10±0.3 | 1.1±0.2 | 1.2±0 |
| phenethyl alcohol | 28.9[a] | 96.8±4.1 | 99.6±3.7 | 60.3±1.6 | 4.5±0.2 | 7.4±0.2 |
| benzyl alcohol | 40.9[a] | 5.4±0.7 | 5.3±0.2 | 8.5±0 | 1.6±0 | 0.3±0 |

[a] Odor threshold was determined in a hydroalcoholic solution at 48% ethanol by volume (25).  [b] Odor threshold was obtained from the literature (26).  * Quantified by HS-GC-FID; ND: Compound was not detected; SD: Standard deviation.

242

**Table 9. Concentration (mg/L ± SD) of Alcohols in Other Aroma Chinese Baijiu**

| Compounds | Threshold (mg/L) | MT | BYB | STJ | HS | GL | GJG6 | GDJJ |
|---|---|---|---|---|---|---|---|---|
| propanol* | 54[a] | 560.9±1.2 | 194.1±2.9 | 219±8 | 64.3±2.2 | 45.9±2.2 | ND | 23.5±0.2 |
| isobutyl alcohol* | 28.3[a] | 48.3±0.8 | 22.2±1 | 28.9±0.3 | 46.1±0.9 | 157.8±4.6 | 7.3±0.3 | 33.3±0.7 |
| isoamyl alcohol* | 179[a] | 125±1.4 | 69.7±0.2 | 122.7±6.8 | 121.5±1.2 | 238.7±6.4 | 22.6±0.2 | 62.8±0.6 |
| 2-methyl-1-butanol | 1.4[b] | 5.8±0.5 | 3.6±0.3 | 6.1±0 | 62.7±4.4 | 55.1±1.1 | 9±0.4 | 3.4±0.1 |
| 2-heptanol | 1.43[a] | 0.01±0 | 0.02±0 | 0.22±0 | 0.26±0.05 | ND | ND | 0.02±0 |
| 1-hexanol | 5.37[a] | 11.8±0.9 | 24.8±1.2 | 43.4±2.4 | 26.3±2.7 | 0.8±0 | 132.8±7.1 | 1±0.1 |
| (Z)-3-hexen-1-ol | 0.07[a] | 0.05±0 | 0.08±0 | 0.09±3 | 0.53±0.1 | 0.15±0 | 0.14±0 | 0.04±0 |
| 1-octen-3-ol | 0.00612[a] | 0.1±0 | 1.8±0 | 0.1±0 | 1.5±0.2 | 0.1±0 | 38.6±3 | ND |
| 2-ethyl-1-hexanol | 0.13[a] | 0.05±0 | 0.03±0 | 0.04±0 | 0.13±0 | 0.05±0 | 0.16±0 | 0.01±0 |
| 1-octanol | 1.1[a] | 1.3±0.1 | 1.5±0 | 1.5±0.1 | 12.2±0.1 | 1.4±0 | 19.9±1.1 | 1.0±0.1 |
| phenethyl alcohol | 28.9[a] | 17.4±0.5 | 6.3±0.2 | 9.4±1 | 62.8±4 | 52.4±1.8 | 10.8±0.4 | 48±2 |
| benzyl alcohol | 40.9[a] | 1.4±0.1 | 14.4±0.8 | 53.3±7 | 7.7±0.1 | 0.2±0 | 55.2±2.1 | 2.1±0.1 |

[a] Odor threshold was determined in a hydroalcoholic solution at 48% ethanol by volume (25). [b] Odor threshold was obtained from the literature (26). * Quantified by HS-GC-FID; ND: Compound was not detected; SD: Standard deviation.

**Table 10. Concentration (± SD) of Terpenoids, Aldehydes, Ketones, and Sulfur-Containing Compounds in Strong-Aroma Chinese Baijiu**

| Compounds | Threshold (µg/L) | GJ | YH | GJGJ | LZLJ | JNC |
|---|---|---|---|---|---|---|
| *Terpenoids (µg/L)* | | | | | | |
| linalool | 13.1[a] | 35±0 | 49±3 | 112±2 | 25±0 | 52±2 |
| nerol | 130[a] | 87±1.9 | 181±18 | 444±10.4 | 494±13.2 | 73±1 |
| geraniol | 120[a] | 2629±36 | 1410±49 | 6659±155 | 13560±154 | 986±11 |
| alpha-terpineol | 680[a] | 63±0 | 63±1 | 68±1 | 72±5 | 63±0 |
| beta-citronellol | 300[a] | ND | ND | ND | ND | ND |
| beta-damascenone | 0.12[a] | ND | 8±1 | 48±12 | 45±1 | 11±1 |
| beta-ionone | 1.3[a] | ND | 2±0 | 17±2 | ND | ND |
| *Aldehydes (mg/L)* | | | | | | |
| acetaldehyde* | 1200[a] | 71.9±0.9 | 57±4.8 | 26.5±0.9 | 62±3.8 | 75.5±3 |
| *Ketones (µg/L)* | | | | | | |
| 2-heptanone | 1330[b] | ND | 1136±12 | ND | ND | 1127±3 |
| 2-nonanone | 483[a] | 32±1 | 69±2 | 144±3 | 51±3 | 33±0 |
| *Sulfur compounds (µg/L)* | | | | | | |
| dimethyl sulfide | 20[a] | ND | ND | ND | 1.82±0.2 | ND |
| methyl thioacetate | 21[a] | ND | 0.72±0.11 | ND | ND | 0.71±0.14 |
| dimethyl disulfide | 10[a] | 0.04±0 | 1.03±0.08 | 0.07±0 | 0.48±0.05 | 0.24±0.02 |
| ethyl thioacetate | 70[a] | ND | ND | 0.03±0 | 0.04±0 | 0.04±0 |
| dimethyl trisulfide | 0.36[a] | 0.61±0.05 | 142.1±2.3 | 25.1±1.6 | 71.9±2.3 | 59.5±2.7 |

[a] Odor threshold was determined in a hydroalcoholic solution at 48% ethanol by volume (25). [b] Odor threshold was obtained from the literature (26). * Quantified by HS-GC-FID; ND: Compound was not detected; SD: Standard deviation.

**Table 11. Concentration (± SD) of Terpenoids, Aldehydes, Ketones, and Sulfur-Containing Compounds in Light-Aroma Chinese Baijiu**

| Compounds | Threshold (µg/L) | CYW | C20 | QK | FJ | HX |
|---|---|---|---|---|---|---|
| *Terpenoids (µg/L)* | | | | | | |
| linalool | 13.1[a] | 154±4 | 184±11 | 325±14 | 51±1 | 46±1 |
| nerol | 130[a] | 90±5 | 139±9 | 130±9 | 38±1 | 9±0 |
| geraniol | 120[a] | 240±40 | 137±14 | 183±14 | 18±2 | 25±2 |
| alpha-terpineol | 680[a] | 75±1 | 77±2 | 92±3 | 62±0 | 62±0 |
| beta-citronellol | 300[a] | ND | ND | 64±19 | ND | 8±2 |
| beta-damascenone | 0.12[a] | 97±3 | 59±5 | 170±0 | 16±0 | 8±1 |
| beta-ionone | 1.3[a] | 40±3 | 20±1 | 42±3 | 1±0 | 1±0 |
| *Aldehydes (mg/L)* | | | | | | |
| acetaldehyde | 1200[a] | 32.1±1.9 | 49.1±2.7 | 30.1±1.8 | 63.2±1.4 | 50.6±0.3 |
| *Ketones (µg/L)* | | | | | | |
| 2-heptanone | 1330[b] | 1161±6 | 1186±11 | 1151±23 | 1128±1 | 1129±0 |
| 2-nonanone | 483[a] | 46±1 | 57±3 | 30±0 | 30±1 | 32±0 |
| *Sulfur compounds (µg/L)* | | | | | | |
| dimethyl sulfide | 20[a] | 0.03±0.01 | ND | ND | ND | ND |
| methyl thioacetate | 21[a] | 0.01±0 | 0.07±0 | 0.60±0.02 | ND | ND |
| dimethyl disulfide | 10[a] | 0.17±0.02 | 1.42±0.1 | 0.41±0.03 | 0.17±0.01 | 0.04±0 |
| ethyl thioacetate | 70[a] | ND | ND | 0.03±0 | 0.03±0 | 0.03±0 |
| dimethyl trisulfide | 0.36[a] | 16.8±0.2 | 126.0±1.3 | 9.4±0 | 7.4±0.3 | 5.3±0.2 |

[a] Odor threshold was determined in a hydroalcoholic solution at 48% ethanol by volume (25).    [b] Odor threshold was obtained from the literature (26).    * Quantified by HS-GC-FID; ND: Compound was not detected; SD: Standard deviation.

**Table 12. Concentration (± SD) of Terpenoids, Aldehydes, Ketones, and Sulfur-Containing Compounds in Other Aroma Style Chinese Baijiu**

| Compounds | Threshold (μg/L) | MT | BYB | STJ | HS | GL | GJG6 | GDJJ |
|---|---|---|---|---|---|---|---|---|
| *Terpenoids (μg/L)* | | | | | | | | |
| linalool | 13.1[a] | 90±8 | 58±2 | 97±8 | 174±8 | 71±3 | ND | 28±0 |
| nerol | 130[a] | 14±2 | 16±3 | 38±2 | 201±18 | 10±1 | 80±7 | 9±1 |
| geraniol | 120[a] | 55±8 | 887±30 | 71±7 | 684±17 | 29±2 | 5819±286 | 42±5 |
| alpha-terpineol | 680[a] | 65±1 | 64±0 | 65±1 | 74±3 | 66±1 | 73±2 | 63±0 |
| beta-citronellol | 300[a] | 65±7 | 24±2 | 97±9 | 44±6 | 20±1 | ND | 5±0 |
| beta-damascenone | 0.12[a] | 12±1 | 9±2 | 27±0 | 86±2 | 31±3 | 37±3 | ND |
| beta-ionone | 1.3[a] | 2±0 | 2±0 | 8±0 | 15±1 | ND | ND | ND |
| *Aldehydes (mg/L)* | | | | | | | | |
| acetaldehyde | 1200[a] | 178.4±0.3 | 86.1±2.4 | 51.6±1.6 | 90.2±3.2 | 2.1±0.1 | 29.3±0.8 | 6.9±0.1 |
| *Ketones (μg/L)* | | | | | | | | |
| 2-heptanone | 1330[b] | 1144±2 | 1138±4 | 1130±2 | 1163±4 | 1126±0 | 1214±60 | 1129±0 |
| 2-nonanone | 483[a] | 45±1 | 162±4 | 61±2 | 49±1 | 73±1 | 417±22 | 30±0 |
| *Sulfur compounds (μg/L)* | | | | | | | | |
| dimethyl sulfide | 20[a] | 0.01±0 | 1.91±0.05 | 4.35±0.2 | 1.15±0 | ND | ND | ND |
| methyl thioacetate | 21[a] | 0.24±0.01 | 0.7±0.04 | 0.74±0.05 | ND | ND | 0.69±0.04 | ND |
| dimethyl disulfide | 10[a] | 0.26±0.02 | 0.18±0 | 0.71±0.03 | 0.05±0 | 0.02±0 | 0.13±0 | 0.02±0 |
| ethyl thioacetate | 70[a] | 0.01±0 | 0.04±0 | 0.03±0 | 0.03±0 | ND | ND | ND |
| dimethyl trisulfide | 0.36[a] | 97.2±1.2 | 94.9±1.6 | 167.9±3.1 | 6.9±0.2 | 0.3±0.0 | 23.1±0.2 | 0.1±0 |

[a] Odor threshold was determined in a hydroalcoholic solution at 48% ethanol by volume (25).   [b] Odor threshold was obtained from the literature (26).   * Quantified by HS-GC-FID; ND: Compound was not detected; SD: Standard deviation.

246

Sulfur-containing compounds are very important to Chinese baijiu due to their powerful characteristic in odors, low sensory thresholds, and high volatility (16, 22, 23). Sulfur-containing odorants are generally transformed from sulfur-containing amino acids. Five sulfur-containing compounds were quantified in this study (see Tables 10–12). Dimethyl disulfide and dimethyl trisulfide were detected in all samples. Dimethyl trisulfide was the highest concentration found in each individual sample. Other researchers demonstrated that both dimethyl disulfide and dimethyl trisulfide were important aroma compounds for strong- and light-aroma style Chinese baijiu (23, 24). Overall, YH, C20, MT, BYB, and STJ had more sulfur-containing compounds with higher concentrations than the other baijiu. GJ (strong-aroma), GL (rice-aroma), and GDJJ (Chi-aroma) contained fewer sulfur-containing compounds. QK, CYW, FJ, and HX fell in the middle in terms of both sulfur-containing compounds and concentration.

## Odor Active Values (OAVs)

To evaluate the potency of quantified odorants, odor activity values (OAVs) were calculated. All the aroma compounds with an OAV of at least 1 are shown in Figure 1 and Figure 2. Because of the wide range of OAVs, they were ranked from high to low by taking all values into account. Also, the OAVs of ethyl hexanoate and β-damascenone in Figure 1a and Figure 2e were divided by 10 in order to fit them into the scale.

Ethyl hexanoate, ethyl butanoate, and ethyl octanoate were identified as three key aroma contributors in Chinese baijiu (Figure 1a). They have OAVs that ranged from 980 to 258166 in strong-aroma (GJ, YH, GJGJ, LZLJ, and JNC), miscellaneous aroma (BYB), Te-aroma (STJ), and Guo Jing-aroma style (GJG6) baijiu, which contributed fruity and floral flavor. Ethyl pentanoate, ethyl isobutanoate, and ethyl 3-methylbutanoate (in Figure 1b) also had higher OAVs (greater than 100) in all strong-aroma and other aroma baijiu (MT, STJ, BYB, HS, and GJG6). YH and JNC had high OAVs for octyl acetate with values of 375 and 228, respectively, suggesting that octyl acetate also played important roles for these two baijiu. The OAV of isopentyl hexanoate was relatively high in JNC (94) and BYB (96) (Figure 1c). Although ethyl lactate was reported as a key aroma compound in some light-aroma baijiu (10), its OAVs were less than 10, except for FJ (13) and STJ (12), due to its high sensory threshold.

Volatile fatty acids are important to the quality and taste of Chinese baijiu (4). Concentrations of hexanoic acid in strong-aroma, miscellaneous aroma (BYB), Te-aroma (STJ), and Guo Jing-aroma baijiu (GJG6) were much higher than their sensory threshold with OAVs ranging from 226–453 (Figure 2a and 2b). Also, QK and HS had relatively high OAVs of butanoic acid with values of 12 and 31, respectively. These acids impart rancid, sweaty, and pungent odors.

Alcohols are mainly formed through fermentation from sugar and amino acid metabolism. In this study, all the Chinese baijiu had relatively lower OAVs for the alcohols quantified with a range from 1 to 46 (Figure 2c and 2d). QK had OAVs for all alcohols quantified that were greater than 1 except benzyl alcohol, which indicate they were important aroma compounds. Among them, 2-methyl-1-butanol showed higher OAVs in HS (44), CYW (43), GL (39), and C20 (34), which produce a roasted flavor, while 1-hexanol with a typical green note showed higher OAVs in LZLJ (45) and GJG6 (24). In addition, GJGJ had the highest OAV for 1-octanol (23) and 1-octen-3-ol (19). The OAV for (Z)-3-hexen-1-ol was higher in HS (7), while benzyl alcohol was higher in GJGJ and LZLJ. QK with CYW and C20 showed slightly higher OAVs for phenethyl alcohol.

For terpenoids, linalool, nerol, β-ionone, geraniol, and β-damascenone were quantified (Figure 2e). Among them, β-damascenone with honey flavor was determined to be the most important compound for all baijiu samples. β-damascenone and β-ionone are important C$_{13}$-norisoprenoids for Chinese baijiu due to their low odor thresholds. OAVs for all five terpenoids in QK were greater than 10 suggesting that they were important aroma contributors. Also, QK had the highest OAVs for β-damascenone and β-ionone with values of 1417 and 32, respectively. The OAVs of β-damascenone for the rest of the baijiu ranged from 67 to 808. The OAVs of β-ionone in CYW, C20, QK, and HS were 31, 16, 13, and 11, respectively, which produced the intensive scent of violets. Geraniol also showed higher OAVs (112) in LZLJ, GJGJ (55), and GJG6 (48), which imparted a floral note.

Acetaldehyde was reported as one of the most important carbonyl compounds in Chinese baijiu. It produces a green apple and dried straw odor and its concentration in Chinese baijiu depend highly on the duration of the fermentation and distillation stage (*16*). The OAV of acetaldehyde for this MT sample was 149, which was the highest among all baijiu samples (Figure 2e).

Several sulfur-containing compounds, including dimethyl sulfide, methyl thioacetate, dimethyl disulfide, ethyl thioacetate, and dimethyl trisulfide, were quantified. Dimethyl trisulfide was the only compound that had a concentration greater than its sensory threshold (an OAV of at least v1) for all the baijiu samples. Dimethyl trisulfide imparts a rotten egg, cabbage, and onion odor that might negatively affect liquor quality (*21*). Its OAVs in STJ, YH, C20, MT, and BYB were 466, 395, 350, 270, and 260, respectively, suggesting it was a potent aroma compound.

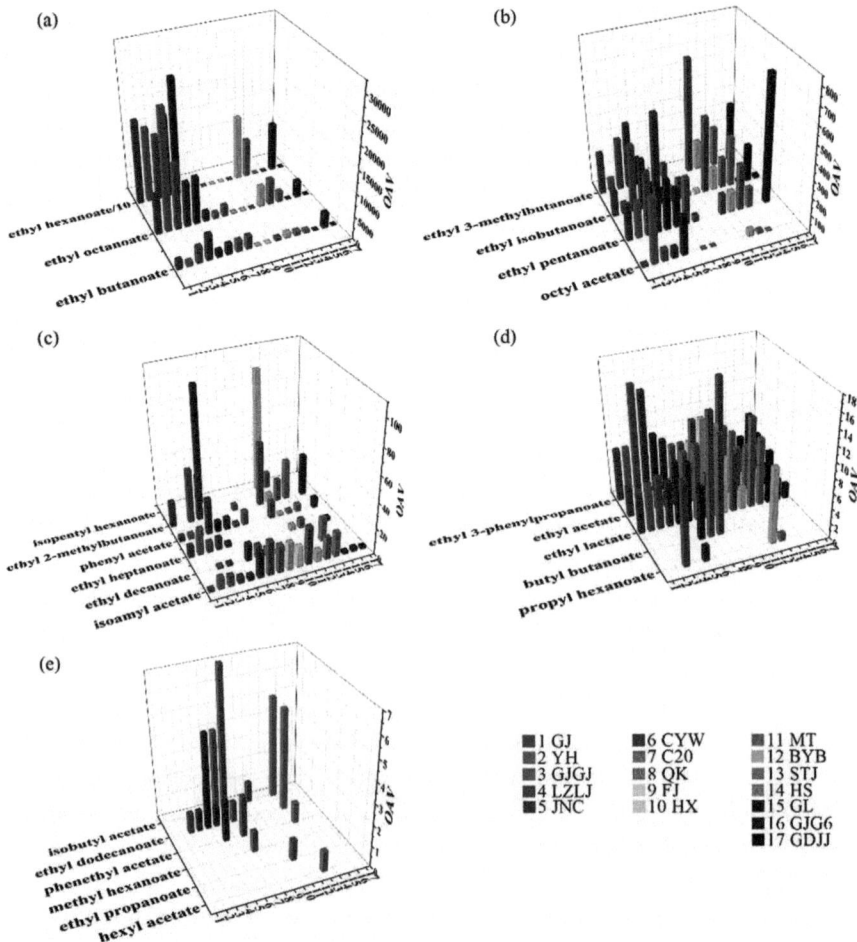

*Figure 1. Comparison of OAVs (≥1) for esters for the 17 Chinese baijiu.*

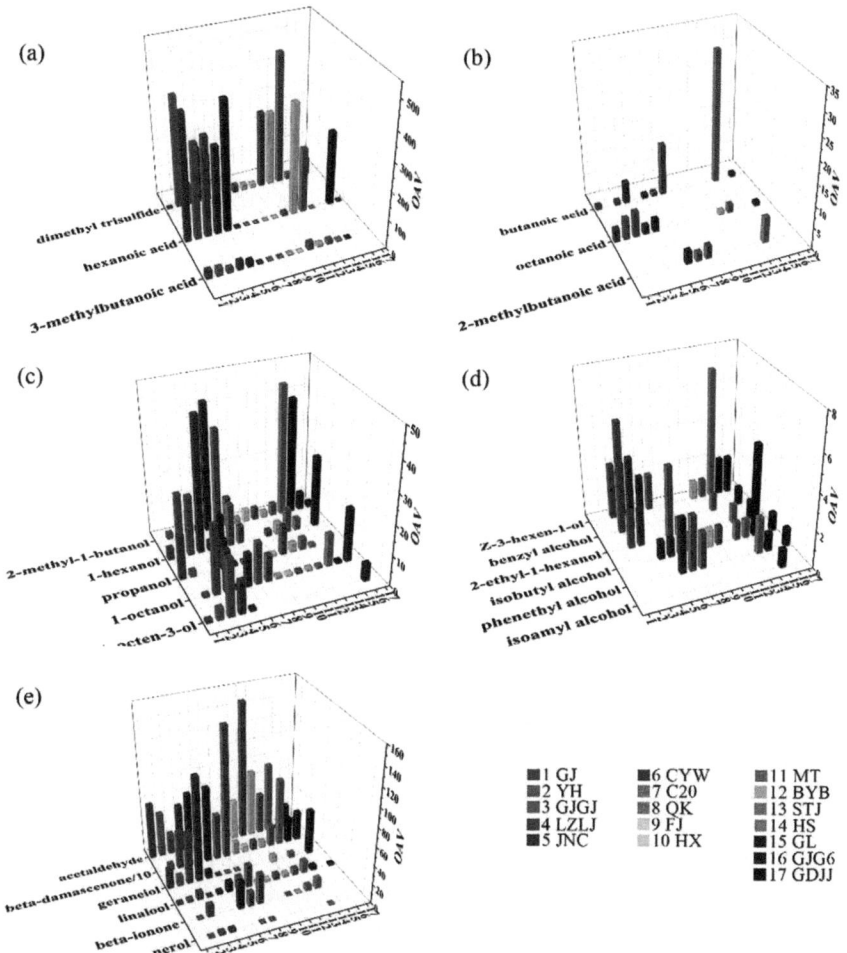

*Figure 2. Comparison of OAVs (≥1) for acids, alcohols, aldehydes, ketones, terpeniods, and sulfur-containing compounds for the 17 Chinese Baijiu.*

## Principal Component Analysis (PCA)

Figure 3 shows the PCA for all the OAVs equal to or greater than 1. Although the first two principal components only described 49% of the total aroma variances, aroma-active compounds can really differentiate the different aroma styles of Chinese baijiu (Figure 3a). Three-dimensional plots of PCA are presented in Figure 3b. Based on the factor loadings (|FL|>0.75) of the first two principal components, 15 compounds, including isobutyl acetate, phenethyl alcohol, 2-heptanol, ethyl butanoate, nerol, 1-octanol, octanoic acid, ethyl hexanoate, hexanoic acid, ethyl heptanoate, phenethyl acetate, ethyl 3-phenylpropanoate, 2-ethyl-1-hexanol, 1-hexanol, and isoamyl alcohol, were selected as the key aroma compounds to differentiate different styles of Chinese baijiu. The distance of the perpendicular projections from the samples onto the vectors of the compound suggested the differences of the samples are affected by that compound. If the perpendicular projections on the vector are further apart, it means that the compound is more important for distinguishing the samples. For example, the perpendicular projections from QK liquor and strong-aroma style baijiu groups (YH, JNC, and GJ) onto phenethyl alcohol vectors were much further than the projections onto 2-heptanol vectors, suggesting that phenethyl alcohol is more important than 2-heptanol for differentiating QK liquor from strong-aroma style baijiu.

All strong-aroma style liquors are on the far right side of Figure 3a, suggesting they have similar aroma composition. However, LZLJ and GJGJ were well separated from the group of JNC, YH, and GJ due to its levels of ethyl butanoate, 1-octanol, and 2-heptanol. This means that within the same aroma style, there were some similarities and differences in aroma composition. STJ (Te-aroma) and BYB (miscellaneous aroma) stay closer with the group of JNC, YH, and GJ, elucidating that they have a similar volatile profile. GJG6 was stand-alone in the middle on the right side. Whereas, light-aroma style baijiu (CYW, C20, and QK) with HS (Lao Bai Gan–aroma style) were on the left side near the vertical axis, which was well distinguished from the strong- and other defined aroma style by isobutyl acetate, phenethyl acetate, phenethyl alcohol, hexanoic acid, ethyl hexanoate, ethyl heptanoate, and octanoic acid. However, the aroma profile of QK had distinct differences from the other two light-aroma baijiu (CYW and C20) but were similar to HS. In addition, MT, GL, GDJJ, FJ, and HX were clustered in the middle of the figure near the horizontal axis but distanced from each other. Figure 3 demonstrated that individual baijiu had its own aroma fingerprints but can be characterized based on aroma composition with PCA analysis.

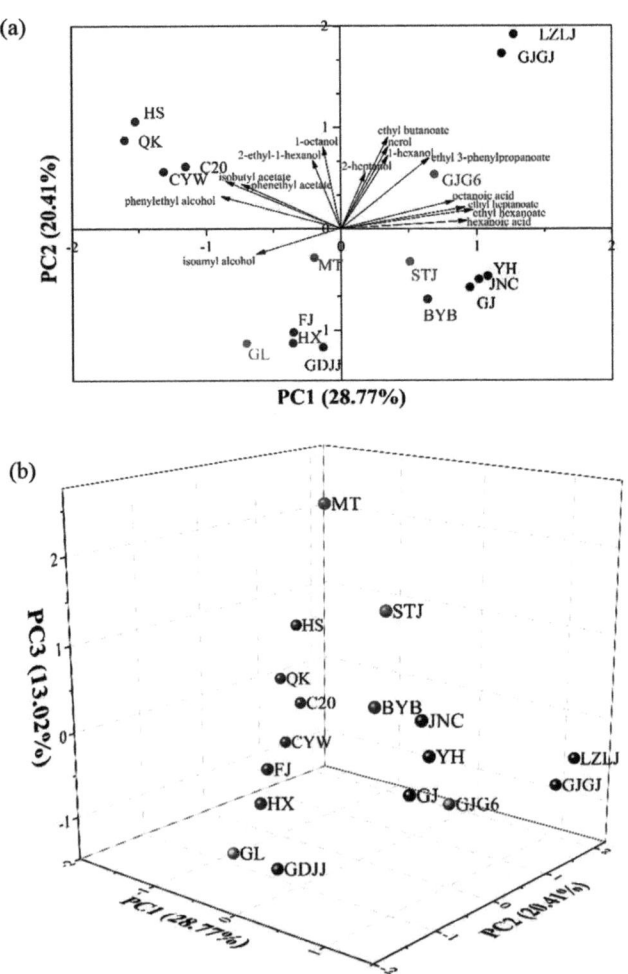

*Figure 3. PCA based on the OAVs (≥1) of aroma compounds detected in the 17 Chinese baijiu. Vector directions of major compounds in Chinese baijiu based on factor loadings (|FL| > 0.75) of the first two principal components were shown in (a), and a three-dimensional plot was shown in (b).*

# Summary

In summary, esters are the most important group of aroma compounds for the 17 Chinese baijiu with ethyl hexanoate, ethyl butanoate, and ethyl octanoate as the top three key aroma contributors. Two $C_{13}$- norisoprenoids, β-damascenone and β-ionone, are also key aroma compounds, especially for the QK liquor. Other potential important aroma compounds are hexanoic acid and dimethyl trisulfide. OAVs are very useful for determining the potency of aroma compounds in baijiu. Different aroma styles of Chinese baijiu can be well differentiated and characterized by PCA based on the OAVs of aroma compounds detected in baijiu.

# References

1. Liu, Z. F.; Yao, Z. J.; Yu, C. Q.; Zhong, Z. M. Assessing Crop Water Demand and Deficit for the Growth of Spring Highland Barley in Tibet, China. *J. Integr. Agric.* **2013**, *12*, 541–551.

2. Dang, B.; Yang, X. J.; Chen, L. S. Analysis on Status of Processing and Utilization of Highland Barley. *Grain Processing* **2009**, *3*, 29.

3. Shen, Y.; Zhang, H.; Cheng, L.; Wang, L.; Qian, H.; Qi, X. In Vitro and In Vivo Antioxidant Activity of Polyphenols Extracted from Black Highland Barley. *Food Chem.* **2016**, *194*, 1003–1012.

4. Wu, Q.; Kong, Y.; Xu, Y. Flavor Profile of Chinese Liquor is Altered by Interactions of Intrinsic and Extrinsic Microbes. *Appl. Environ. Microbiol.* **2016**, *82*, 422–430.

5. Wang, H. Y.; Xu, Y. Effect of Temperature on Microbial Composition of Starter Culture for Chinese Light Aroma Style Liquor Fermentation. *Lett. Appl. Microbiol.* **2015**, *60*, 85–91.

6. Xu, Y.; Wang, D.; Fan, W. L.; Mu, X. Q.; Chen, J. Traditional Chinese Biotechnology. In *Biotechnology in China II*; Springer: Berlin, Germany, 2009; pp 189–233.

7. Fan, W.; Qian, M. C. Identification of Aroma Compounds in Chinese 'Yanghe Daqu' Liquor by Normal Phase Chromatography Fractionation Followed by Gas Chromatography [Sol] Olfactometry. *Flavour Fragrance J.* **2006**, *21*, 333–342.

8. Wu, Q.; Chen, L.; Xu, Y. Yeast Community Associated with the Solid State Fermentation of Traditional Chinese Maotai-Flavor Liquor. *Int. J. Food Microbiol.* **2013**, *166*, 323–330.

9. Wang, C.; Shi, D.; Gong, G. Microorganisms in Daqu: A Starter Culture of Chinese Maotai-Flavor Liquor. *World J. Microbiol. Biotechnol.* **2008**, *24*, 2183–2190.

10. Wang, X.; Fan, W.; Xu, Y. Comparison on Aroma Compounds in Chinese Soy Sauce and Strong Aroma Type Liquors by Gas Chromatography–Olfactometry, Chemical Quantitative and Odor Activity Values Analysis. *Eur. Food Res. Technol.* **2014**, *239*, 813–825.

11. Gao, W.; Fan, W.; Xu, Y. Characterization of the Key Odorants in Light Aroma Type Chinese Liquor by Gas Chromatography–Olfactometry, Quantitative Measurements, Aroma Recombination, and Omission Studies. *J. Agric. Food Chem.* **2014**, *62*, 5796–5804.

12. Fan, W.; Shen, H.; Xu, Y. Quantification of Volatile Compounds in Chinese Soy Sauce Aroma Type Liquor by Stir Bar Sorptive Extraction and Gas Chromatography-Mass Spectrometry. *J. Sci. Food Agric.* **2011**, *91*, 1187–1198.

13. Yao, F.; Yi, B.; Shen, C.; Tao, F.; Liu, Y.; Lin, Z.; Xu, P. Chemical Analysis of the Chinese Liquor Luzhoulaojiao by Comprehensive Two-Dimensional Gas Chromatography/time-of-Flight Mass Spectrometry. *Sci. Rep.* **2015**, *5*, 9553.

14. Fan, W.; Qian, M. C. Characterization of Aroma Compounds of Chinese "Wuliangye" and "Jiannanchun" Liquors by Aroma Extract Dilution Analysis. *J. Agric. Food Chem.* **2006**, *54*, 2695–2704.

15. Pang, X.; Han, B.; Huang, X.; Zhang, X.; Hou, L.; Cao, M.; Gao, L.; Hu, G.; Chen, J. Effect of the Environment Microbiota on the Flavour of Light-Flavour Baijiu During Spontaneous Fermentation. *Sci. Rep.* **2018**, *8*, 3396.

16. Niu, Y.; Yao, Z.; Xiao, Q.; Xiao, Z.; Ma, N.; Zhu, J. Characterization of the Key Aroma Compounds in Different Light Aroma Type Chinese Liquors by GC-Olfactometry, GC-FPD, Quantitative Measurements, and Aroma Recombination. *Food Chem.* **2017**, *233*, 204–215.

17. Zheng, X. W.; Han, B. Z. Baijiu, Chinese liquor: History, Classification and Manufacture. *J. Ethnic Foods* **2016**, *3*, 19–25.

18. Liu, H.; Sun, B. Effect of Fermentation Processing on the Flavor of Baijiu. *J. Agric. Food Chem.* **2018**, *66*, 5425–5432.

19. Zhou, Q.; Zhang, S.; Li, Y.; Xie, C.; Li, H.; Ding, X. A Chinese Liquor Classification Method Based on Liquid Evaporation with One Unmodulated Metal Oxide Gas Sensor. *Sens. Actuators, B* **2011**, *160*, 483–489.

20. Fan, W.; Qian, M. C. Headspace Solid Phase Microextraction and Gas Chromatography−Olfactometry Dilution Analysis of Young and Aged Chinese "Yanghe Daqu" Liquors. *J. Agric. Food Chem.* **2005**, *53*, 7931–7938.

21. Wu, Q.; Zhu, W.; Wang, W.; Xu, Y. Effect of Yeast Species on the Terpenoids Profile of Chinese Light-Style Liquor. *Food Chem.* **2015**, *168*, 390–395.

22. Sun, J.; Zhao, D.; Zhang, F.; Sun, B.; Zheng, F.; Huang, M.; Sun, X.; Li, H. Joint Direct Injection and GC–MS Chemometric Approach for Chemical Profile and Sulfur Compounds of Sesame-Flavor Chinese Baijiu (Chinese Liquor). *Eur. Food Res. Technol.* **2018**, *244*, 145–160.

23. Sha, S.; Chen, S.; Qian, M.; Wang, C.; Xu, Y. Characterization of the Typical Potent Odorants in Chinese Roasted Sesame-Like Flavor Type Liquor by Headspace Solid Phase Microextraction–Aroma Extract Dilution Analysis, with Special Emphasis on Sulfur-Containing Odorants. *J. Agric. Food Chem.* **2016**, *65*, 123–131.

24. Zheng, J.; Liang, R.; Wu, C.; Zhou, R.; Liao, X. Discrimination of Different Kinds of Luzhou-Flavor Raw Liquors Based on Their Volatile Features. *Food Res. Int.* **2014**, *56*, 77–84.

25. Fan, W.; Xu, Y. Determination of Odor Thresholds of Volatile Aroma Compounds in Baijiu by a Forced-Choice Ascending Concentration Series Method of Limits [J]. *Liquor Making* **2011**, *4*, 035.

26. Burdock, G. A. *Fenaroli's Handbook of Flavor Ingredients*, Fourth ed.; CRC Press: Boca Raton, FL, 2016; pp 798–1739.

# Chapter 15

# Research Progress on Aroma Compounds in Wuliangye

Dong Zhao and Jia Zheng[*]

**Wuliangye Yibin Co. Ltd., 150# Minjiang West Road, Cuiping District, Yibin, Sichuan, China 644007**
[*]E-mail: zhengwanqi86@163.com.

Wuliangye liquor is one of the premium liquors in China. According to the development history, Wuliangye is a well-known brand (King of Chinese liquor) because making technologies such as fermentation and blending technologies for Wuliangye are widely applied in the Chinese liquor-making industry. In this review, the progress of flavor research on Wuliangye liquor is summarized, higlighting factors such as raw materials (five grains), auxiliary materials (rice husk), and starters (*baobaoqu*) that can significantly influence the flavor feature of Wuliangye liquor. The identification of aroma compounds and aroma active compounds from raw materials to the liquor was discussed as the topic issue.

## Introduction

Chinese liquor (*baijiu*) is a distilled spirit with an annual production of 11,981,000 kL in 2017 (*1*). Based on annual production levels, the strong aroma liquor (Chinese named *nongxiang*) is the most prominent liquor in China contributing to more than 80% of total production (9,855,000 kL in 2017). Wuliangye liquor is one of the most famous liquor brands in the world, and it is also considered to be the most famous *nongxiang* liquor in China due to its large consumption levels as well as its specific aroma, taste features, and the technology used to produce it (*2, 3*). In this history of China national liquor tasting conference in 1952, 1963, 1979, 1984, and 1989, respectively, Wuliangye liquor was awarded the distinction of "National Famous Liquor" in 1963, 1979, 1984, and 1989 at China's National Liquor Tasting Conference (*2*). Its annual sales reached 30.186 billion Chinese Yuan and annual liquor consumption reached 180,007 tons in 2017 ranking it as the most popular *nongxiang* liquor. In 2015, its fermentation technology was listed in the Intangible Cultural Heritage in China. In addition, the old fermentation pits (known as *laojiao*, the oldest pits were built in 1368 A.D.) and related liquor-making workshops have been listed in the National Industry Heritage Lists of China in 2018.

Wuliangye liquor is fermented from five grains (Figure 1A), steamed rice husk (Figure 1B), and the starter (*baobaoqu* [Figure 1C]) under the anaerobic condition of the fermentor (*laojiao* [Figure

1D]). The core ideas for Wuliangye making are recycling fermentation and transferring fermentation. The former is the main technical process and the latter is the strategy for quality stabilization.

The production process for Wuliangye liquor is a multi-step process described as follows. In the main fermentation process (recycling process) (Figure 2A), the fresh five grains are milled and mixed with the fermented grains from the previous batch and the steamed rice husk. The mixed grains are then layered in the distillation apparatus (known as *zengtong*). In the distillation process, raw liquors including head, heart, and tail fraction are separately collected depending on their sensory judgment. After distillation is finished, the distilled mixed grains are further cooled and mixed with *baobaoqu* powder. The mixtures are then placed in a *laojiao* to ferment for 70 days. The fermented grains will then undergo the same procedure again once the fermentation is finished. In order to keep the mass balance, the top layer of fermented grains (Figure 2B) is only mixed with steamed rice husk for liquor distillation and is then discarded after liquor distillation. The heart raw liquor is then selected for aging in pottery jars (500–1000 L) for less than three years, blended, and bottled accordingly.

The transferring fermentation was first used in the production of Wuliangye to stabilize the liquor quality (Figure 2C). Whole fermented grains in *laojiao* are divided into 10–11 parts, and each part weighs 1000 kg filling the *zengtong*. The mixtures, as previously described for fermentation, are usually moved from one *laojiao* to another one nearby. As a result, the fermented grains from different *laojiao* in the same department can affect each other. After long-term uninterrupted fermentation, aroma compounds could be accumulated into the fermented grains. Aroma compounds in fresh raw liquors could come from various sources, such as the fresh five grains, the steamed rice husk, the *baobaoqu*, or the fermented grains.

This review describes the current understanding of the effects of the raw materials, auxiliary materials, and starters on liquor aroma. Studies identifying compounds and aroma active compounds that contribute to the aroma of Wuliangye liquor, usually determined by gas chromatography (GC) and gas chromotography-olfactometry (GC-O), are summarized as well.

Figure 1. (A) Five grains, (B) steamed rice husk, (C) baobaoqu, and (D) laojiao for Wuliangye liquor fermentation.

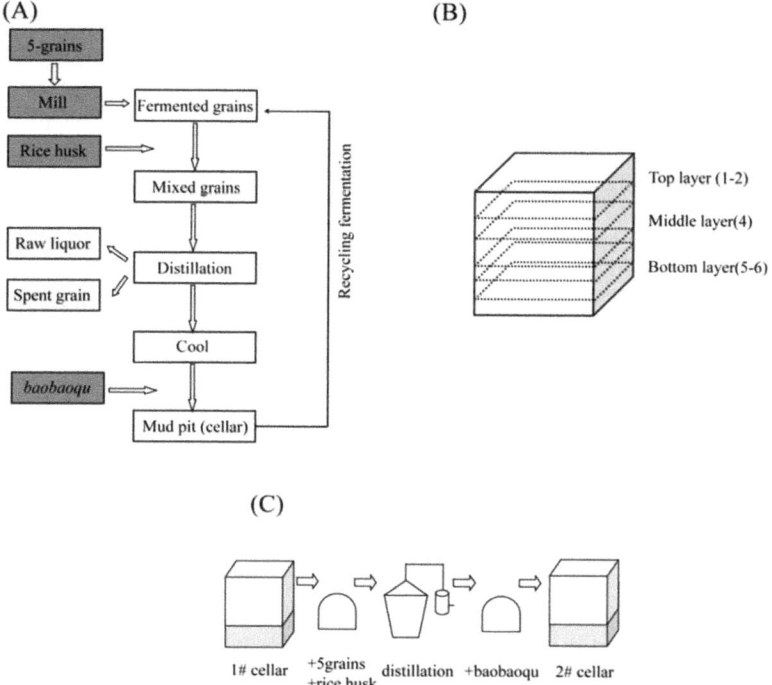

Figure 2. (A) Production process flowchart of Wuliangye liquor, (B) layer distribution, and (C) flowchart of transferring fermentation.

## Aroma Compounds in Five Grains

The raw material is one of most important aroma sources for Chinese liquor. For the *nongxiang* liquor, the amount of raw material added to the liquor production mixture ranges from 20%–25% (4). Normally, the cooking process of raw cereals is always combined with the liquor distillation in *nongxiang* liquor, which is also called concomitant of distillation and cooking.

Cereals are the basis for liquor fermentation. Five grains are used in the fermentation process of Wuliangye liquor: sorghum, rice, glutinous rice, wheat, and corn (Figure 1A). The combination of these five grains forms the specific formula fermentation, which is also called "Chen's mystery formula." Raw materials provide abundant nutrients, such as sugar and protein, for the metabolism of microorganisms in the liquor fermentation process, and these cereals also contain a large number of aroma compounds. In the liquor distillation process, the free form aroma compounds are usually concentrated in raw liquors. Numerous studies have been conducted to identify the free form aroma compounds of distilled five grains. Aroma compounds in sorghum (5), rice (6), glutinous rice (6), wheat (7), and corn (8) were identified by simultaneous distillation extraction and head space solid phase microextraction (HS-SPME). However, the identification of each compound is limited when comparing the mass spectrum of each compound with the standard mass spectrum in the NIST database.

In order to better understand potential aroma active compounds, GC-O was used to determine free form aroma active compounds in the five grains. Wu *et al.* applied solid phase microextraction (SPME) GC-O to identify aroma active compounds in five different grains (sorghum, rice, glutinous rice, wheat, and corn); the results showed that ethyl hexanoate, 2-phenylethyl alcohol, nonanal, (*E*)-octenal, and 2-pentylfuran were the most important aroma compounds (9), while cooked sorghum had the strongest flavor intensity and esters, aldehydes, and ketones were main compounds.

Although various potential aroma active compounds in each grain were detected above, the pretreatment method for the five grains in previous study did not coincide with the commercial scale. In this way, the distillates from the different distilling time points of the five grains in the commercial scale were collected, and aroma active compounds were determined using aroma extract dilution analysis (AEDA) in our latest study (10). The results showed that a total of 49 aroma compounds were identified as having flavor dilution (FD) values greater than 3. The most important aroma active compounds were 2-methylbutanal, hexanal, (E)-2-heptenal, benzeneacetaldehyde, guaiacol, o-cresol, p-cresol, ethyl hexanoate, dimethyl trisulfide, and β-damascenone (FD of at least 81, see Table 1), with the odor descriptions of green and grassy, pungent, grassy and fatty, nutty, apricot, medicinal, pineapple-like, rotten cabbage, and tobacco notes.

Meanwhile, in the fermentation process, the aroma precursors in the five grains could be easily hydrolyzed by enzymes, organic acids, and other factors. Aroma precursors in different grains were extracted by solid phase extraction ($C_{18}$ column), hydrolyzed by HCl, and quantified by GC-MS, sequentially (11). According to the heat map of aroma precursors, diethyl succinate, hexanoic acid, nonanal, and decanal were found to be the main compounds in sorghum; benzeneacetaldehyde, tetramethylpyrazine, and dehydro-β-ionone were the main compounds in corn; ethyl hexanoate, 2-phenylethyl alcohol and phenol were the main compounds in wheat; octanoic acid, nonanoic acid, and (E)-2-octenoic acid were the main compounds in rice; 1-octen-3-ol, (E,E)-2,4-decadienal, and cis-linaloloxide were the main compounds in glutinous rice. However, very little is known about the degree of hydrolyzing and the contribution of aroma precursors to the overall aroma profile in commercial scale.

**Table 1. Aroma Active Compounds in Five Grains, *Baobaoqu*, Rice Husk, and Wuliangye Liquor**

| Materials | Aroma Active Compounds |
| --- | --- |
| five grains (10) | 3-methylbutanal, hexanal, (E)-2-heptenal, ethyl hexanoate, ethyl octanoate, 2-phenylethyl acetate, 4-methylguaiacol, benzeneacetaldehyde, guaiacol, o-cresol, p-cresol, dimethyl trisulfide, pyrrole, β-damascone, furfural, 3,5-dimethyl-2-butylpyrazine |
| baobaoqu (3) | butanal, 3-methylbutanal, dimethyl disulfide, heptanal, (E)-2-octenal, 3-methylbutanol, 1-octen-3-ol, 2,6-dimethylpyrazine, trimethylpyrazine, tetramethylpyrazine, 2-ethnyl-6-methylpyrazine, acetic acid, furfural, benzenacetaldehyde, guaiacol, 2-phenylethyl alcohol, phenol, γ-pantolactone, 4-vinylguaiacol |
| rice husk | butanal, 2-propenal, 2-methylbutanal, 3-methylbutanal, diacetyl, pentanal, hexanal, heptanal, octanal, nonanal, decanal, (E)-2-octenal, 3-octen-2-one, furfural, 3,7-dimethyl-1,6-octadien-3-ol, 3-methyl-2(5H)-furanone, isophorone, γ-pentalactone, β-butyrolactone, benzeneacetaldehyde, naphthalene, β-damascenone, 2-methylnapthalene, 1-methylnapthalene, β-ionone, (Z)-whiskey lactone, phenol, p-cresol, 4-vinylguaiacol |
| liquor (26) | 2-methylpropanal, 1,1-diethoxy-3-methylbutane, butanol, isoamyl alcohol, ethyl acetate, ethyl propanoate, ethyl butanoate, ethyl hexanoate, ethyl isovalerate, diethyl succinate, hexyl acetate, ethyl octanoate, ethyl phenylacetate, 3-methylbutanoic acid, 2-methylbutanoic acid, butanoic acid, pentanoic acid, hexanoic acid, octanoic acid, dimethyl trisulfide, 2-butylfuran, dihydro-2(3H)-furanone, 4-methylguaiacol, benzeneacetaldehyde, phenol, γ-nonalactone, 4-ethylguaiacol |

## Aroma Compounds in *Baobaoqu*

The brick-shaped starters (*daqu*) could provide crude enzymes and microorganisms (mold, bacterium, and yeasts) for fermentation of Chinese liquor (*12–15*). Aroma compounds, or its precursors produced during the starter-making period, can also influence the quality of the liquor. The *daqu* for Chinese liquor is always prepared using pure raw wheat or raw wheat mixed with other grains or beans, and their classification is based on their highest incubation temperature, including high temperature (60–70 °C), middle to high temperature (50–60 °C), and medium temperature (40–50 °C). The incubation strains for *daqu* usually come from the wild microorganisms in the surrounding environment. Thus, the quality and aroma profile of each kind of starter are also influenced by local geological parameters because the microbial community in the air changes from place to place.

*Baobaoqu* (Figure 1B), prepared using pure wheat, is the necessary medium to high temperature starter for Wuliangye liquor. The incubation process of *baobaoqu* was introduced in previous studies (*3*). Initially, the raw pure wheat particle is ground and adjusted to a moisture content of about 38%. The wheat mash is then pressed into a specific brick-shaped rough starter with a big bulge on one side. The rough *baobaoqu* is then placed into an incubation room for about 30 days. During this period, the rough *baobaoqu* experiences three major stages: the temperature-rise phase, the high-temperature phase, and the temperature-fall phase. In the temperature-rise phase, the body temperature of *baobaoqu* increases from an ambient temperature to its highest temperature (58–60 °C). In the high-temperature phase, the body temperature will be kept at the highest temperature for one week. In the temperature-fall phase, the body temperature will slowly decrease until the end of the incubation process.

In 2006, HS-SPME was first applied to determine the aroma compounds in *baobaoqu* (*16*), and a total of 68 volatile compounds were identified including 30 pyrazines and other similar chemical substances in liquor (i.e, acetic acid, hexanoic acid, and ethyl hexanoate). However, the result of this report was only based on the comparison between compound mass spectrum and NIST standard mass spectrum. Recently, in order to get a more credible result, a detailed identification of aroma compounds was conducted in which 19 aldehydes, 40 esters, 8 ketones, 15 alcohols, 22 aromatic compounds, 12 pyrazines, 7 furans, 3 acids, 3 lactones, 6 terpenes, and 2 other compounds were determined on the basis of their linear retention indices (RIs), pure chemical standards, and a total of 26 compounds were identified in middle to high temperature *daqu* for the first time (*17*).

However, there has not been enough published information on aroma active compounds in *baobaoqu*. Zheng et al. (*3*) systematically analyzed that the changing aroma profile of *baobaoqu* in the fermentation process using flavor chemistry analyses including sensory analysis, Osme analysis, quantification, and odor active value. The sensory property of starters significantly changed from the green, fatty, and raw bready notes on Day 1 to strong moldy and nutty notes on Day 27. Acetic acid, methional, ethyl acetate, ethyl 2-phenylacetate, 2-phenylethyl alcohol, ethyl 2-phenylacetate, and 4-vinylguaiacol are potential aroma active compounds based on their Osme values (Table 1). On the basis of quantification analysis, (*E*)-2-hexenal, ethyl acetate, isoamyl alcohol, 2-phenylethyl alcohol, 2-pentylfuran, acetic acid, and lactic acid represented the highest concentration at Day 7, and (*E,E*)-2,4-hexadienal, 2,6-dimethylpyrazine, trimethylpyrazine, and tetramethylpyrazine presented the highest level at the temperature-fall phase.

## Aroma Compounds in Rice Husk

The rice husk is the necessary auxiliary material for Chinese liquor; for *nongxiang* liquor, it accounts for 18%–25% of total weight of raw materials (*18*). Traditionally, prior to the liquor distillation, the rice husk is mixed with the fermented grains in order to improve the specific surface area and increase the rectification of aroma compounds in the fermented grains during liquor distillation. However, the raw rice husk has to be steamed prior to the liquor distillation; otherwise, it will produce unpleasant odor notes including pungent, dusty, and earthy notes after long-term storage and transportation. The steaming time is traditionally dependent on the operator's experience, and there have only been two papers focused on rice husk flavor.

In order to better understand the off-flavor compounds in raw rice husk and its variation during steaming, a series of experiments on identification and quantification of aroma compounds in raw rice husk was conducted. A total of 180 compounds were identified: 28 aldehydes, 38 ketones, 36 aromatic compounds, 21 alcohols, 12 furans, 12 terpenes, 9 lactones and furanones, 6 pyrazines, 4 sulfurs, 3 esters, and 11 other compounds (*19*). In addition, variations in aroma compounds were further detected in collections with different distillation time points (*20*). In Figures 3 and 4, concentrations of total aroma compounds, alcohols, pyrazines, and terpenoids decreased with distillation; the concentration of aldehydes increased in the initial 5 min and decreased after that; and concentrations of lactones, furfural, phytone, and vanillin increased with the distillation. Additionally, *trans*-1,10-dimethyl-*trans*-9-decalol was confirmed as one of earthy note compounds in steamed rice husk (*21*). Recently, results of aroma active compounds in rice husk suggested that butanal, 2-methylbutanal, hexanal, heptanal, octanal, nonanal, 1-octen-3-ol, 3,7-dimethyl-1,6-octadien-3-ol, 3-methyl-2(5H)-furanone, $\beta$-butyrolactone, $\beta$-damascenone, and 1-methylnapthalene were potent aroma active compounds based on their FD values (more than 1024, Table 1).

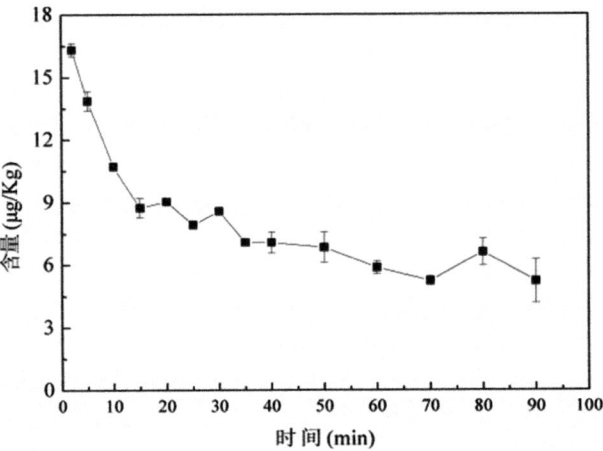

*Figure 3. Variation of total concentration of aroma compounds in raw rice husk distillation process. (Reproduced with permission from reference (20). Copyright 2017 Food and Fermentation Industries).*

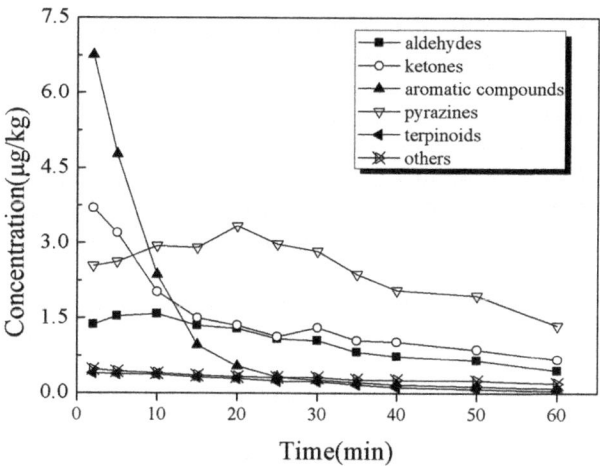

*Figure 4. Variation of different types of aroma compounds in the raw rice husk distillation process. (Reproduced with permission from reference (20). Copyright 2017 Food and Fermentation Industries).*

## Aroma Compounds in Wuliangye Liquor

The classification of Chinese liquors is primarily based on their aroma profile. Moutai liquor and Wuliangye liquor, two of the better known liquors in China, are classified as soy–sauce aroma liquor and *nongxiang* liquor, respectively. With the development of chromatography, more complex analytical instruments such as GC-O, GC-MS, multidimensional GC, and time-of-flight MS have been used to identify aroma compounds in Chinese liquor (*22–24*).

To date, GC with a flame ion detector is the main instrument for quantifying the main aroma compounds in Wuliangye liquor and other *nongxiang* liquors; these main compounds included ethyl acetate, ethyl butanoate, ethyl hexanoate, ethyl lactate, isoamyl alcohol, 1-propanol, and isobutanol among others. Fan and Qian (*25*) first compared aroma active compounds in Jiannanchun and Wuliangye liquors. According to the AEDA results, ethyl butanoate, ethyl pentanoate, ethyl hexanoate, ethyl octanoate, butyl hexanoate, ethyl 3-methylbutanoate, hexanoic acid, and 1,1-diethoxy-3-methylbutane were the most important aroma compounds with FD values greater than 1024 and aroma notes of fruity, floral, apple, pineapple, and sweaty. Several pyrazines including 2,5-dimethyl-3-ethylpyrzine, 2-ethyl-6-methylpyrazine, 2,6-dimethylpyrazine, 2,3,5-trimethylpyrazine, and 3,5-dimethyl-2-pentylpyrazine were also potent aroma active compounds.

Recently, the difference in aroma profiles between low-proof (35% and 39% vol) and high-proof (45% and 52% vol) Wuliangye liquors was analyzed using Osme analysis (*26*). It was found that main aroma compounds (such as ethyl hexanoate and ethyl butanoate) with the fruity notes presented similar odor intensities between low-proof and high-proof liquors, while the acids and aromatic compounds presented different ranking sequences, especially for ethyl 2-phenylacetate, 2-phenylethyl alcohol, and 4-methylguaiacol with significant sweet, floral, and woody notes.

## Summary

In summary, the research discussed in this review covered the identification of aroma compounds and potential aroma active compounds. The studies provide information on each factor (five grains, rice husk, and *baobaoqu*) that can significantly influence liquor quality. For example, the presence of pyrazines in *baobaoqu* is associated with roasted, nutty, and chocolate notes; aldehydes especially straight–chain saturated aldehydes with green and grassy notes and geosmin with earthy note were main off-flavor compounds for raw rice husk; and various lactones and vanillin in steamed rice husk are associated with sweet, coconut-like, and vanilla notes. However, aroma identification only represents the first step in Wuliangye flavor research. The aroma profile of liquor is significantly different depending on the raw materials (Table 1) and the combination of the five grains and the rice husk is an important condition for liquor fermentation. Until now, the aroma active compounds in other key components of liquor-making (such as cellar mud, fermented grains, and raw liquor) are still unknown. Therefore, running a series of detailed modern flavor chemistry experiments such as aroma active compound identification, strict quantification, and aroma recombination and omission tests on Wuliangye liquor (both high- and low-quality) and related aroma sources are topical tasks. Solving these problems would allow objective characterization of the quality of Wuliangye liquor, link liquor with its aroma sources, and provide more reference for improving the liquor quality.

## Acknowledgments

We would like to thank Kangzhuo Yang, Jianmin Zhang, Xia Zhang, Qingchun Luo, and Fang Liu for the supplemental materials.

## References

1. Jun, L. In 2017, The Output of Chinese Liquor was 11.98 Million k=Kiloliters, An Increase of 6.9%. *Chinese Liquor Network* [Online], Jan 22, 2018. http://www.baijw.com/html/xwpd/xysj/23856.html (accessed Mar 14, 2019).

2. Shen, Y. F. *Manual of Chinese Liquor Manufactures Technology*; Light Industry Publishing House of China: Beijing, 2007.

3. Zheng, J.; Zhao, D.; Peng, Z. F.; Yang, K. Z.; Zhang, Q.; Zhang, Y. K. Variation of Aroma Profile in Fermentation Process of Wuliangye *Baobaoqu* Starter. *Food Res. Inter.* **2018**, *114*, 64–71.

4. Lai, D. Y.; Ding, Z. X. Study on the Change Rules of Seven Factors for Pit Entry and Their Correlations (VII): Pit Entry Fermenting Grains. *Liquor-Making Sci. Technol.* **2013** (5), 30–32.

5. Lian, S. C.; Xie, Z. M.; Ye, H. X.; Li, Y. H.; Wang, F.; Wu, Y. H. Research on the Flavouring Compositions of Sorghum. *Liquor-Making Sci. Technol.* **2012** (3), 40–42.

6. Peng, Z. F.; Li, Y. H.; Lian, S. C.; Xie, Z. M.; Ye, H. X. Research on the Flavoring Compositions of Rice and Sticky Rice by SDE and SPME. *Liquor-Making Sci. Technol.* **2014** (12), 42–46.

7. Ye, H. X.; Lian, S. C.; Xie, Z. M.; Li, Y. H.; Liao, Q. J. Analysis of the Volatile Flavoring Components of Wheat in Cooking and Steaming. *Liquor-Making Sci. Technol.* **2014** (1), 38–42.

8. Xie, Z. M.; Lian, S. C.; Ye, H. X.; Li, Y. H.; Liao, Q. J.; Wang, X. Q. Research on the Flavouring Compositions of Maize. *Liquor-Making Sci. Technol.* **2012** (9), 68–71.

9. Wu, Y. R.; Liu, S. Y.; Fan, X. L.; Yang, J. W.; Jiang, W.; Wang, D. L.; Li, N. Analysis of Aroma Components of Five Different Cooked Grains Used for Chinese Liquor Production by GC-O-MS. *Food Sci.* **2016**, *37*, 94–98.

10. Peng, Z. F.; Zhao, D.; Zheng, J.; Peng, Z. Y.; Yang, K. Z.; Zhang, J. M.; Lv, X. L. Comparison of Odor-Active Compounds in Distillates of Five Grains Between First Time and Second Time Distillation Using AEDA. *Food Ferment. Ind.* **2017**, *43*, 1–8.

11. Zhu, W. A.; Wu, Q.; Li, J. M.; Xu, Y. Isolation and Analysis of Bound Aroma Compounds in Different Raw Brewing Materials. *J. Food Sci. Biotechnol.* **2015**, *34*, 456–462.

12. Jin, G. Y.; Zhu, Y.; Xu, Y. Mystery Behind Chinese Liquor Fermentation. *Trends Food Sci. Technol.* **2017**, *63*, 18–28.

13. Zheng, X. W.; Tabrizi, M. R.; Nout, M. J. R.; Han, B. Z. *Daqu*—A Traditional Chinese Liquor Fermentation Starter. *J. Inst. Brew.* **2012**, *117*, 82–90.

14. Zheng, X.; Yan, Z.; Nout, M. J. R.; Boekhout, T.; Han, B. Z.; Zwietering, M. H.; Smid, E. J. Characterization of the Microbial Community in Different Types of *Daqu* Samples as Revealed by 16S rRNA and 26S rRNA Gene Clone Libraries. *World J. Microbiol. Biotechnol.* **2015**, *31*, 199–208.

15. Wu, C. D.; Deng, J. C.; He, G. Q.; Zhou, R. Q. Metaproteomic Characterization of Daqu, a Fermentation Starter Culture of Chinese Liquor. *J. Prot. Bioinfor.* **2016**, *9*, 49–52.

16. Zhao, D.; Li, Y. H.; Xiang, S. Q.; Wu, Y. H.; Wang, F.; Lan, S. R. Study on Flavouring Components of Daqu by Headspace Solid-Phase Microextraction Gas Chromatography-Mass Spectrometry. *Liquor-Making Sci. Technol.* **2006** (5), 92–94.

17. Yang, K. Z.; Zhao, D.; Zheng, J.; Zhang, J. M.; Liu, F. Identification of Volatile Compounds in Fermentation Process of Wuliangye Wrapped Starter. *Liquor-Making Sci. Technol.* [Online] Nov 27, 2018, http://kns.cnki.net/kcms/detail/52.1051.TS.20181127.1556.003.html (accessed Mar 18, 2019).

18. Yin, X. M.; Zhang, S. Y.; Ao, Z. H.; Sui, L. Y.; Liu, Y. H.; Wang, F.; He, J. Investigation of the Effects of Different Use Level of Bran on the Fermentation of Non-Flavor Liquor by Fuzzy Mathematics Evaluation. *Liquor-Making Sci. Technol.* **2013** (3), 50–53.

19. Zhao, D.; Zheng, J.; Peng, Z. Y.; Yang, K. Z.; Zhang, J. M.; Lv, X. L.; Yang, R. Identification of Volatile Compounds in Rice Husk by Head–Space Solid-Phase Microextraction, Liquid-Liquid Extraction and Flavor Fractionation. *Liquor-Making Sci. Technol.* **2016** (12), 31–39.

20. Zheng, J.; Zhao, D.; Peng, Z. Y.; Yang, K. Z.; Zhang, J. M.; Lv, X. L. Variation of Aroma Compounds in Rice Husk During Distillation Process. *Food Ferment. Ind.* **2017**, *43*, 8–15.

21. Xie, Z. M.; Lian, S. C.; Li, Y. H.; Ye, H. X. Research on Earthy-Musty Substances in Liquor-Making Bran. *Liquor-Making Sci. Technol.* **2016** (1), 47–50.

22. Zhu, S. K.; Lu, S.; Ji, K. L.; Guo, K. L.; Li, Y. L.; Wu, C. Y.; Xu, G. W. Characterization of Flavor Compounds in Chinese Liquor Moutai by Comprehensive Two-Dimensional Gas Chromatography/Time-of-Flight Mass Spectrometry. *Anal. Chim. Acta* **2007**, *597*, 340–348.

23. Fan, W. L.; Qian, M. C. Identification of Aroma Compounds in Chinese "Yanghe Daqu" Liquor by Normal Phase Chromatography Fractionation Followed by Gas Chromatography Olfactometry. *Flav. Fragr. J.* **2006**, *21*, 333–342.

24. Fan, W. L.; Tang, K.; Zhang, Y. H. Characterization of Pyrazines in Some Chinese Liquors and Their Approximate Concentrations. *J. Agric. Food Chem.* **2007**, *24*, 9956–9962.

25. Fan, W. L.; Qian, M. C. Characterization of Aroma Compounds of Chinese "Wuliangye" and "Jiannanchun" Liquors by Aroma Extract Dilution Analysis. *J. Agric. Food Chem.* **2006**, *54*, 2695–2704.

26. Peng, Z. F.; Zhao, D.; Zheng, J.; Yuan, J. B.; Cao, H. Y.; Peng, Z. Y. Comparison of Flavor Characteristics Between Low-Alcohol and High-Alcohol Wuliangye by Using Modern Flavor Chemistry Technology. *Liquor-Making Sci. Technol.* **2018** (12), 17–22.

# Chapter 16

# Aroma Profile of *Folium isatidis* Leaf as a Raw Material of Making *Bingqu* for *Chixiang* Aroma- and Flavor-Type Baijiu

Zhanglan He,[1] Wenlai Fan,[*,1] Yan Xu,[1] Songgui He,[2] and Xinyi Liu[2]

[1]State Key Laboratory of Food Science & Technology, Key Laboratory of Industrial Biotechnology of Ministry of Education & School of Biotechnology, Jiangnan University, 1800 Lihu Blvd., Wuxi, Jiangsu, China, 214122

[2]Center of Technology in Guangdong Jiujiang Distillery Co., Ltd., Foshan, Guangdong, China, 528203

[*]E-mail: wenlai.fan@163.com.

*Folium isatidis* leaf is one of the main raw materials used in the production of chixiang *bingqu*, a type of *xiaoqu* for *chixiang* aroma-type *baijiu* (Chinese liquor). To better understand the aroma profile of *F. isatidis*, aroma compounds of the leaf were identified using gas chromatography-olfactometry (GC–O) coupled with gas chromatography-mass spectrometry (GC–MS) following liquid-liquid extraction (LLE). A total of 45 aroma compounds in *F. isatidis* were detected using the DB-FFAP column. On the basis of Osme values, the most important aroma compounds (Osme value of at least 3.5) were 4-vinylguaiacol (smoky), guaiacol (phenolic, smoky, and herbal), acetic acid (sour), (*E*)-cinnamaldehyde (cinnamon), 2-phenylethanol (floral and rose), phenol (phenolic, plastic, and medicinal), and L-(−)-borneol (camphoreous). Terpenoids are the most diverse class of compounds and contribute to herbal, mentholic, floral, sweet, camphoreous, and fruity odors. The limits of detection (LOD) of the quantitative method are greater than 0.0873 μg/L ([−]-*trans*-pinocarveol). Linearity was satisfied in all cases. Quantification studies showed that benzoic acid is the most abundant (2.46 mg/g [DW]) constituent, followed by benzyl alcohol, 4-vinylguaiacol, acetic acid, 2-phenylethanol, and 3,4-dimethoxyacetophenone. The results of this study might provide useful information for flavor research involving *chixiang bingqu* and *baijiu*.

# Introduction

The *bingqu* (a kind of *xiaoqu*) used for producing the *chixiang* aroma- and flavor-type *baijiu* (liquor) is the key ingredient in the production of ethanol and flavor during liquor-making, as is mentioned in the old-saying "*Qu* is the spirit of liquor." The main raw materials required for making *bingqu* are rice, soybeans, and Chinese herbs (*1*).

The *xiaoqu* pellet is the starter material required for making *chixiang bingqu*. The raw materials required for making the *xiaoqu* pellet include rice, Chinese herbs, and *xiaoqu* pellets (Chinese name *qumu*) from a previous lot. The raw materials are milled, sifted, and then mixed well with water (rice: water is 1:0.45–0.5, w/w). The mixture is then pressed into a rectangular cake-like mould and fermented for five days at 30–35 °C in a dedicated room. After fermentation, the *xiaoqu* pellet is dried and crushed into powder before use.

The *chixiang bingqu* is made from cooked rice, soybeans (20% by weight of rice), Chinese herbs (0.9% by weight), and *xiaoqu* pellet powder (2.2% by weight) (Figure 1). The mixture is pressed into rectangular cake-like shapes (27 × 21 × 3 cm). The cakes are then fermented in a dedicated room at 30–35 °C for seven days (*2, 3*).

Raw materials of chixiang bingqu

*Figure 1. Raw materials used in the production of bingqu and different stages in the production of bingqu.*

*Folium isatidis* and *Cinnamomum cassia* Presl leaves are the chief Chinese herbs used in the preparation of *chixiang bingqu*. Previous studies have characterized the volatile compounds of *C. cassia* Presl leaf. A total of 41 aroma compounds were identified from *C. cassia* Presl leaf using liquid-liquid extraction (LLE) and gas chromatography-olfactometry (GC–O) coupled with gas chromatography-mass spectrometry (GC–MS). (*E*)-cinnamaldehyde was identified as the most important aroma compound (Osme value of 4.2) and was the most abundant (3.71 mg/g [DW]). This was followed by acetic acid with an Osme vale of 4.0 and then by 4-vinylguaiacol, acetophenone, guaiacol, 1,2-dimethoxybenzene, and $\beta$-ionone, all of which had Osme values of at least 3.0. Among them, 1,2-dimethoxybenzene was detected for the first time in *C. cassia* Presl leaf and had an intense flowery aroma (*4*).

*F. isatidis* is the dried leaf of *Isatis indigotica* Fort and is a well-known and valuable traditional Chinese medicinal herb. It is used in the treatment of sore throat and redness of the skin and as an antipyretic (*5*). Most studies on *F. isatidis* have focused on its antibacterial, antiviral, antipyretic, anti-inflammatory, and immuno-stimulatory properties (*6*), but there has yet to be a study on its aroma profile.

GC–O coupled with GC–MS is widely used to study the aroma profiles of multiple alcoholic beverages (7–9), food products (10), and the raw materials used in the production of alcohol and food (11, 12). In this study, we aimed to identify the important aroma compounds in *F. isatidis* using GC–O and to quantify these aroma compounds. The findings of this research can help us to better understand the aroma chemistry of *F. isatidis* and its contribution to the flavor of *chixiang bingqu*.

## Material and Methods

### Chemicals

All chemicals were of analytical reagent grade with at least 95% purity. Eucalyptol, *m*-cymene, (–)-fenchone, D-camphor, linalool, 4-terpinenol, (–)-*trans*-pinocarveol, (–)-*α*-terpineol, verbenol, L-(–)-borneol, verbenone, myrtenol, caryophyllene oxide, *β*-eudesmol, benzaldehyde, methyl benzoate, benzyl alcohol, 2-phenylethanol, (*E*)-cinnamaldehyde, 2-methoxybenzyl alcohol, cinnamyl alcohol, dihydroactinidiolide, benzoic acid, 3,4-dimethoxyacetophenone, 4-hydroxybenzaldehyde, guaiacol, phenol, methyl eugenol, 4-vinylguaiacol, syringol, isobutanoic acid, butanoic acid, hexanoic acid, heptanoic acid, 1-methyl-(1*H*)-pyrrole, methyl heptanone, (*Z*)-3-hexen-1-ol, and methyl hexanoate (internal standard [IS]) were purchased from Sigma–Aldrich Co., Ltd. (Shanghai, China). Globulol, *p*-cymenol, and acetic acid were obtained from J&K Scientific Ltd. (Shanghai, China).

Diethyl ether, anhydrous sodium sulfate, and sodium chloride were of analysis-grade and purchased from Shanghai Pharmaceuticals Holding Co., Ltd (Shanghai, China). Ethanol was obtained from Sigma–Aldrich Co., Ltd. (Shanghai, China). Water used was obtained from a Milli-Q purification system (Millipore, Bedford, MA, USA). Diethyl ether was redistilled, and Milli-Q water was boiled for 5 min and cooled to 20 °C before use.

### Samples

*F. isatidis* leaves were provided by Guangdong Jiujiang Distillery Co., Ltd. (Foshan, China).

### LLE for *F. isatidis*

*F. isatidis* (8 g) was ground with liquid nitrogen in a ceramic mortar and then soaked in 150 mL of Milli-Q water overnight. The solution was filtered using filter paper (Whatman, No.102, 18 cm in diameter, Shanghai, China) and the filtrate was then saturated with sodium chloride and extracted three times with 20 mL of redistilled diethyl ether in a separatory funnel. The combined extracts were centrifuged at 10,000 rpm for 5 min. The supernatant was dried over anhydrous sodium sulfate overnight, concentrated to 250 µL under a gentle stream of nitrogen, and then stored at –20 °C until GC analysis.

### GC–O Analysis

GC–O analysis was performed using an Agilent 6890 GC equipped with an Agilent 5975 mass-selective detector (MSD, Agilent Technologies, Palo Alto, CA, USA) and an olfactometer (ODP 2, Gerstel, Mülheim an der Ruhr, Germany). The column carrier gas was helium held at constant pressure (2 mL/min). Half of the column flow was directed to the MSD, while the other half was directed to the olfactometer. The samples were analyzed on a DB-FFAP column (60 m × 0.25 mm i.d., 0.25 µm film thickness; J&W Scientific, Folsom, CA, USA). Each concentrated fraction (1 µL) was injected with a split ratio of 1:1. The oven temperature was held at 50 °C for 2 min, then raised to 230 °C at a rate of 6 °C/min, and held at 230 °C for 15 min. Injector and detector temperatures were 250 °C.

Three panelists (one male and two females) from the Laboratory of Brewing Microbiology and Applied Enzymology at Jiangnan University were selected for GC–O analysis. The panelists responded to the aroma intensity of the stimulus by using a 6-point scale ranging from 0 to 5 where 0 denoted no aroma, 3 denoted moderate aroma, and 5 denoted an extreme aroma. The retention time, intensity, and aroma descriptor were recorded. The extraction was replicated five times by each panelist. The aroma intensities were averaged for the 15 analyses. When a panelist could not detect an aroma compound, the intensity was considered was recorded as 0 for the averaging process (13).

### Identification of Aroma Compounds

The aroma compounds were identified by comparing the mass spectra, aromas, and retention indices (RIs) of the compounds with reference compounds. RIs were determined using a series of standard linear $C_5$–$C_{30}$ alkanes under the same chromatographic conditions (13).

### Quantitation of Aroma Compounds from *F. isatidis*

The aroma compounds were isolated as previously described. Methyl hexanoate as an internal standard (IS) (6.792 mg/L final concentration) was added to the sample before LLE. The results are expressed as the mean value of three replicates.

The response ratio of standard compounds to the IS was plotted against the concentration ratio. The selected ion monitor (SIM) for the IS was $m/z$ 74. The limit of detection (LOD) was calculated as the analyte concentration of a standard that produced a signal-to-noise ratio of 3.

Because some aroma compounds' chemical references (camphene hydrate, 2,4-dimethylbenzaldehyde, [E]-3-hexenoic acid, spathulenol, *cis*-isoeugenol) are not available, their calibration curves cannot be established as they were analyzed by a semi-quantitative method using the following equation (14):

$$C(\mu g/g) = \frac{A_c}{A_{IS}} C_{IS}(\mu g/g) \tag{1}$$

where $C$ is the relative concentration of the analyte (expressed in $\mu g/g$), $C_{IS}$ is the final concentration of IS in the sample, $A_c$ is the peak area of the analyte, and $A_{IS}$ is the peak area of the IS.

### GC–MS

GC–MS analysis was performed using an Agilent 6890 GC equipped with an Agilent 5975 MSD. Samples were analyzed on a DB-FFAP column (60 m × 0.25 mm i.d., 0.25 μm film thickness, J&W Scientific, Folsom, CA, USA). The column carrier gas was helium held at a constant flow rate of 2 mL/min. The column effluent was split 1:1 into the MSD and olfactometer (250 °C). The injector temperature was set at 250 °C. The oven temperature was held at 50 °C for 2 min, then programmed to 230 °C at a rate of 6 °C/min, and held for 15 min.

The ion source and quadrupole temperatures were set at 230 °C and 150 °C, respectively. The electron impact (EI) energy was 70 eV, and mass spectra were recorded in the 35–350 amu range.

# Results and Discussion

## Selection of Soaking Solvent

To extract the aroma compounds, the *F. isatidis* powder was soaked in different solvents including Milli-Q water, ethanol-water (10:90, by volume), and 100% ethanol (Figure 2).

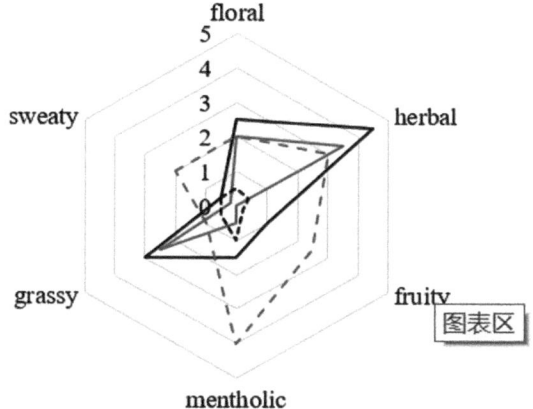

—leaf —water – – –ethanol-water (10:90, volume) ---ethanol

*Figure 2. Comparison of overall odor profiles from different solvents.*

When comparing the overall odor profiles of the extracts extracted using different solvents, it was found that the aroma profile of the Milli-Q water extract was the most similar to that of the *F. isatidis* leaf sample. It had a strong herbal and floral aroma. The ethanol-water mixture (10:90, by volume) showed a characteristically strong mentholic aroma, while the 100% ethanol extract had a very weak aroma profile.

## Identification of Aroma Compounds of *F. isatidis*

The *F. isatidis* extracts were analyzed by the GC–O and Osme methods (Figure 3). A total of 45 aroma compounds were detected: 17 terpenoids, 13 aromatics, 6 phenolics, 6 fatty acids, and 3 other aroma compounds (Table 1). Fourteen of the compounds were important aroma compounds based on their Osme values (≥3.0); 4-Vinylguaiacol (smoky), guaiacol (phenolic, smoky, and herbal), and acetic acid (sour) had the highest Osme values (4.0). This was followed by (*E*)-cinnamaldehyde (cinnamon), 2-phenylethanol (floral and rose), phenol (phenolic, plastic, and medicinal), and L-(–)-borneol (camphoreous) with Osme values ranging from 3.5–3.8.

Terpenoids are the most diverse class of aroma compounds in *F. isatidis*. This class is widely found in nature and has special odors. Of the terpenoids, L-(–)-borneol has the highest Osme value (3.5) and has a camphoreous aroma. It is an important aroma compound in herbaceous aroma-type liquor (*15*). In addition, verbenone (mentholic and sweet), (–)-*trans*-pinocarveol (mentholic), linalool (floral and sweet), and eucalyptol (herbal and mentholic) also had high Osme values (≥3.0).

A total of 13 aromatic compounds were identified in the *F. isatidis* leaf. (*E*)-cinnamaldehyde (Osme value of 3.8), 2-phenylethanol (Osme value of 3.6), and benzyl alcohol (Osme value of 3.0) had the highest Osme values and contributed to the cinnamon and floral aromas. 2-Phenylethanol is also an important aroma compound in *chixiang* aroma-type liquor (*7*), giving rosy and honey aromas. This compound can be produced by yeast (*16*).

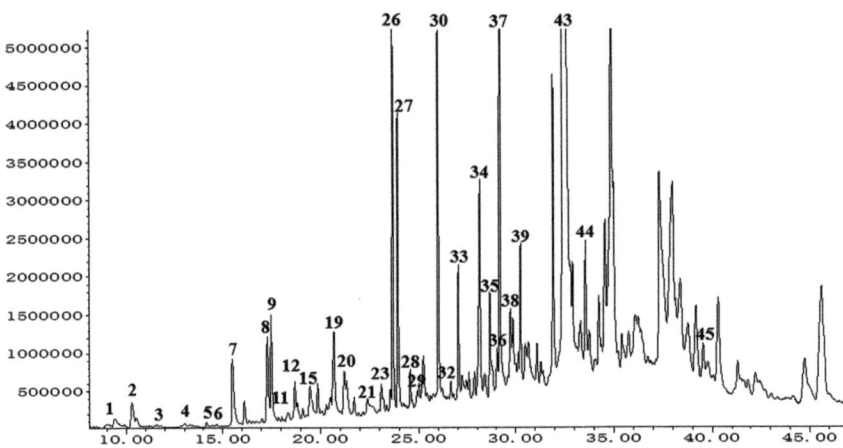

*Figure 3. The Total Ion Chromatogram of aroma compounds in F. isatidis generated by GC-MS. 1:1-methyl-(1H)-pyrrole, 2: eucalyptol, 3: m-cymene, 4: methyl heptanonea, 5: (Z)-3-hexen-1-ol, 6: (−)-fenchone, 7: acetic acid, 8: D-camphor, 9: linalool, 11: isobutanoic acid, 12: camphene hydrate, 15: (−)-trans-pinocarveol, 19: L-(−)-borneol, 20: verbenone, 21: myrtenol, 23: heptanoic acid, 26: guaiacol, 27: benzyl alcohol, 28: 2-phenylethanol, 29: (E)-3-hexenoic acid, 30: phenol, 32: methyl eugenol, 33: (E)-cinnamaldehyde, 34: spathulenol, 35: 2-methoxybenzyl alcohol, 36: globulol, 37: 4-vinylguaiacol, 38: β-eudesmol, 39: syringol, 43: benzoic acid, 44: 3,4-dimethoxyacetophenone, 45: 4-hydroxybenzaldehyde.*

Although only a few phenolic compounds were detected, they were highly abundant. All phenolic compounds other than methyl eugenol and syringol had Osme values greater than 3.0. These included phenol, 4-vinylguaiacol, guaiacol, and *cis*-isoeugenol (temporary identification), and were responsible for phenolic, smoky, herbal, and floral odors. Among these, 4-vinylguaiacol is an important aroma compound in wheat qu of Chinese rice wine (*17*) and can be produced by decarboxylation from the ferulic acid present in grains. During fermentation, 4-vinylguaiacol gets converted to vanillin (*18*).

Six fatty acids and three other aroma compounds were detected in this study. Acetic acid and (Z)-3-hexen-1-ol (Osme value of 3.2) show strong intensity peaks. Acetic acid also has a high FD factor in *chixiang* aroma-type liquor (*7*), suggesting that it contributes important aromas to the liquor. (Z)-3-hexen-1-ol is found in many plants and is responsible for the green odor.

### Table 1. Aroma Compounds of *F. isatidis* Leaf by GC–O

| No. | $RI_{DB\text{-}FFAP}$ | Aroma Compounds | Odor Description | Osme | Basis of Identification[a] |
|---|---|---|---|---|---|
| **Terpenoids** | | | | | |
| 2 | 1159 | eucalyptol | herbal, mentholic | 3.2 | MS, aroma, RI |
| 3 | 1230 | *m*-cymene | citrus | 2.4 | MS, aroma, RI |
| 6 | 1357 | (−)-fenchone | herbal, woody | 1.6 | MS, aroma, RI |
| 8 | 1491 | *D*-camphor | camphoreous | 2.5 | MS, aroma, RI |
| 9 | 1501 | linalool | floral, sweet | 3.2 | MS, aroma, RI |
| 12 | 1565 | camphene hydrate | mentholic | 2.0 | MS, aroma, RIL |
| 13 | 1571 | 4-terpinenol | woody, musty | 2.2 | MS, aroma, RI |

268

Table 1. (Continued). Aroma Compounds of *F. isatidis* Leaf by GC–O

| No. | $RI_{DB\text{-}FFAP}$ | Aroma Compounds | Odor Description | Osme | Basis of Identification[a] |
|---|---|---|---|---|---|
| 15 | 1627 | (–)-*trans*-pinocarveol | mentholic | 3.3 | MS, aroma, RI |
| 16 | 1630 | (–)-$\alpha$-terpineol | floral, soapy | 2.5 | MS, aroma, RI |
| 18 | 1650 | verbenol | fresh, herbal | 2.5 | MS, aroma, RI |
| 19 | 1671 | L-(–)-borneol | camphoreous | 3.5 | MS, aroma, RI |
| 20 | 1707 | verbenone | mentholic, sweet | 3.4 | MS, aroma, RI |
| 21 | 1770 | myrtenol | camphoreous, medicinal | 2.4 | MS, aroma, RI |
| 31 | 1958 | caryophyllene oxide | woody | 1.2 | MS, aroma, RI |
| 34 | 2091 | spathulenol | fruity | 1.5 | MS, aroma, RIL |
| 36 | 2146 | globulol | floral | 1.0 | MS, aroma, RI |
| 38 | 2233 | $\beta$-eudesmol | green | 2.0 | MS, aroma, RI |
| **Aromatics** | | | | | |
| 10 | 1529 | benzaldehyde | bitter | 2.0 | MS, aroma, RI |
| 16 | 1610 | methyl benzoate | phenolic, leathery | 2.0 | MS, aroma, RI |
| 24 | 1818 | 2,4-dimethylbenzaldehyde | sweet, chemical | 1.6 | MS, aroma, RIL |
| 25 | 1826 | *p*-cymenol | sweet, fruity | 1.0 | MS, aroma, RI |
| 27 | 1860 | benzyl alcohol | floral | 3.0 | MS, aroma, RI |
| 28 | 1896 | 2-phenylethanol | floral, rosy | 3.6 | MS, aroma, RI |
| 33 | 2031 | (*E*)-cinnamaldehyde | cinnamon | 3.8 | MS, aroma, RI |
| 35 | 2133 | 2-methoxybenzyl alcohol | medicinal | 2.5 | MS, aroma, RI |
| 40 | 2282 | cinnamyl alcohol | cinnamon | 1.2 | MS, aroma, RI |
| 42 | 2432 | dihydroactinidiolide | fruity, berry | 2.6 | MS, aroma, RI |
| 43 | 2427 | benzoic acid | fruity | 1.2 | MS, aroma, RI |
| 44 | 2502 | 3,4-dimethoxyacetophenone | sweet, floral | 1.2 | MS, aroma, RI |
| 45 | 2942 | 4-hydroxybenzaldehyde | woody | 1.0 | MS, aroma, RI |
| **Phenolics** | | | | | |
| 26 | 1845 | guaiacol | phenolic, smoky, herbal | 4.0 | MS, aroma, RI |
| 30 | 1974 | phenol | phenolic, plastic, medicinal | 3.6 | MS, aroma, RI |
| 32 | 1985 | methyl eugenol | clove | 1.6 | MS, aroma, RI |
| 37 | 2185 | 4-vinylguaiacol | smoky | 4.0 | MS, aroma, RI |
| 39 | 2262 | syringol | sweet, floral | 2.4 | MS, aroma, RI |

**Table 1. (Continued). Aroma Compounds of *F. isatidis* Leaf by GC–O**

| No. | $RI_{DB\text{-}FFAP}$ | Aroma Compounds | Odor Description | Osme | Basis of Identification[a] |
|---|---|---|---|---|---|
| **41** | 2351 | *cis*-isoeugenol | clove | 3.5 | MS, aroma, RIL |
| **Fatty acids** | | | | | |
| **7** | 1394 | acetic acid | sour | 4.0 | MS, aroma, RI |
| **11** | 1521 | isobutanoic acid | acid, rancid | 1.2 | MS, aroma, RI |
| **14** | 1585 | butanoic acid | cheesy | 1.5 | MS, aroma, RI |
| **22** | 1805 | hexanoic acid | cheesy | 1.2 | MS, aroma, RI |
| **23** | 1812 | heptanoic acid | sweat | 2.8 | MS, aroma, RI |
| **29** | 1939 | (*E*)-3-hexenoic acid | honey | 0.9 | MS, aroma, RIL |
| **Others** | | | | | |
| **1** | 1098 | 1-methyl-(1*H*)-pyrrole | smoky, herbal | 2.8 | MS, aroma, RI |
| **4** | 1295 | methyl heptanone | fruity, citrus | 1.8 | MS, aroma, RI |
| **5** | 1333 | (*Z*)-3-hexen-1-ol | fresh, green | 3.2 | MS, aroma, RI |

[a] MS: compounds were identified by MS spectra; aroma: compounds were identified by the aroma descriptors; RI: compounds were identified by comparison to the RIs of pure standard; RIL: compounds were identified by comparison with the RIs from the literature.

## Quantitative Analysis of Aroma Compounds in *F. isatidis*

To gain a deeper insight into the aroma of *F. isatidis*, a total of 40 odorants were quantified using LLE. As shown in Table 2, the 13 aromatic compounds accounted for 2.90 mg/g (DW) and were the most prominent class of compounds found in the leaf. This was followed by the phenolics (6 compounds; 299 µg/g [DW]), the terpenoids (17 compounds; 117 µg/g [DW]), and the fatty acids (6 compounds, 75.1 µg/g [DW]).

Of the aroma compounds, benzoic acid was the most abundant (2.46 mg/g [DW]), followed by benzyl alcohol (243 µg/g [DW]), 4-vinylguaiacol (148 µg/g [DW]), acetic acid (69.3 µg/g [DW]), 2-phenylethanol (57.7 µg/g [DW]), and 3,4-dimethoxyacetophenone (55.0 µg/g [DW]). All were important aroma compounds (Osme values of at least 3.0), except 3,4-dimethoxyacetophenone (Osme value of 1.2).

In the *bingqu*-making process, benzoic acid can inhibit yeast and mold growth. It is also effective against a wide range of bacteria (*19*). In food processing, benzoic acid is used as a preservative. The Food and Agriculture Organization (FAO) and World Health Organization (WHO) evaluated acceptable daily intake (ADI) of benzoic acid is in the order of 5 mg/kg of body weight (*20*). The WHO conducted a risk assessment of benzoic acid and sodium benzoate in its Concise International Chemical Assessment Document 26 and identified that benzoic acid is considered to be a natural ingredient in food existing in many plants and animals (*21*).

Phenolics were the second most abundant class of compounds in *F. isatidis*. Although the concentrations of 4-vinylguaiacol, phenol, syringol, and guaiacol were more than 40 µg/g (DW), they were not detected in the *chixiang* aroma-type liquor suggesting that they we-re consumed by the microorganisms or converted into other compounds during fermentation.

**Table 2. Standard Curves and Concentrations of Aroma Compounds in *F. isatidis* Leaf**

| No. | Aroma Compounds | $m/z^c$ | Slope | Intercept | $R^2$ | n | LOD (µg/L) | Concentration (µg/g(DW)) |
|---|---|---|---|---|---|---|---|---|
| 1 | 1-methyl-(1*H*)-pyrrole | 81 | 0.660 | 0.0876 | 0.9965 | 8 | 0.490 | 0.868±0.259 |
| 2 | eucalyptol | 81 | 2.32 | -0.0170 | 0.9992 | 9 | 1.07 | 0.944±0.567 |
| 3 | *m*-cymene | 119 | 3.08 | 2.83 | 0.9973 | 11 | 0.470 | 0.732±0.135 |
| 4 | methyl heptanone[a] | 43 | 2.01 | 0.328 | 0.9982 | 8 | 0.980 | <ql. [d] |
| 5 | (Z)-3-hexen-1-ol | 67 | 2.58 | 0.0110 | 0.9990 | 10 | 0.720 | 1.78±0.0562 |
| 6 | (−)-fenchone | 81 | 0.500 | 0.0193 | 0.9986 | 11 | 0.560 | 2.09±0.0118 |
| 7 | acetic acid | 60 | 12.4 | -0.132 | 0.9969 | 11 | 3.47 | 69.3±8.81 |
| 8 | D-camphor | 95 | 1.73 | -0.0310 | 0.9848 | 11 | 0.0300 | 4.34±2.62 |
| 9 | linalool | 71 | 1.68 | -0.0440 | 0.9915 | 10 | 0.780 | 26.8±20.4 |
| 10 | benzaldehyde | 105 | 1.77 | 0.106 | 0.9994 | 9 | 0.130 | 18.6±2.84 |
| 11 | isobutanoic acid | 73 | 4.43 | 0.552 | 0.9959 | 5 | 5.31 | <ql. [d] |
| 12 | camphene hydrate[a] | 71 | | | | | | 1.87±0.324 |
| 13 | 4-terpinenol | 71 | 0.623 | 0.0727 | 0.9912 | 9 | 0.570 | 1.71±0.483 |
| 14 | butanoic acid | 60 | 1.94 | 0.556 | 0.9915 | 8 | 12.0 | 1.10±0.553 |
| 15 | (−)-*trans*-pinocarveol | 92 | 2.99 | -0.00520 | 0.9946 | 10 | 0.0900 | 1.86±0.677 |
| 16 | methyl benzoate | 105 | 0.309 | 0.0567 | 0.9881 | 9 | 0.190 | 2.41±0.267 |
| 17 | (−)-α-terpineol | 121 | 1.55 | 0.0110 | 0.9999 | 10 | 0.550 | 6.65±3.27 |
| 18 | verbenol | 94 | 2.01 | 0.403 | 0.9965 | 11 | 0.550 | 0.495±0.280 |
| 19 | L-(−)-borneol | 95 | 0.654 | -0.131 | 0.9890 | 10 | 1.10 | 16.7±1.83 |
| 20 | verbenone | 107 | 1.28 | -0.0600 | 0.9661 | 6 | 0.740 | 0.844±0.235 |
| 21 | myrtenol | 79 | 0.443 | 0.0361 | 0.9959 | 7 | 0.480 | 0.898±0.270 |
| 22 | hexanoic acid | 60 | 0.682 | 1.20 | 0.9896 | 6 | 4.02 | 3.88±2.24 |
| 23 | heptanoic acid | 60 | 0.103 | -0.00460 | 0.9992 | 9 | 0.0900 | <ql. [d] |

Table 2. (Continued). Standard Curves and Concentrations of Aroma Compounds in *F. isatidis* Leaf

| No. | Aroma Compounds | $m/z^c$ | Slope | Intercept | $R^2$ | n | LOD ($\mu g/L$) | Concentration ($\mu g/g(DW)$) |
|---|---|---|---|---|---|---|---|---|
| 24 | 2,4-dimethylbenzaldehyde[a] | 133 | | | | | | 1.29±0.652 |
| 25 | p-cymenol | 135 | 0.335 | 0.0561 | 0.9917 | 10 | 0.160 | <ql.[d] |
| 26 | guaiacol | 109 | 1.02 | -0.0510 | 0.9857 | 10 | 0.250 | 46.2±21.5 |
| 27 | benzyl alcohol | 79 | 1.93 | 0.230 | 0.9959 | 10 | 3.34 | 243±21.8 |
| 28 | 2-phenylethanol | 91 | 3.27 | 0.00460 | 0.9968 | 8 | 0.470 | 57.7±4.63 |
| 29 | (E)-3-hexenoic acid[a] | 55 | | | | | | 0.776±0.131 |
| 30 | phenol | 94 | 0.944 | -0.0490 | 0.9915 | 10 | 0.110 | 48.5±2.89 |
| 31 | caryophyllene oxide | 135 | 4.32 | 0.0369 | 0.9941 | 8 | 2.28 | 44.3±18.0 |
| 32 | methyl eugenol | 178 | 0.410 | 0.0201 | 0.9907 | 10 | 0.0900 | 0.635±0.321 |
| 33 | (E)-cinnamaldehyde | 131 | 1.35 | 0.000700 | 0.9968 | 10 | 0.200 | 6.18±3.58 |
| 34 | spathulenol[a] | 205 | | | | | | 3.79±2.05 |
| 35 | 2-methoxybenzyl alcohol | 138 | 2.43 | 0.198 | 0.9871 | 10 | 1.08 | 27.0±5.58 |
| 36 | globulol | 109 | 2.10 | 0.0932 | 0.9924 | 6 | 6.18 | 0.399±0.0842 |
| 37 | 4-vinylguaiacol | 150 | 1.84 | -0.301 | 0.9883 | 10 | 1.98 | 148±9.05 |
| 38 | β-eudesmol | 59 | 1.65 | -0.0140 | 0.9762 | 6 | 0.650 | 2.45±1.23 |
| 39 | syringol | 154 | 1.48 | 0.337 | 0.9950 | 5 | 2.78 | 47.2±5.04 |
| 40 | cinnamyl alcohol | 92 | 0.931 | 0.0329 | 0.9907 | 9 | 0.370 | 33.1±2.85 |
| 41 | cis-isoeugenol[a] | 164 | | | | | | 8.92±5.14 |
| 42 | dihydroactinidiolide | 111 | 0.605 | 0.0329 | 0.9966 | 9 | 0.250 | <ql.[d] |
| 43 | benzoic acid[b] | 105 | 3.12 | -0.656 | 0.9798 | 9 | 7.24 | 2.46±1.58 |

**Table 2. (Continued). Standard Curves and Concentrations of Aroma Compounds in *F. isatidis* Leaf**

| No. | Aroma Compounds | $m/z^c$ | Slope | Intercept | $R^2$ | $n$ | LOD $(\mu g/L)$ | Concentration $(\mu g/g(DW))$ |
|-----|-----------------|---------|-------|-----------|-------|-----|------------------|-------------------------------|
| 44 | 3,4-dimethoxyacetophenone | 165 | 1.04 | 0.372 | 0.9972 | 10 | 4.49 | 55.0±18.3 |
| 45 | 4-hydroxybenzaldehyde | 122 | 0.685 | 0.122 | 0.9957 | 9 | 0.770 | <ql. [d] |

[a] Aroma compounds were quantitively measured by semi-quantification. [b] Concentration (mg/g [DW]). [c] m/z, quantifying, and qualifying ions. [d] The concentration was below the quantification limit.

Terpenoids were the most diverse class of aroma compounds in *F. isatidis*, but their total concentration was low. Our results showed that terpenoids were only present in trace concentrations in *F. isatidis*. However, it is well-known that the higher concentrations of aroma compounds present in a food product do not always contribute to its aroma, and that only those compounds with concentrations higher than their odor thresholds can contribute to the aroma (*10*). Further research should be conducted to determine the odor thresholds of these terpenoids.

## Conclusion

A total of 45 aroma compounds were detected in *F. isatidis*, 14 of which were important aroma compounds (Osme value of at least 3.0). The most important aroma compounds were 4-vinylguaiacol, guaiacol, and acetic acid. Terpenoids were the most diverse class of aroma compounds contributing mentholic, floral, camphoreous, and herbal aromas to the leaf.

The results of quantitation showed that the total concentration of aromatic compounds was highest (2.90 mg/g [DW]), followed by phenolics (299 µg/g [DW]), terpenoids (117 µg/g [DW]), and fatty acids (75.1 µg/g [DW]). Of the aroma compounds, benzoic acid had the highest concentration (2.46 mg/g [DW]), followed by benzyl alcohol, 4-vinylguaiacol, acetic acid, 2-phenylethanol, and 3,4-dimethoxyacetophenone.

Further work on the fate of the aroma compounds during fermentation is needed to understand the biochemical effects of *F. isatidis* in the *bingqu*-making process.

## Acknowledgments

Financial support from the Ministry of Science and Technology, P. R. China under Nos. National Key R&D Program of China (2016YFD0400503) is gratefully acknowledged.

## References

1.  Li, D. Discussion on the Formation of Specific Style of Soybean-Flavor Liquor. *Liquor-Making Sci. Technol.* **2004**, *1*, 24–27.

2.  Shen, Y. *Manual of Chinese Liquor Manufactures Technology*; Light Industry Publishing House of China: Beijing, China, 1996.

3.  Wang, C.; Fan, W.; Xu, Y.; He, S.; Liu, Y. Sugars and Sugar Alcohols in Traditional Bingqu and Mechanical-Making Moldy Bran (Fuqu) for Chixiang Aroma Type Liquor. *Food Fermentation Ind.* **2018**, *44*, 235–239.

4.  He, Z.; Fan, W.; Xu, Y.; He, S.; Liu, X. Characterization of Odorants in *Cinnamomum cassia* Presl Leaf by GC-Olfactometry. In *The 2nd International Flavor and Fragrance Conference*; Qian, M., Xu, Y. , Eds.; American Chemical Society: Wuxi, China, 2018; pp 127.

5.  Muluye, R. A.; Bian, Y.; Alemu, P. N. Anti-Inflammatory and Antimicrobial Effects of Heat-Clearing Chinese Herbs: A Current Review. *J. Tradit. Complement Med.* **2014**, *4*, 93–98.

6.  Liu, Z.; Yang, Z.; Xiao, H. Antiviral Activity of the Effective Monomers from *Folium isatidis* Against Influenza Virus in Vivo. *Virol. Sin.* **2010**, *25*, 445–451.

7.  Fan, H.; Fan, W.; Xu, Y. Characterization of Key Odorants in Chinese Chixiang Aroma-Type Liquor by Gas Chromatography-Olfactometry, Quantitative Measurements, Aroma Recombination, and Omission Studies. *J. Agric. Food Chem.* **2015**, *63*, 3660–3668.

8. Fan, W.; Qian, M. C. Characterization of Aroma Compounds of Chinese "Wuliangye" and "Jiannanchun" Liquors by Aroma Extract Dilution Analysis. *J. Agric. Food Chem.* **2006**, *54*, 2695–2704.

9. Culleré, L.; Escudero, A.; Cacho, J.; Ferreira, V. Gas Chromatography-Olfactometry and Chemical Quantitative Study of the Aroma of Six Premium Quality Spanish Aged Red Wines. *J. Agric. Food Chem.* **2004**, *52*, 1653–1660.

10. Grosch, W. Evaluation of the Key Odorants of Foods by Dilution Experiments, Aroma Models and Omission. *Chem. Senses* **2001**, *26*, 533–545.

11. Fan, W.; Xu, Y.; Jiang, W.; Li, J. Identification and Quantification of Impact Aroma Compounds in 4 Nonfloral *Vitis vinifera* Varieties Grapes. *J. Food Sci.* **2010**, *75*, S81–S88.

12. Zhang, C.; Ao, Z.; Chui, W.; Shen, C.; Tao, W.; Zhang, S. Characterization of the Aroma-Active Compounds in Daqu: A Tradition Chinese Liquor Starter. *Eur. Food Res. Technol.* **2012**, *234*, 69–76.

13. Qian, M. C.; Fan, W. Identification of Aroma Compounds in Chinese "Yanghe Daqu" Liquor by Normal Phase Chromatography Fractionation Followed by Gas Chromatography/ Olfactometry. *Flavour Fragr. J.* **2006**, *21*, 333–342.

14. Luo, T.; Fan, W.; Xu, Y. Characterization of Volatile and Semi-Volatile Compounds in Chinese Rice Wines by Headspace Solid Phase Microextraction Followed by Gas Chromatography Mass Spectrometry. *J. Inst. Brew.* **2012**, *114*, 172–179.

15. Fan, W.; Hu, G.; Xu, Y.; Jia, Q.; Ran, X. Analysis of Aroma Components in Chinese Herbaceous Aroma Type Liquor. *J. Food Sci. Biotechnol.* **2012**, *8*, 810–819.

16. Peinado, R. A.; Moreno, J. J.; Ortega, J. M.; Mauricio, J. C. Effect of Gluconic Acid Consumption During Simulation of Biological Aging of Sherry Wines by a Flor Yeast Strain on the Final Volatile Compounds. *J. Agric. Food Chem.* **2003**, *51*, 6198–6203.

17. Mo, X.; Xu, Y.; Fan, W. Characterization of Aroma Compounds in Chinese Rice Wine Qu by Solvent-Assisted Flavor Evaporation and Headspace Solid-Phase Microextraction. *J. Agri. Food Chem.* **2010**, *58*, 2462–2469.

18. Mo, X.; Xu, Y. Ferulic Acid Release and 4-Vinylguaiacol Formation During Chinese Rice Wine Brewing and Fermentation. *J. Inst. Brew.* **2012**, *116*, 304–311.

19. Tfouni, S. A. V.; Toledo, M. C. F. Determination of Benzoic and Sorbic Acids in Brazilian food. *Food Control* **2002**, *13*, 117–123.

20. Sieber, R.; Butikofer, U.; Bosset, J. O. Benzoic Acid as a Natural Compound in Cultured Dairy Products and Cheese. *Int. Dairy J.* **1995**, *5*, 227–246.

21. *Concise International Chemical Assessment Documents 26: Benzoic Acid and Sodium Benzoate*; World Health Organization (WHO): Geneva, 2000.

# Chapter 17

# Influence of Ethanol on Flavor Perception in Distilled Spirits

**Zhuzhu Wang,[1] Chelsea M. Ickes,[2] and Keith R. Cadwallader[\*,1]**

[1]**Department of Food Science and Human Nutrition, University of Illinois, 1302 W. Pennsylvania Ave., Urbana, Illinois 61801, United States**
[2]**Silesia Flavors Inc., 5250 Prairie Stone Parkway, Hoffman Estates, Illinois 60192, United States**
[\*]**E-mail: cadwlldr@illinois.edu.**

People often enjoy drinking distilled spirits on the rocks or with a splash of water to open up the flavors. From a physicochemical perspective, the addition of water (dilution of ethanol) affects the spirit's matrix in several ways. These include a decrease in surface tension, a change in the structure of the liquid water and ethanol matrix, and a change in volatile compound partitioning and release from the bulk solution. From a physiological perspective, the dilution of ethanol partially reduces the pungency associated with high ethanol content, and more importantly, it increases olfactory sensitivity to the aroma compounds. However, there is not enough evidence to prove which effect dominates nor are there any systematic studies to link sensory perception alteration to quantitative data. This chapter discusses the influence of ethanol on flavor perception from both physicochemical and physiological perspectives and provides additional insights into the physiological effects of ethanol on individual aroma compound perception based on our own recent studies.

## Introduction

Whiskey, rum, gin, and tequila occupy the majority of the global spirits market. According to the Alcoholic Drinks Report 2018-Spirits (*1*), in North America, revenue in the spirits market amounted to \$56.148 million in 2018, and the market is expected to grow annually by 3.5%. In recent years, China has been the greatest source of revenue (\$190.234 million in 2018). With respect to spirit type, whiskey showed the most growth in the past few years, and its popularity is expected to increase in the coming years due to increasing consumption in countries such as China and India (*2*).

The various types of spirits are characterized by their own unique flavors, due mainly to different starting materials used in their manufacture. This includes grains for whiskeys, molasses for rum, juniper berries for gin, grapes for cognac, and agave for tequila. The subsequent post-distillation treatment (i.e., unaged versus aged in oak barrels) also plays an important role. Some key aroma components in selected distilled spirits are listed in Table 1. Volatiles including fusel alcohols, acetates, and esters formed during fermentation of starting materials impart fruity and solvent-like characteristics on the final product. Raw distilled spirits have a harsh flavor; however, aging in the wood barrels creates a product with much more desirable flavor characteristics that come from ethanolysis of wood components, lignin pyrolysis from the charring of the barrel, and by direct extraction of wood volatiles.

Most distilled spirits are bottled at around 40% alcohol by volume (ABV). While some consumers drink spirits neat (unmixed), others prefer to drink them "on the rocks" (with ice cubes). The ice both cools and dilutes the spirit, which ultimately affects flavor perception. This effect is well recognized since it is a common practice for master blenders in the whiskey industry to dilute the spirit to 20% or 23% ABV prior to nosing and tasting. A common explanation for this practice is that it reduces the pungency of the alcohol, but others argue that the pungency is not solely dependent on the ethanol concentration and may be partially attributed to other effects such as the activity of ethyl esters in the headspace or the oak extractives in solution (3).

**Table 1. Key Aromas from Selected Distilled Spirits**

| Spirits | Reported Representative Aromas |
|---|---|
| Bourbon (18) | (E)-β-damascenone, γ-nonalactone, γ-decalactone, eugenol, vanillin, isoamyl alcohol, α-damascone, whisky lactone, 2-phenylethanol, guaiacol, ethyl 2-methylbutyrate, β-ionone, (E)-2-heptenal |
| Rum (19, 20) | (E)-β-damascenone, 3-methylbutanal, 2,3-butanedione, ethyl 2-methylbutyrate, vanillin, ethyl butanoate, (S)-2-methylbutanal, guaiacol, 1,1-diethoxyethane (acetal), 3-hydroxy-4,5-dimethylfuran2(5H)-one, ethyl octanoate, (S)-2-methyl-1-butanol, ethyl hexanoate, 4-ethylphenol, ethyl hexanoate, 2-methoxy-4-propylphenol |
| Gin (21) | α-pinene, β-pinene, β-myrcene, limonene, γ-terpinene, ρ-cymene, linalool, bornyl acetate, α-terpineol, valencene, β-citronellol, τ-geraniol, caryophyllene oxide |
| Cognac (22, 23) | (E)-β-damascenone, methylpropanal, ethyl 2-methylbutyrate, ethyl 2-methylpropanoate, ethyl 3-methylbutanoate, ethyl hexanoate, 3-methylbutanal, 2,3-butanedione, ethyl butanoate, ethyl octanoate, vanillin, acetal, 3-methoxybutanol, nerolidol, linalool |
| Tequila (24, 25) | Ethyl acetate, acetal, (E)-β-damascenone, ethyl octanoate, ethyl hexanoate, α-terpineol, diacetyl, linalool, (Z)-linalool oxide, isoamyl acetate, citronellol |

Research on distilled spirits has shown that the presence of ethanol in a water solution decreases surface tension and changes the structure of the liquid water-ethanol matrix (4–7). This impacts volatile compound partition and release from the bulk solution, which ultimately results in an alteration of the headspace constituents. However, physicochemical properties of the aroma compounds and ethanol only determine the likelihood that the aroma compounds will be released from the matrix into the headspace. It is known that the overall odor of a mixture is not simply the summation of the individual odors of the constituent chemicals. In odor psychophysics, the rated intensity of suprathreshold mixtures generally falls below the sum of the intensities of the unmixed components (8–13). Some earlier studies have reported that ethanol exposure (i.e., during ingestion) results in alteration of olfactory sensitivity (14, 15). This may be partially due to the depressant effect of ethanol inhibition of synaptic excitation in the olfactory bulb (16, 17). Consequently, the numbers and concentrations of odorants present in the headspace of the spirit are not necessarily the same patterns being perceived ortho- or retronasally. Therefore, there must been some physiological effects involved, but the question is which effect dominates: physiochemistry or physiology?

## Physicochemistry of Flavor Release

### Ethanol Concentration Impacts the Structure of the Liquid Water-Ethanol Matrix

Structurability of ethanol-water mixtures at different alcoholic strengths was thought to be closely related to aroma release (26–28).

Coccia et al. (29) reported that in water-ethanol mixtures, the hydrogen-bonding structure of water is strengthened by the presence of small amounts of ethanol accompanied by the formation of new water-ethanol hydrogen bonds or the increased water-water association in the region of the ethanol mole fraction (f), from 0–0.07 (f is defined as moles of alcohol divided by the total moles of the mixture, i.e., alcohol and water).

Ethanol molecules at all concentrations are surface active and thus preferentially bind at the solution-vapor interface (30). At low ethanol concentrations, ethanol preferentially moves to the surface layer while in the bulk of the liquid phase, it is accommodated in existing cavities in the three-dimensional, hydrogen-bonded, low density arrangement of water (31). Around f = 0.016, 40% of the surface area is already occupied by alcohol molecules (31). In the bulk solution, the hydroxyl ion of ethanol is hydrogen-bonded to the enclosing water molecules such that the hydrophobic alkyl group of the alcohol is masked in a fluctuating clathrate hydrate structure ($C_2H_5OH.17H_2O$) in order to reinforce the overall arrangement. This effect is maximized around f = 0.055, where presumably the cavity sites in the water are saturated with alcohol (31). Under these conditions, the excess alcohol molecules fill up the second layer under the surface and are in mutual contact (28, 31). Stewardt et al. (30) reported that for ethanol concentrations where f ≤ 0.059, the spatial distribution of ethanol molecules at the interface is not entirely uniform, and the interfacial ethanol molecules tend to cluster into small strands with methylene carbon atoms of the nearest neighbor ethanol molecules being approximately 5Å apart. The surface thickness stays consistent with having a single monolayer of ethanol adsorbed at the solution liquid-vapor interface until f = 0.11 (30). The surface tension of the ethanol-water mixture decreases sharply as the ethanol mole fraction in the liquid reaches ~f = 0.2, beyond where the variation is marginal (30, 32, 33). The rapid fall in surface tension is predominantly correlated with the decrease in the water-water hydrogen bonds and is indicative of rapid coverage (surface adsorption) of ethanol at the solution-air interface (4). At f = 0.2, the so-called "critical micelle concentration (CMC)" has been achieved, the point beyond which any further increment in ethanol concentration fails to lower the surface tension further (28). A 20% ABV at 25 °C corresponds to an f = 0.069, a point in which the water structure is most strengthened relative to all ranges of ethanol contents (34).

### Ethanol Impacts Aroma Compound Partitioning between Bulk Solution and Headspace

Tsachaki et al. (35) reported that under static headspace conditions, an increase in the ethanol content of an aqueous matrix (from 4% to 40% ABV) will cause a general decrease in the headspace concentration for most aroma compounds. This has been attributed to increases in the solubility of aroma compounds upon addition of ethanol. The concentration of a volatile compound in the headspace above an ethanolic solution is governed largely by its LogP value (a partition coefficient equivalent to the concentration of a compound in an octanol phase versus a water phase).

Hydrophillic odorants with LogP values that were less than 1 were not affected by 12% ABV, but the effect became stronger as LogP values increased toward 3 (i.e., for more hydrophobic odorants). Thereafter, there was some evidence that the trend may reverse around a LogP value of 4. At high LogP values (very hydrophobic molecules), the presence of ethanol in the solution did not decease the concentration of volatiles consistently (35).

Surprisingly, under dynamic headspace conditions, the effect of increasing the ethanol content is the reverse of what is observed under static headspace conditions because for most volatiles, the absolute volatile concentration above an ethanolic solution was greater than the concentration above a pure water solution (35–37). Taylor et al. (37) demonstrated that other than furfuryl alcohol and phenylacetaldehyde (which showed no difference in release between water and 12% ethanol), all other odorants studied (with a LogP range from -1.34 to 4.47) demonstrated an increased release rate in a matrix containing higher ethanol content under dynamic headspace conditions with the magnitude of the effect varying by a factor from 1.05 to 13 depending on the compound.

In fact, how ethanol impacts odorant release under both static and dynamic headspace is compound dependent, and the level at which the effect ceases is also ethanol concentration dependent (35, 36, 38, 39). For example, Conner et al. (38) reported that the static headspace concentration of ethyl esters (C2–C12) followed a similar pattern with little change in the gradient in matrices from 5%, 10%, and 17% ABV. However, there was a decrease at 23% ABV and another marked decrease at 40% ABV. Boothroyd et al. (40) have also observed an exaggerated increase of headspace concentration for hydrophobic compounds (LogP > 2) upon dilution from 40% to 23% ABV, while dilution from 23% to 5% ABV had a progressive but lesser impact. On the contrary, the headspace concentration of hydrophilic compounds (LogP < 2) were minimally affected by a change to the ethanol concentration.

According to Conner et el. (38), from 20% to 57% ABV, there is a progressive aggregation of ethanol molecules resulting in a reduction of hydrophobic hydration. Therefore, a critical point must exist between this concentration range where ethanol starts to self-associate to form "pseudo-micelles." An increased proportion of small hydrophobic volatile molecules move into these aggregates, which cause a more rapid decrease of concentration in the headspace. For amphiphic volatiles with large hydrophobic surface areas (i.e., long carbon chain ethyl esters), the surface-active ethanol molecules can also be adsorbed onto the surface to lower the interfacial tension between the aqueous phase and ethyl esters. This reduces the free energy of mixing which increases the solubility of esters (38). Following up with Conner's conclusion, it is reasonable to believe that after diluting the spirits (over ice cubes), the concentration of ethanol drops below this critical point and interrupts aggregation of ethanol molecules, leading to the collapse of ethanol "pseudo-micelles", thus resulting in the release of hydrophobic volatiles from ethanol clusters. Moreover, it is also possible that adding water increases the surface tension by creating and strengthening the hydrogen bonding between water molecules so that more clathrate structures of water molecules become available to incorporate ethanol molecules. This causes less ethanol to be available for adsorbtion on the surface of esters, which increases the interfacial tension between esters and the solution resulting in a decrease in ester solubility.

Under dynamic conditions, ethanol molecules evaporate faster at lower ethanol concentrations (31, 41). The increased ethanol evaporation rate at the interface will create an even greater surface tension gradient. This causes the molecules of the adjacent low surface tension regions to move toward the higher surface tension region, carrying with them an appreciable volume of underlying liquid (42). The resulting interfacial turbulence (Marangoni effects) can significantly accelerate the mass transfer process (5, 42), and, as a result, less soluble hydrophobic compounds gain access to the interfacial layer for partition (27). In addition, surface evaporation causes cooling of the interfacial layer. The difference in temperature between the surface layer and the bulk solution sets up a Rayleigh-Berbards convection (27–31, 36, 37, 43), whereby the colder surface layer descends through the bulk phase toward the bottom of the glass while the warmer bulk solution moves toward the surface. Consequently, a current is created to effectively "stir" the bulk phase. When odorants are present, this "stirring" effect assists in their mass transfer from bulk phase to the interfacial layer for partition (27). However, not all of the amphiphilic compounds gain access to the interfacial layer. With respect to soluble amphiphilic compounds (such as long-chain alcohols, aldehydes, and esters), the solubility lowered by reducing ethanol content can induce "structuring" of the solution via the formation of agglomerates. These agglomerates serve as reservoirs for small hydrophobic volatile compounds (LogP > 2), which lower their concentrations in the headspace potentially altering the balance of volatile compounds (27, 40, 44).

## Physiological Effect on Flavor Perception

### Ethanol Affects Olfactory Sensitivity

Ethanol is an effective olfactory and trigeminal stimulus, contributing to the warming and burning perception of alcoholic drinks (45–48). The rank order of ethanol threshold is taste > nasal irritancy > olfaction, and the predominant taste attribute near the threshold is bitterness (49, 50). Males have significantly higher (1.68% ± 0.18% v/v) ethanol taste thresholds than females (1.19% ± 0.09% v/v) (49). A person's ability to discriminate alcoholic strength by taste has an upper threshold; this is usually at an alcoholic strength above 40% ABV (51). In a ranking trial where participants received rum samples at 30%, 40%, 50%, and 60% ABV (Samples were standardized to contain the same amount of flavor compounds and color so the only substantial difference in the test samples was alcoholic strength), they were unable to detect a significant taste difference between the four alcoholic strength levels (52).

Research on the alteration of flavor perception caused by the presence of alcohol can be dated back to the 1950's when Margulies et al. (53) first reported that substituting a meal at lunchtime with 75 mL of a 12% ABV solution increased the recognition threshold (i.e., it decreased sensitivity) to the odor of ground coffee. In the 1970's, Engen et al. (14) reported that ingesting ethanol in orange juice enhanced odor detection performance of guaiacol, which contradicted an earlier report showing a decrease in olfactory sensitivity (53). Nobel (54) reported that the bitterness of quinine (10 ppm), epicatechin (1000 ppm), and catechin (1000 ppm) were more intense in 5% ethanol than in water. However, Hirsch and Bissell (55) later reported a decreased sensitivity to thiophane after drinking a series of alcoholic beverages. Patel et al. (15) determined that ethanol ingestion markedly decreases olfactory threshold sensitivity to ethanol, but not to the rose-like smelling substance, 2-phenylethanol. Mattes and coworkers (49) reported that rinsing with ethanol (beer) prior to tasting suppressed the initial bitterness of quinine, but enhanced its bitter aftertaste. Matin et al. (56) studied four concentration of ethanol (4%, 8%, 12%, and 24% ABV) combined with four suprathreshold concentrations of sucrose, sodium chloride, citric acid, and quinine. They found that ethanol enhanced the intensity of quinine in all four concentration, but only enhanced the sweetness of sucrose in the lower two concentrations of 4% and 8% ABV, had no effect at 12% ABV, and slightly depressed sweetness at the highest concentration of 24% ABV. Moreover, ethanol significantly depressed the saltiness of sodium chloride and the sourness of citric acid at the higher concentrations of 12% and 24% ABV. Fontoin et al. (57) reported in alcoholic drink models (0%, 7%, 11%, and 15% ethanol) that bitterness intensity was higher when the ethanol level increased. However, to their surprise, an increased ethanol concentration significantly decreased astringency of grape seed tannin.

Recently, Ickes et al. (58) performed the first sensory study to demonstrate how flavor perception in spirits is altered as a result of its ethanol dilution. In their study, they diluted rum 1:2 (v/v) with either pure water to a final ABV of 20% or with a pure aqueous 40% ABV solution to account for the flavor dilution effect while keeping the ethanol concentration the same as the original liquor. To their surprise, the flavor profiles of the original Ron Abuelo rum (RA) and the sample that was diluted with water (RA20) were very similar, with the RA20 generally having only slightly lower attribute intensities. In contrast, the rum diluted with 40% ABV (RA40) had a significantly different flavor profile than the RA (Table 2).

**Table 2. Mean Intensity Rating for Significant Aroma, Mouthfeel, Taste, Aftertaste and Aroma-by-Mouth Attributes of the Ron Abuelo (RA) 7-Year Rum Dilution Series[*]. (Data from (58)).**

| Modality | Attribute | RA | RA20 | RA40 |
|---|---|---|---|---|
| Aroma | Alcohol | 9.63[a] | 7.94[b] | 4.88[c] |
| | Caramel | 8.13[a] | 7.31[a] | 5.81[b] |
| | Vanilla | 8.44[a] | 7.50[a,b] | 6.19[b] |
| | Dark Fruit | 7.13[a] | 5.13[b] | 5.50[b] |
| Aroma-by-Mouth | Alcohol | 10.81[a] | 9.38[a] | 3.94[b] |
| | Caramel | 8.44[a] | 7.06[b] | 5.06[c] |
| | Maple | 8.06[a] | 7.19[b] | 5.38[c] |
| | Vanilla | 8.63[a] | 7.19[b] | 5.75[c] |
| Taste | Bitter | 9.06[a] | 8.38[a] | 4.13[b] |
| Aftertaste | Bitter | 10.25[a] | 9.06[a] | 6.00[b] |
| | Brown Spice | 7.56[a] | 7.13[a] | 5.63[b] |
| | Vanilla | 7.06[a] | 6.25[a,b] | 5.31[b] |
| | Plastic | 4.94[a] | 4.50[a.b] | 3.44[b] |
| Mouthfeel | Astringent | 10.25[a] | 8.81[b] | 4.75[c] |
| | Slick | 6.63[a] | 5.75[a] | 3.13[b] |
| | Warming | 9.63[a] | 9.06[a] | 3.63[b] |

[*] Superscripts of the same letter within an attribute indicate no significant difference by Fisher's least significant difference (LSD) test at $\alpha = 0.05$. "RA20" is 1:2 dilution of Ron Abuelo aged 7 years with water to achieve 20% ABV, "RA40" is 1:2 dilution of Ron Abuelo aged 7 years with 40% ethanol to achieve 40% ABV.

## Ethanol Affects Aroma Detection Threshold

We believe that the alteration of flavor perception at different ethanolic strengths can be partly explained by the alteration of the detection threshold of each aroma compound.

The odor threshold of ethanol is 84 ppm (59) and odorants have detection thresholds in ethanol that are 10–312 times higher than in water (60). Some researchers have studied the effect of ethanol concentration on aroma perception in wine (61–65) and beer (66). To the best of our knowledge, there are few studies that have systematically investigated the effect of ethanol concentration on aroma perception during consumption of distilled spirits. Ickes and Cadwallader (27) reviewed and summarized aroma detection thresholds for well-known whiskey aroma compounds at different ethanolic strengths (Table 3), but there appeared to be no consistent trend in how ethanol concentration affects odor thresholds. This may be attributed to the fact that these thresholds were determined by different labs where testing may have been performed under different conditions (27). In addition, individuals can have sensitivities to a stimulus that differ by a factor of 1,000 to 10,000 (66). Consequently, published group threshold values for a given stimulus are highly variable from one sensory panel to another, often varying by a factor of more than 100-fold (66, 67).

**Table 3. Odor Detection Thresholds (ppb)[a] for Selected Aroma Compounds in Various Ethanol-Water Matrices**

| Compound | Water | 10% ABV | 23% ABV | 40% ABV |
|---|---|---|---|---|
| β-Damascenone | 0.0004 | 0.05 | | 0.1 |
| Ethyl Butanoate | 1 | 20 | 100 | 9.5 |
| Ethyl Hexanoate | 5 | 5 | 70 | 30 |
| Ethyl Octanoate | 70 | 2 | | 147 |
| 2-Phenethyl Acetate | 356 | 250 | | 108 |
| 1,1-Diethoxethane | 25 | 50 | | 719 |
| 3-Methyl butanol | 1000 | 30000 | | 56100 |
| 2-Phenylethanol | 1000 | 10000 | 8000 | 2600 |
| Acetaldehyde | 25 | 500 | 2800 | 19200 |
| Acetic Acid | 180000 | 200000 | 28000 | 75521 |
| Linalool | 0.14 | 15 | | 23 |
| Geraniol | 3.2 | 30 | 30 | |
| Guaiacol | 2.5 | 10 | 90 | 9.2 |
| 4-Vinyl Guaiacol | 100 | 40 | 700 | 7.1 |
| Vanillin | 4.9 | 200 | 600 | 22 |

[a] Values from published literature, sources cited in (27).

We recently performed a systematic study utilizing a sensory panel and appropriate protocols (68) to determine the detection threshold of 2-phenylethanol in different %ABV matrices. Results clearly showed that the detection threshold of the rose-smelling compound increased when the ethanol concentration was increased (Table 4). However, under these circumstances, it is possible that the beverage matrix (ethanol and water) played a major role where the 2-phenylethanol would be retained to a greater extent in a higher ethanol matrix, thereby resulting in a higher threshold value.

**Table 4. Odor Threshold of 2-phenylethanol in Different Ethanol Concentrations**

| Matrix | Best Estimate Threshold (BET)[a] (ppm) | Reported Threshold (ppm) |
|---|---|---|
| 0% ABV | 0.41 | 0.48 (69) |
| 5% ABV | 0.99 | |
| 10% ABV | 1.99 | 10 (70) |
| 23% ABV | 2.64 | 8 (71) |
| 30% ABV | 7.23 | 7.5 (72)[b] |
| 40% ABV | 18.74 | 2.6 (73) |

[a] Determined using the method described in reference (68).    [b] Threshold was determined at 34% (w/w).

In order to provide a clearer picture of the physiological effect of ethanolic strengths on the detection threshold of aroma compounds, a modified gas chromatography-olfactometry (GC-O) system was built so that the panel was able to sniff the GC effluent in a background of constant ethanol vapor of differencing concentrations (generated from 0% ABV, 20% ABV, or 40% ABV solutions). In this way, the solvating effect of ethanol was totally excluded, and any experienced alteration of aroma detection thresholds could be attributed solely to the physiological effect of ethanol from the vapor background.

## Materials and Methods

A stock solution of each aroma compound was prepared in diethyl ether and then stepwise diluted 1:3 (v/v) in the same solvent. Each dilution was analyzed using GC-O. The GC-O system consisted of a 6890 GC (Agilent Technologies, Inc., Palo Alto, CA, USA), equipped with an FID, olfactory detection port (Datu, Geneva, NY, USA), and cool on-column injector. Each extract (2 uL) was separated using an RTX®-Wax column (15 m length x 0.53 mm i.d. x 1.0 μm film thickness; Restek Corp, Bellefonte, PA, USA). Helium was used as the carrier gas at a constant flow rate of 5.0 mL/min. The oven temperature was programmed as follows: initial temperature is set at 35 °C (5 min hold); ramp rate is 10 °C/min; final temperature is 225 °C (30 min hold). All dilutions were analyzed by GC-O under the same conditions, except for the humidified air (background vapor) supplied to the olfactometry detection port (ODP) that was passed through either pure water, 20% ABV, or 40% ABV. To avoid prolonged exposure to the background vapor, panelists were asked to only sniff at a specific retention time (RT) region (±0.5 min of the RT expected for a specific odorant).

The highest dilution at which an odorant was detected was defined as its flavor dilution (FD) factor. For example, an odorant with an FD factor of 9 was detected in the original stock solution, the 1:3 dilution, and the 1:9 dilution, but not in the 1:27 dilution. This means that the odorant is present at or above its odor detection threshold concentration (in air) in the 1:9 dilution but below it in the 1:27 dilution.

## Results and Discussion

The experimental results clearly demonstrated the physiological effect of ethanol on detection thresholds for all aroma compounds (Table 5). The FD factors of each aroma compound differed at each ethanolic background strengh and tended to decrease with increasing ethanol vapor concentration. These results also indicated that decreases in aroma detection thresholds at different ethanolic strengths is independent of the physicochemistry related to partition and release from the bulk solution. Ethanol molecules from the vapor must have interacted with aroma compounds differently somewhere in the nasal cavity; results from Table 5 have also exhibited that the degree of physiological suppression effect of ethanol on aromas were not universal with some compounds more affected than the others, suggesting a potential relationship between ethanol aroma suppression and aroma molecular structures. For example, syringol and vanillin both had FD factors that were 3 times lower in ethanol vapor of 20% ABV and 27 times lower in 40% ABV, when compared to 0% ABV. However, linalool had an FD factor that was 9 times lower in ethanol vapor of 20% ABV and 81 times lower in 40% ABV, when compared to 0% ABV. Ethyl butyrate had an FD factor that was 27 times lower in ethanol vapor of 20% ABV and more than 243 times lower in 40% ABV, when compared to 0% ABV. As a result, it is worthwhile to repeat this experiment by expanding the number of aroma compounds with varying molecular structures. By generating more sensory data and relating it to structural and physiocochemical features of the aroma compounds, we aim to provide additional insights into the structure-function understanding at the molecular level in an effort to better explain the physiological suppression effect of ethanol on flavor perception.

**Table 5. FD Factors of Selected Whiskey Aroma Compounds in 0%, 20%, and 40% ABV Vapor Environments**

| Aroma Compound | LogP | Vapor Environment | | |
|---|---|---|---|---|
| | | 0% ABV | 20% ABV | 40% ABV |
| β-Damascenone (2.04 ppm)[a] | 4.402 | 81 | 27 | 9 |
| 2-Phenylethanol (12.48 ppm) | 1.360 | 27 | 9 | 3 |
| Linalool (2.56 ppm) | 2.970 | 243 | 27 | 3 |
| Whiskey Lactone (13.3 ppm) | 1.968 | 243 | 81 | 27 |
| Eugenol (1536 ppm) | 2.270 | 81 | 27 | 3 |
| Ethyl Butyrate (2034 ppm) | 1.804 | 243 | 9 | 1[b] |
| 3-Methyl-1-butanol (1223 ppm) | 1.160 | 27 | <1[c] | <1 |
| Guaicol (82 ppm) | 1.320 | 27 | 3 | 1 |
| γ-Nonalactone (16.56 ppm) | 1.942 | 729 | 243 | 81 |
| Syringol (4338 ppm) | 1.150 | 27 | 9 | 1 |
| Vanillin (40 ppm) | 1.210 | 729 | 243 | 27 |

[a] Numbers in parentheses represent the concentration of the aroma compound in stock solution.   [b] FD = 1, odorant was only detected in stock solution.   [c] FD < 1, odorant was not even detected in the stock solution.

## Conclusion

Other than water, ethanol is the most abundant volatile compound in any alcoholic beverage. Studies have shown that the concentration of ethanol has a significant impact on the aroma released from a bulk solution due to physicochemical effects on the structure of ethanol-water mixtures, particularly in terms of molecular packing and hydrogen-bonding networks. Additional attention has also been drawn to the physiological impact of ethanol on aroma perception as it alters olfactometry sensitivity and suppresses the detection thresholds of aroma compounds. This chapter provides an up-to-date review of the published literature on the physicochemical and physiological aspects of this subject and shares new data from our lab to further support the physiological suppression effects of ethanol on aroma detection thresholds in air, suggesting a possible relationship between the aroma compound structure and the degree of ethanol aroma suppression. There is still not enough evidence to illustrate which effect actually dominates flavor perception during consumption of spirits. It is true that aroma compounds do need to be released into the headspace before they can be perceived. However, the perception of aroma compounds may also be suppressed by the presence of ethanol in the nasal cavity. Consequently, a number of systematic sensory and analytical studies are needed to provide more insights into both the physicochemical and physiological effects. For example, if we are able to compare the degree of aroma detection threshold suppression above ethanol-water matrixes and in an equivalent ethanol-water vapor environment then we will be able to better understand which of the the fundamental driving forces is dominant in the flavor perception alteration observed upon dilution of ethanol during consumption of distilled spirits.

## References

1.  Spirits North America. *Statista: The Statistics Portal*. https://www.statista.com/outlook/ 10020000/104/spirits/ north-america (accessed Nov 1, 2018).

2.  *Bourbon/Tennesse Whiskey*. Distilled Spirits Council. https://www.distilledspirits.org/ products/bourbon-tennessee-whiskey/ (accessed Nov 1, 2018).

3.  Withers, S. J.; Piggott, J. R.; Leroy, G.; Conner, J. M.; Paterson, A. Factors Affecting Pungency of Malt Distillates and Ethanol-Water Mixtures. *J. Sens. Stud.* **1995**, *10*, 273–283.

4.  Biscay, F.; Ghoufi, A.; Malfreyt, P. Surface Tension of Water-Alcohol Mixtures from Monte Carlo Simulations. *J. Chem. Phys.* **2011**, *134*, 44709–44718.

5.  Alvarez, E.; Correa, A.; Correa, J. M.; Garcia-Rosello, E.; Navaza, J. M. Surface Tension of Three Amyl Alcohol + Ethanol Binary Mixtures from (293.15 to 323.15) K. *J. Chem. Eng. Data* **2011**, *56*, 4235–4238.

6.  Cahn, J. W. Critical point wetting. *J. Chem. Phys.* **1977**, *66*, 3667–3672.

7.  Raina, G.; Kulkarni, G. U.; Rao, C. N. Mass Spectrometric Determination of the Surface Compositions of Ethanol-Water Mixtures. *Int. J. Mass Spectrom.* **2001**, *212*, 267–271.

8.  Jones, F. N.; Woskow, M. H. On the Intensity of Odor Mixtures. *Ann. N.Y. Acad. Sci.* **1964**, *116*, 484–494.

9.  Berglund, B.; Berglund, U.; Lindvall, U.; Svensson, L. T. A Quantitative Principle of Perceived Intensity Summation in Odor Mixtures. *J. Exp. Psychol.* **1973**, *100*, 29–38.

10. Cain, W. S. Odor Intensity: Mixtures and Masking. *Chem. Senses* **1975**, *1*, 339–352.

11. Laing, D. G.; Panhuber, H.; Willcox, M. E.; Pittman, E. A. Quality and Intensity of Binary Odor Mixtures. *Physiol. Behav.* **1984**, *33*, 309–319.

12. Cain, W. S.; Schiet, F. T.; Olsson, M. J.; de Wijk, R. A. Comparison of Models of Odor Interaction. *Chem. Senses* **1995**, *20*, 625–637.

13. Miyazawa, T.; Gallagher, M.; Preti, G.; Wise, P. M. Synergistic Mixture Interactions in Detection of Periththreshold Odors by Humans. *Chem. Senses* **2008**, *33*, 363–369.

14. Engen, T.; Kilduff, R. A.; Rummo, N. J. The Influence of Alcohol on Odor Detection. *Chem. Senses* **1975**, *1*, 323–329.

15. Patel, S. J.; Bollhoefer, A. D.; Richard, L. D. Influences of Ethanol Ingestion on Olfactory Function in Humans. *Psychopharmacology* **2004**, *171*, 429–434.

16. Nicoll, R. A. The Effect of Anaesthetics on Synaptic Excitation and Inhibition in the Olfactory Bulb. *J. Physiol.* **1972**, *223*, 803–814.

17. Austin, A.; Scholfield, C. N. Interaction Between Phorbol Dibutyrate and Anaesthetics on Synaptic Responses from Olfactory Cortex of Rat. *Neuropharmacology* **1991**, *30*, 1113–1118.

18. Poisson, L.; Schieberle, P. Characterization of the Most Odor-Active Compouds in an American Bourbon Whisky by Application of the Aroma Extract Dilution Analysis. *J. Agric. Food Chem.* **2008**, *56*, 5813–5819.

19. Franitza, L.; Granvogl, M.; Schieberle, P. Influence of the Production Process on the Key Aroma Compounds of Rum: From Molasses to the Spirit. *J. Agric. Food Chem.* **2016**, *64*, 9041–9053.

20. Pino, J. A.; Tolle, S.; Gok, R.; Winterhalter, P. Characterisation of Odour-Active Compounds in Aged Rum. *Food Chem.* **2012**, *132*, 1436–1441.

21. Vichi, S.; Riu-Aumatell, M.; Mora-Pons, M.; Buxaderas, S.; Lopez-Tamames, E. Characterization of Volatiles in Different Dry Gins. *J. Agric. Food Chem.* **2005**, *53*, 10154–10160.

22. Ferrari, G.; Lablanquie, O.; Cantagrel, R.; Ledauphin, J.; Payot, T.; Fournier, N.; Guichard, E. Determination of Key Odorant Compounds in Freshly Distilled Gognac Using GC-O, GC-MS, and Sensory Evaluation. *J. Agric. Food Chem.* **2004**, *52*, 5670–5676.

23. Uselmann, V.; Schieberle, P. Decoding the Combinatorial Aroma Code of a Commercial Cognac by Application of the Sensonics Concept and First Insights into Differences from a German brandy. *J. Agric. Food Chem.* **2015**, *63*, 1948–1956.

24. Gonzalez-Robles, I. W.; Cook, D. J. The Impact of Maturation on Concentrations of Key Odour Active Compounds Which Determine the Aroma of Tequila. *J. Inst. Brew.* **2016**, *122*, 369–380.

25. Lahne, J.; Cadwallader, K. R. Streamlined Analysis of Potent Odorants in Distilled Alcoholic Beverages: The Case of Tequila. In *Flavor Chemistry of Wine and Other Alcoholic Beverages*; Qian, M. C.; Shellhammer, T. H., Eds.; ACS Symposium Series 1104; American Chemical Society: Washington, DC, 2012; pp 37–53.

26. D'Angelo, M.; Onori, G.; Santucci, A. Self-Associattion of Monohydric Alcohols in Water: Compressibility and Infrared Absorption Measurements. *J. Chem. Phys.* **1994**, *100*, 3107–3113.

27. Ickes, C. M.; Cadwallader, K. R. Effects of Ethanol on Flavor Perception in Alcoholic Beverages. *Chemosens. Percept.* **2017**, *10*, 119–134.

28. Onori, G.; Santucci, A. Dynamical and Structual Properties of Water/Alcohol Mixtures. *J. Mol. Liq.* **1996**, *69*, 161–181.

29. Coccia, A.; Indovina, P. L.; Podo, F.; Viti, V. PMR Studies on the Structures of Water-Ethyl Alcohol Mixtures. *Chem. Phys.* **1975**, *7*, 30–40.

30. Stewardt, E.; Shields, R. L.; Taylor, R. S. Molecular Dynamics Simulations of the Liquid/Vapor Interface of Aqueous Ethanol Solutions as a Function of Concentration. *J. Phys. Chem. B* **2003**, *107*, 2333–2343.

31. Spedding, P. L.; Grimshaw, J. Abnormal Evaporation Rate of Ethanol from Low Concentration Aqueous Solution. *Langmuir* **1993**, *9*, 1408–1413.

32. Ghoufi, A.; Artzner, F.; Malfreyt, P. Physical Properties and Hydrogen-Bonding Network of Water-Ethanol Mixtures from Molecular Dynamics Simulations. *J. Phys. Chem. B* **2016**, *120*, 793–803.

33. Bagheri, A.; Fazil, M.; Bakhshaei, M. Surface Properties and Surface Thickness of Aqueous Solutions of Alcohols. *J. Mol. Liq.* **2016**, *224*, 442–451.

34. Nose, A.; Hojo, M. Hydrogen Bonding of Water-Ethanol in Alcoholic Beverages. *J. Biosci. Bioeng.* **2006**, *102*, 269–280.

35. Tsachaki, M.; Aznar, M.; Linforth, R. T.; Taylor, A. J. Dynamics of Flavor Release from Ethanolic Solutions. *Dev. Food Sci.* **2006**, *43*, 441–444.

36. Tsachaki, M.; Linforth, R. S.; Taylor, A. Dynamic Headspace Analysis of the Release of Volatile Organic Compounds from Ethanolic Systems by Direct APCI-MS. *J. Agric. Food Chem* **2005**, *53*, 8328–8333.

37. Taylor, A. J.; Tsachaki, M.; Lopez, R.; Morris, C.; Ferreira, V.; Wolf, B. Odorant Release from Alcoholic Beverages. In *Flavors in Noncarbonated Beverages*; Da Costa, N. C.; Cannon R. J. , Eds.; ACS Symposium Series 1036; American Chemical Society: Wahington, DC, 2010; pp 161–175.

38. Conner, J. M.; Birkmyre, L.; Pasterson, A.; Piggott, J. Headspace Concentrations of Ethyl Esters at Different Alcoholic Strengths. *J. Sci. Food Agric.* **1998**, *77*, 121–126.

39. Tsachaki, M.; Gady, A. L.; Kalopesas, M.; Linforth, R. S.; Athes, V.; Marin, M.; Taylor, A. J. Effect of Ethanol, Temperature, and Gas Flow Rate on Volatile Release from Aqueous Solutions Under Dynamic Headspace Dilution Conditions. *J. Agric. Food Chem* **2008**, *56*, 5308–5315.

40. Boothroyd, E. L.; Linforth, R. T.; Cook, D. J. Effects of Ethanol and Long-Chain Ethyl Ester Concentrations on Volatile Partitioning in a Whisky Model System. *J. Agric. Food Chem* **2012**, *60*, 9959–9966.

41. O'Hare, K. D.; Spedding, P. L.; Grimshaw, J. Evaporation of the Ethanol and Water Components Comprising a Binary Liquid Mixture. *Dev. Chem. Eng. Miner. Process.* **1993**, *1*, 118–128.

42. Marangoni, C. Ueber die Ausbreitung der Tropfen Einer Flüssigkeit auf der Oberfläche Einer Anderen (On the Expansion of a Drop of Fluid on the Surface of Another). *Anna. Phys.* **1871**, *219*, 337–354.

43. Hosoi, A. E.; Bush, J. M. Evaporative Instabilities in Climbing Films. *J. Fluid Mech.* **2001**, *442*, 217–239.

44. Conner, J. M.; Paterson, A.; Piggott, J. R. Agglomeration of Ethyl Esters in Model Spirit Solutions and Malt Whiskies. *J. Sci. Food Agric* **1994**, *66*, 45–53.

45. Green, B. G. The Sensitivity of the Tongue to Ethanol. *Ann. N.Y. Acad. Sci.* **1987**, *510*, 315–317.

46. Cometto-Muniz, J. E.; Cain, W. S. Thresholds for Odor and Nasal Pungency. *Physiol. Behav.* **1990**, *48*, 719–725.

47. Laska, M.; Distel, H.; Hudson, R. Trigeminal Perception of Odorant Quality in Congenitally Anosmic Subjects. *Chem. Senses* **1997**, *22*, 456–477.

48. Ramsey, I.; Ross, C.; Ford, R.; Fisk, I.; Yang, Q.; Gomez-Lopez, J.; Hort, J. Using a Combined Temporal Approach to Evaluate the Influence of Ethanol Concentration on Liking and Sensory Attributes of Larger Beer. *Food Qual. Prefer.* **2018**, *68*, 292–303.

49. Mattes, R. D.; DiMeglio, D. Ethanol Perception and Ingestion. *Physiol. Behav.* **2001**, *72*, 217–229.

50. Cretin, B. N.; Dubourdieu, D.; Marchal, A. Influence of Ethanol Content on Sweetness and Bitterness Perception in Dry Wines. *Food Sci. Technol.* **2018**, *87*, 61–66.

51. Lachenmeier, D. W.; Kanteres, F.; Rehm, J. Alcoholic Beverage Strength Discrimination by Taste May Have an Upper Threshold. *Alcohol. Clin. Exp. Res.* **2014**, *38*, 2460–2467.

52. Lachenmeier, D. W.; Kanteres, F.; Rehm, J. Is It Possible To Distinguish Vodka by Taste? Comment on Structurability: A Collective Measure of the Structural Differences in Vodkas. *J. Agric. Food Chem.* **2011**, *59*, 464–465.

53. Margulies, N. R; Goetzl, F. R. The Effect of Alcohol upon the Acuity of the Sense of Taste for Sucrose and the Sensation Complex of Appetite and Satiety. *Perm. Found Med. Bull.* **1950**, *8*, 102–106.

54. Noble, A. C. Bitterness in Wine. *Physiol. Behav.* **1994**, *56*, 1251–1255.

55. Hirsch, A. R.; Bissell, G. Effects of Acute Alcohol Inebriation and Human Olfaction: A Preliminary Report. *J. Neurol. Orthop. Med. Surg.* **1998**, *18*, 114–121.

56. Matin, S.; Pangborn, R. M. Taste Interaction of Ethyl Alcohol with Sweet, Salty, Sour and Bitter Compounds. *J. Sci. Food Agric.* **1970**, *21*, 653–655.

57. Fontoin, H.; Saucier, C.; Teissedre, P. L.; Glories, Y. Effect of pH, Ethanol and Acidity on Astringency and Bitterness of Graphe Seed Tannin Oligomers in Model Wine Solution. *Food Qual. Prefer.* **2008**, *19*, 286–291.

58. Ickes, C. M.; Cadwallader, K. R. Effect of Ethanol on Flavor Perception of Rum. *Food Sci. Nutr.* **2018**, *6*, 912–924.

59. Amoore, J. E.; Hautala, E. Odor as an Ald to Chemical Safety: Odor Threshold Compared with Threshold Limit Values and Volatilities for 214 Industrial Chemicals in Air and Water Dilution. *J. Appl. Toxicol.* **1983**, *3*, 272–290.

60. Grosch, W. Evaluation of the Key Odorants of Foods by Dilution Experiments, Aroma Models and Omission. *Chem. Senses* **2001**, *26*, 533–545.

61. Guth, H. Comparison of Different White Wine Varieties in Odor Profiles by Instrumental Analysis and Sensory Studies. In *Chemistry of Wine Flavor*; Waterhouse, A. L.; Ebeler, S., Eds.; ACS Symposium Series 714; American Chemical Society: Washington, DC, 1998; pp 39–52.

62. Escudero, A.; Campo, E.; Farina, L.; Cacho, J.; Ferreira, V. Analytical Characterization of the Aroma of Five Premium Red Wines. Insights into the Role of Odor Families and Concept of Fruitiness of Wines. *J. Agric. Food Chem.* **2007**, *55*, 4501–4510.

63. Le Berre, E.; Atanasova, B.; Langlois, D.; Etievant, P.; Thomas-Danguin, T. Impact of Ethanol on the Perception of Wine Odorant Mixtures. *Food Qual. Prefer.* **2007**, *18*, 901–908.

64. Jones, P. R.; Gawel, R.; Francis, I. L.; Waters, E. J. The Influence of Interactions Between Major White Wine Components on the Aroma, Flavor and Texture of Model White Wine. *Food Qual. Prefer.* **2008**, *19*, 596–607.

65. Goldner, M. C.; Zamora, M. C.; Lira, P. D.; Gianninoto, H.; Bandoni, A. Effect of Ethanol Level in the Perception of Aroma Attributes and the Detection of Volatile Compounds in Red Wine. *J. Sens. Stud.* **2009**, *24*, 243–257.

66. Peltz, M.; Shellhammer, T. Ethanol Content has Little Effect on the Sensory Orthonasal Detection Threshold of Hop Compounds in Beer. *J. Am. Soc. Brew. Chem.* **2017**, *75*, 221–227.

67. Meilgaard, M. C. Testing for Sensory Threshold of Added Substances. *J. Am. Soc. Brew. Chem.* **1991**, *49*, 128–135.

68. Watcharananun, W.; Cadwallader, K. R.; Huangrak, K.; Kim, H.; Lorjaroenphon, Y. Identification of Predominant Odorants in Thai Desserts Flavored by Smoking with "Tian Op", a Traditional Thai Scented Candle. *J. Agric. Food Chem.* **2009**, *57*, 996–1005.

69. Tandon, K.; Baldwin, E.; Shewfelt, R. Aroma Perception of Individual Volatile Compounds in Fresh Tomatoes (*Lycopersicon esculentum*, Mill.) as Affected by the Medium of Evaluation. *Postharvest Biol. Technol.* **2000**, *20*, 261–268.

70. Guth, H. Quantitation and Sensory Studies of Character Impact Odorants of Different White Wine Varieties. *J. Agric. Food Chem.* **1997**, *45*, 3027–3032.

71. Lee, K.; Paterson, A.; Piggott, J. Measurement of Thresholds for Reference Compounds for Sensory Profiling of Scotch Whiskey. *J. Inst. Brew.* **2000**, *106*, 287–294.

72. Salo, P.; Nykanen, L.; Suomalainen, H. Odor Thresholds and Relative Intensities of Volatile Aroma Components in an Artificial Beverage Imitating Whiskey. *J. Food. Sci.* **1972**, *37*, 394–398.

73. Poisson, L.; Schieberle, P. Characterization of the Key Aroma Compounds in an American Bourbon Whisky by Quantitative Measurements, Aroma Recombination, and Omission Studies. *J. Agric. Food Chem.* **2008**, *56*, 5820–5826.

# Characterization of the Key Aroma Compounds in Rum Made from Sugar Cane Juice by Means of the Sensomics Approach

Laura Franitza,[1] Peter Schieberle,[1] and Michael Granvogl[*,1,2,3]

[1]Lehrstuhl für Lebensmittelchemie, Technische Universität München, Department für Chemie, Lise-Meitner-Straße 34, D-85354 Freising, Germany

[2]Lehrstuhl für Analytische Lebensmittelchemie, Technische Universität München, Wissenschaftszentrum Weihenstephan für Ernährung, Landnutzung und Umwelt, Maximus-von-Imhof-Forum 2, D-85354 Freising, Germany

[3]Institut für Lebensmittelchemie, Fachgebiet für Lebensmittelchemie und Analytische Chemie (170a), Universität Hohenheim, Fakultät Naturwissenschaften, Garbenstrasse 28, D-70599 Stuttgart, Germany

[*]E-mail: Michael.Granvogl@ch.tum.de.

The sensomics approach was used to characterize the key odorants in rum made from sugar cane juice. Aroma extract dilution analysis (AEDA) revealed 39 aroma-active compounds in the flavor dilution (FD) factor range from 8 to 2048. Per this analysis, 2- and 3-methyl-1-butanol, *cis*-whisky lactone, 2-phenylethanol, decanoic acid, and vanillin were identified with the highest FD factors. In total, 36 of the identified aroma compounds showed FD factors of at least 32 and were analyzed using stable isotope dilution analysis (SIDA) or an enzymatic reaction (acetic acid and ethanol). With the quantitative data at hand, odor activity values (OAVs), the ratios of concentration to odor threshold, were calculated; (*E*)-$\beta$-damascenone, ethyl (*S*)-2-methylbutanoate, and ethanol had the highest OAVs. The aroma was successfully simulated by a recombination experiment using reference compounds in their naturally occurring concentrations. Finally, formation pathways for the key odorants are discussed.

## Introduction

Rum is a traditional and popular spirit originating from the Carribean and American countries. Traditionally, rum is produced from molasses, which is the remainder after sugar production. After fermentation, the molasses is used for distillation. As a final step, the distillate is stored in oak barrels leading to the brown, sometimes slightly golden color of the spirit. In contrast, white rum is stored in steel cylinders or filtered to remove any colorings. Before the rum is bottled, it is diluted to the desired alcohol content and sometimes blended to keep a constant quality. In addition, rum can also be produced directly from fresh or cooked sugar cane juice, which is common in Haiti, Martinique, and Guadeloupe. If the production process complies with the local regulations called "Appelation d'Origine Contrôlée," the rum can be labeled as "Rhum Agricole."

The first studies on rum were performed in 1963 by Bober and Haddaway (1). Using gas chromatography, characteristic chromatograms were determined for rum, brandy, and whisky and showed 1-propanol and methylpropanol as the main components. Further studies were done by Maarse et al. (2) identifying more than 100 volatile compounds. Pino et al. (3) suggested 20 compounds (e.g., ethyl hexanoate, 2-phenylethanol, and vanillin) as key aroma compounds in aged rum using aroma extract dilution analysis (AEDA), quantitation experiments, and the calculation of odor activity values (OAVs).

Nowadays, various rum qualities with different aroma profiles are commercially available. The key aroma compounds of two different rums made from molasses with variation in pricing were analyzed by Franitza et al. (4), showing a clear difference in the amounts of their respective key odorants (e.g., in (E)-$\beta$-damascenone, vanillin, and 2,3-butanedione). In general, a less pronounced overall aroma was shown in the lower priced rum. A further study by Franitza et al. (5) proved the strong influence of the production process on the formation of the key aroma compounds in rum.

In a third study, Franitza et al. (6) analyzed the volatile fraction of different rums by means of a metabolomics approach and showed a successful authentication of sugar cane rums for the first time. A targeted analysis was performed and revealed 12 marker compounds (e.g., 3-methyl-1-butanol, 1-hexanol, and $\beta$-ionone), assuming that there is also a difference in the key aroma compounds of rums that were either produced from molasses or sugar cane juice.

Therefore, the aim of the present study is the characterization of the key aroma compounds in rum produced from sugar cane juice by the Molecular Sensory Science Concept (7) consisting of the identification of the odorants with a high aroma impact by AEDA in combination with gas chromatography-mass spectrometry, quantitation experiments by means of stable isotope dilution assays, calculation of OAVs, and a recombination experiment.

## Experimental Section

### Rum Samples

A high-priced rum produced from cooked sugar cane juice (40% alcohol by volume [ABV]) was bought from an internet supplier. The rum was labeled as "Rhum Vieux Agricole," manufactured in Martinique, and stored in oak barrels for at least four years. Samples were stored at room temperature in the dark prior to analysis.

### Chemicals

The following compounds were used as references for identification and quantitation experiments and were obtained from commercial sources: acetic acid; 4-allyl-2-methoxyphenol; 2,3 butanedione; (E,E)-2,4-decadienal; decanoic acid; 1,1-diethoxyethane; ethyl butanoate; ethyl hexanoate; ethyl (S)-2-methylbutanoate; ethyl 3-methylbutanoate; ethyl pentanoate; 4-ethylphenol; ethyl 3-phenylpropanoate; hexanal; (Z)-3-hexenol; 3-hydroxy-4,5-dimethyl-2(5H)-furanone (sotolon); $\beta$-ionone; 2-isobutyl-3-methoxypyrazine; linalool; 2-methoxyphenol; 3-methylbutanal; (S)-2-methylbutanoic acid; 3-methylbutanoic acid; (S)-2-methyl-1-butanol; 3-methyl-1-butanol; 3-methylbutyl acetate; 4-methylphenol; methylpropanol; (E,E)-2,4-nonadienal;

γ-nonalactone; (E)-2-nonenal; octanoic acid; 2-phenylethanol; 4-propyl-2-methoxyphenol; cis- and trans-whisky lactone (Sigma-Aldrich Chemie, Taufkirchen, Germany); 2-methylbutanal (Alfa Aesar, Karlsruhe, Germany); ethyl cyclohexanoate and 4-ethyl-2-methoxyphenol (Lancaster, Mühlheim/Main, Germany); butanoic acid; ethanol; and 4-hydroxy-3-methoxybenzaldehyde (vanillin) (Merck, Darmstadt, Germany). 1-(2,6,6-Trimethyl-1,3-cyclohexadien-1-yl)-2-buten-1-one ((E)-β-damascenone) was kindly provided by Symrise (Holzminden, Germany).

The diethyl ether (Merck) was freshly distilled prior to use. Pentane, hydrochloric acid, sodium chloride, sodium sulfate, and sodium carbonate were also from Merck. Liquid nitrogen was obtained from Linde (Munich, Germany). All chemicals were at least of analytical grade.

## Stable Isotopically Labeled Standards

$[^2H_{12}]$-Hexanal and $[^2H_9]$-methylpropanol were purchased from Sigma Aldrich. $[^2H_9]$-2-Methylbutanoic acid, $[^2H_7]$-4-methylphenol, and $[^2H_5]$-2-phenylethanol were obtained from C/D/N Isotopes (Quebec, Canada).

The following stable isotopically labeled standards were prepared as previously described: $[^{13}C_4]$-2,3-butanedione, $[^2H_3]$-ethyl butanoate, and $[^2H_3]$-ethyl cyclohexanoate (8); $[^2H_2]$-butanoic acid (9); $[^2H_{4-7}]$-(E)-β-damascenone (10); $[^2H_{2-4}]$-(E,E)-2,4-decadienal (11); $[^2H_2]$-decanoic acid (12); $[^{13}C_2]$-1,1-diethoxyethane, $[^2H_3]$-ethyl hexanoate, and $[^2H_2]$-3-ethylphenol (13); $[^2H_{2-4}]$-4-ethyl-2-methoxyphenol and $[^2H_3]$-vanillin (14); $[^2H_3]$-ethyl 2-methylbutanoate (15); $[^2H_5]$-ethyl pentanoate and $[^2H_5]$-ethyl 3-phenylpropanoate (16); $[^{13}C_2]$-3-hydroxy-4,5-dimethyl-2(5H)-furanone (17); $[^2H_3]$-2-methoxyphenol (18); $[^2H_3]$-2-methoxy-4-(1-propenyl)phenol (19); $[^2H_2]$-3-methylbutanal and analog $[^2H_2]$-2-methylbutanal (20); $[^2H_2]$-3-methylbutanoic acid (21); $[^2H_2]$-3-methyl-1-butanol (22); $[^2H_{11}]$-3-methylbutyl acetate (23); $[^2H_{2-4}]$-4-propyl-2-methoxyphenol (24); and $[^2H_2]$-cis-whisky lactone (25).

$[^2H_9]$-Ethyl 3-methylbutanoate was obtained by esterifying $[^2H_9]$-3-methylbutanoic acid with ethanol following the procedure described for $[^2H_3]$-ethyl 2-methylbutanoate (15).

Concentrations of the isotopically labeled standards were determined using a ThermoQuest Trace 2000 gas chromatograph (Egelsbach, Germany) equipped with a flame ionization detector (FID). Methyl octanoate was used as internal standard and the FID response factor was determined for each unlabeled reference. Afterwards, the peak areas of the labeled compound and methyl octanoate as well as the FID response factor were used to calculate the concentration of the labeled standard.

## Isolation of Volatiles

Rum (100 mL) was extracted by liquid-liquid extraction using diethyl ether (3 x 100 mL) at room temperature. To remove the majority of the ethanol, the combined organic phases were washed with aqueous NaCl solution (1 mol/L; 3 x 300 mL), dried over anhydrous sodium sulfate, and filtered. To separate the volatiles from the non-volatiles, high vacuum distillation using the solvent assisted flavor evaporation (SAFE) technique (26) was applied and the distillate obtained was concentrated using a Vigreux column (50 cm x 1 cm) and microdistillation (27) to a final volume of ~200 μL.

### High-Resolution Gas Chromatography-Olfactometry (HRGC-O)

HRGC-O was performed using a Carlo Erba Instruments-type 5610 gas chromatograph (Hofheim, Germany). The samples were manually injected using a cold-on column technique at 40 °C, and two fused silica capillaries were used for separation: DB-FFAP (30 m x 0.32 mm i.d.; 0.25 μm film thickness; 1.9 mL/min flow rate of the carrier gas helium) or DB-5 (30 m x 0.25 μm i.d.; 0.25 μm film thickness; 1.2 mL/min flow rate) (both J&W Scientific; Agilent, Waldbronn, Germany). After injection, the initial temperature was held for 2 min, raised at 6 °C/min to 230 °C, and then held for 5 min. At the end of the column, the effluent was split into two equal parts by means of a Y-type quick-seal glass splitter (Chrompack, Frankfurt, Germany) and two deactivated fused silica capillaries of the same length (25 cm x 0.32 mm i.d.). One part was directed to an FID held at 250 °C and the other part was directed to a sniffing-port held at 220 °C. To determine linear retention indices (RIs) for each compound, a series of n-alkanes (C6 – C26 [FFAP] and C6-C18 [DB-5]) was used as previously described (28).

### AEDA

The extract obtained after SAFE distillation and concentration was used for AEDA. Therefore, the distillate was diluted stepwise 1+1 (v+v) with diethyl ether, and the original distillate and each dilution were analyzed using HRGC-O to determine the flavor dilution (FD) factors. The original distillate was analyzed by at least three experienced panelists to avoid overlooking odor-active compounds.

### High Resolution Gas Chromatography-Mass Spectrometry (HRGC-MS)

HRGC-MS was performed using a Hewlett-Packard gas chromatograph 5890 series II (Waldbronn) coupled to a Finnigan sector field mass spectrometer type MAT 95 S (Bremen, Germany). Mass spectra were generated in the electron ionization mode (MS-EI) at 70 eV and in the chemical ionization mode (MS-CI) at 115 eV using isobutane as reactant gas.

### Quantitation by Stable Isotope Dilution Analysis (SIDA)

Quantitation of the odor-active compounds was performed using SIDA. Therefore, the diethyl ether (1+1 by vol., but a minimum of 10 mL) and the internal standards (dissolved in diethyl ether or ethanol; amounts were determined in preliminary experiments) were added to the rum (0.1–400 mL, depending on the concentration of each analyte). For equilibration, the mixture was stirred for 30 min at room temperature, and the work-up procedure was performed as described above to isolate volatiles.

To determine the respective response factors (R$_f$), mixtures of known amounts of the unlabeled and labeled compounds in five different ratios (5:1, 3:1, 1:1, 1:3, 1:5) (Table 1) were analyzed using the respective systems.

For quantitation of butanoic acid, 2- and 3-methylbutanoic acid, and decanoic acid, the distillate was separated into an acidic fraction (AF) and a neutral/basic fraction (NBF). As a result, liquid-liquid extraction was performed using an aqueous $Na_2CO_3$ solution (0.5 mol/L; 1+1 by vol., three times, pH 10.0). The combined aqueous phases were adjusted to pH 2 using hydrochloric acid and the acidic volatiles were extracted with diethyl ether (1+1 by vol., three times). The combined organic phases were dried over anhydrous $Na_2SO_4$ and concentrated to ~100 μl.

**Table 1. Stable Isotopically Labeled Standards, Selected Ions, and Response Factors ($R_f$) used in the Stable Isotope Dilution Assays**

| Compound | Isotope Label | Ion (m/z)[a] | | $R_f$[b] |
|---|---|---|---|---|
| | | Analyte | Internal Standard | |
| 4-allyl-2-methoxyphenol[c] | _[c] | 165 | 168[c] | 0.86 |
| 2,3-butanedione | $^{13}C_4$ | 87 | 91 | 0.99 |
| (E)-β-damascenone | $^2H_{2-5}$ | 192 | 194-197[d] | 0.91 |
| (E,E)-2,4-decadienal | $^2H_{2-4}$ | 153 | 155-157[d] | 0.60 |
| decanoic acid | $^2H_2$ | 187 | 189 | 0.89 |
| 1,1-diethoxyethane | $^{13}C_2$ | 73 | 75 | 0.90 |
| ethyl butanoate | $^2H_3$ | 117 | 120 | 0.98 |
| ethyl cyclohexanoate | $^2H_3$ | 157 | 160 | 1.00 |
| ethyl hexanoate | $^2H_3$ | 145 | 148 | 0.98 |
| 4-ethyl-2-methoxyphenol | $^2H_{2-4}$ | 153 | 155-157[d] | 0.66 |
| ethyl 2-methylbutanoate | $^2H_3$ | 131 | 134 | 0.99 |
| ethyl 3-methylbutanoate | $^2H_9$ | 131 | 140 | 1.00 |
| ethyl pentanoate | $^2H_5$ | 131 | 136 | 0.96 |
| 3-ethylphenol | $^2H_2$ | 123 | 125 | 0.87 |
| 4-ethylphenole | _[e] | 123 | 125[e] | 0.87 |
| ethyl 3-phenylpropanoate | $^2H_5$ | 179 | 184 | 1.00 |
| hexanal | $^2H_{12}$ | 101 | 113 | 0.94 |
| 2-methoxyphenol | $^2H_3$ | 125 | 128 | 0.98 |
| 2-methoxy-4-(1-propenyl)phenol | $^2H_3$ | 165 | 168 | 0.86 |
| 2-methoxy-4-propylphenol | $^2H_{2-4}$ | 167 | 169-171[d] | 0.98 |
| 2-methylbutanal | $^2H_2$ | 87 | 89 | 0.88 |
| 3-methylbutanal | $^2H_2$ | 87 | 89 | 0.98 |
| 2-methylbutanoic acid | $^2H_9$ | 103 | 112 | 0.88 |
| 3-methylbutanoic acid | $^2H_2$ | 103 | 105 | 0.98 |
| 2-methylbutanol[f] | _[f] | 71 | 73[f] | 0.84 |
| 3-methylbutanol | $^2H_2$ | 71 | 73 | 0.99 |
| 3-methylbutyl acetate | $^2H_{11}$ | 131 | 142 | 0.90 |

**Table 1. (Continued). Stable Isotopically Labeled Standards, Selected Ions, and Response Factors ($R_f$) used in the Stable Isotope Dilution Assays**

| Compound | Isotope Label | Ion (m/z)[a] Analyte | Ion (m/z)[a] Internal Standard | $R_f^b$ |
|---|---|---|---|---|
| 4-methylphenol | $^2H_7$ | 109 | 116 | 0.79 |
| methylpropanol | $^2H_9$ | 57 | 66 | 0.62 |
| phenylacetic acid | $^{13}C_2$ | 137 | 139 | 0.94 |
| 2-phenylethanol | $^2H_5$ | 105 | 110 | 0.71 |
| sotolon | $^{13}C_2$ | 129 | 131 | 0.70 |
| cis-whisky lactone | $^2H_2$ | 157 | 159 | 0.79 |
| trans-whisky lactone[g] | -[g] | 157 | 159[g] | 1.00 |
| vanillin | $^2H_3$ | 153 | 156 | 1.00 |

[a] Ions used for quantitation. [b] Response factor determined by analyzing defined mixtures of analyte and internal standard. [c] 4-Allyl-2-methoxyphenol was quantitated using [$^2H_2$]-2-methoxy-4-(1-propenyl)phenol as internal standard. [d] Internal standard was used as a mixture of isotopologues. [e] 4-Ethylphenol was quantitated using [$^2H_2$]-3-ethylphenol as internal standard. [f] 2-Methylbutanol was quantitated using [$^2H_2$]-3-methylbutanol as internal standard. [g] trans-Whisky lactone was quantitated using [$^2H_2$]-cis-whisky lactone as internal standard.

Quantitation experiments were performed by means of a Varian 431 gas chromatograph (Darmstadt, Germany) equipped with a DB-FFAP (30 m x 0.25 mm i.d., 0.25 μm film thickness; J&W Scientific) coupled to a Varian 220 ion trap mass spectrometer.

If a trace compound (e.g., 3-methylbutanal) was overlapped by a major volatile or the solvent, two-dimensional gas chromatography-mass spectrometry was used. A Trace 2000 series gas chromatograph (ThermoQuest) equipped with a DB-FFAP column (30 m x 0.32 mm i.d., 0.25 μm film thickness; J&W Scientific) was coupled to a Varian CP 3800 GC equipped with an OV-1701 column (30 m x 0.25 μm i.d., 0.25 μm film thickness; J&W Scientific). Injection was performed by a Combi PAL autosampler (CTC Analytics, Zwingen, Switzerland). By means of a moving capillary stream switching system, volatiles were transferred onto the second column, and the volatiles were cryo-focused before separation using a cold trap cooled to -100 °C with liquid nitrogen. Mass spectra were generated by means of a Varian Saturn 2000 ion trap mass spectrometer running in CI mode using methanol as reactant gas at 70 eV.

### Separation of Ethyl (R)-2- and Ethyl (S)-2-Methylbutanoate, (R)-2- and (S)-2-Methylbutanoic Acid, and (R)-2- and (S)-2-Methylbutanal

It w as recently reported that chiral odorants were separated by GC/GC-MS using a DB-FFAP column (30 m x 0.32 mm i.d., 0.25 μm film thickness) in the first dimension and an appropriate chiral column (BGB Analytik, Böckten, Switzerland) in the second dimension (29). Simultaneously, a separation of 2- and 3-methyl-1-butanol as well as 2- and 3-methylbutanoic acid was achieved.

## Quantitation of Ethanol and Acetic Acid

Ethanol and acetic acid were determined using a commercial enzyme kit (R-Biopharm, Darmstadt, Germany). Analyses were performed using a Shimadzu UV-2401PC photometer (Duisburg, Germany) according to the instructions of the manufacturer.

## Determination of Orthonasal Odor Threshold

Orthonasal odor thresholds were determined in an aqueous solution containing 40% of pure ethanol (by vol.) as previously described (30).

## Aroma Profile Analysis (APA)

For APA, a sensory panel rated the intensities of selected odor attributes (butter-like, clove-like, ethanolic, fruity, malty, and vanilla-like) on a seven-point linear scale ranging from zero (not perceivable) to three (strongly perceivable). The panel consisted of 20 experienced assessors participating in weekly sensory training sessions intended to train their abilities to recognize and describe different aroma qualities. Sensory analyses were performed in a sensory room at $21 \pm 1\,°C$ equipped with single booths. The samples (15 mL) were presented in covered glass vessels (40 mm i.d., total volume of 45 mL).

## Aroma Recombination

For aroma recombination, the rum (200 mL) was extracted with pentane (3 x 200 mL) and diethyl ether (3 x 200 mL) until the remaining liquid was odorless. The liquid was freeze-dried and the lyophilisate was used as a matrix. All analyzed aroma compounds with an OAV of at least 1 were prepared in aqueous solutions containing 40% ABV and were added to the matrix in the naturally occurring concentrations determined in the rum. The recombinate and the original rum were evaluated by the sensory panel as explained above for the APA.

## Results and Discussion

To get an intial idea of the overall rum aroma, aroma profile analyses were performed (Figure 1). In general, the rum elicited a pleasant and intense aroma. The ethanolic odor quality was rated with the highest intensity followed by the fruity and vanilla-like attributes, while the butter-like note showed the lowest intensity. To elucidate which odorants were responsible for the overall aroma impression, the sensomics concept was applied (7).

### Identification of Key Aroma Compounds

To identify key aroma compounds, the rum was extracted with diethyl ether. The volatile fraction was separated from the non-volatiles by means of high vacuum distillation using the SAFE technique (26). The distillate obtained revealed the typical rum aroma when it was evaluated on a strip of filter paper and was, therefore, subsequently used for HRGC-O and AEDA. For AEDA, the distillate was diluted stepwise 1+1 (v+v) with diethyl ether and each dilution was analyzed by HRGC-O. The FD factor was determined for each odor-active compound and corresponds to the highest dilution in which the respective compound was last recognized.

Altogether, 39 odor-active regions were determined in the FD factor range between 8 and 2048 (Figure 2). The highest FD factor of 2048 was determined for compounds **11a, b** (malty odor impression) and **29** (coconut-like); followed by **27** (flowery), **38** (soapy and musty), and **39** (vanilla-like) with an FD factor of 1024; **8** (malty), **24** (smoky and gammon-like), **31** (smoky and gammon-like), and **34** (phenolic and clove-like) with an FD factor of 512; and **3** (fruity), **5** (fruity), **14** (fruity and sweet), and **37** (seasoning-like) with an FD factor of 256.

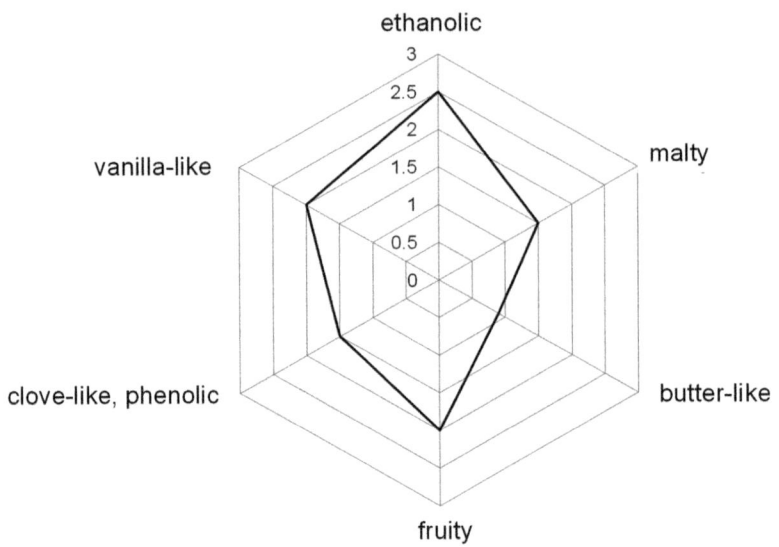

*Figure 1. Aroma profile of the rum.*

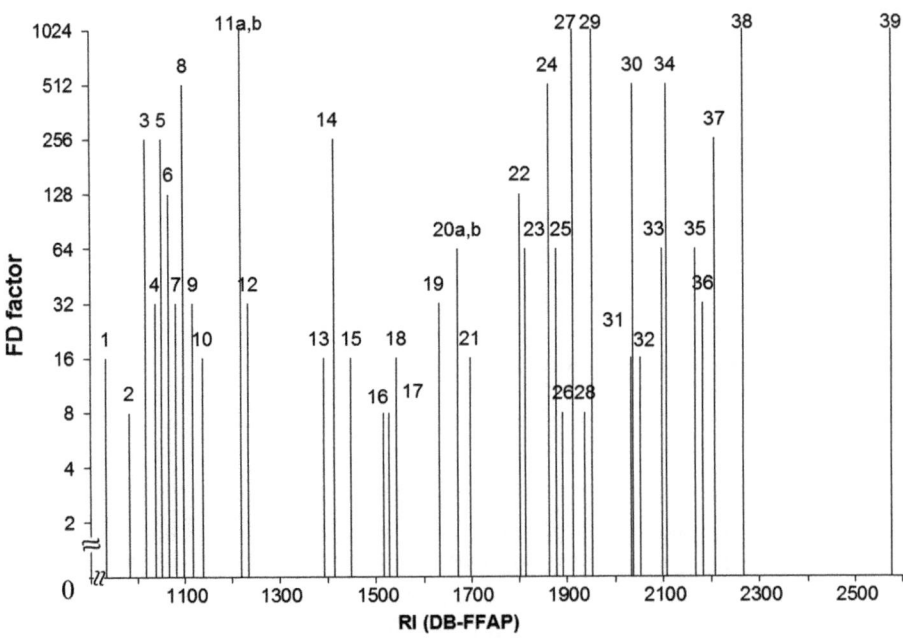

*Figure 2. FD chromatogram obtained using AEDA on the rum distillate. Odorants with an FD factor ≥ 8 are displayed. Numbering is identical to Table 2.*

298

To identify the compounds responsible for the single odor impressions, RIs on two GC columns of different polarities were compared to data available in an in-house database containing ~1000 volatiles. In addition, comparing the aroma quality and intensity of the odor impressions detected in the rum distillate to possible reference compounds at similar concentration ranges was performed. For an unequivocal identification, mass spectra (MS-EI, MS-CI) were recorded for both the odorants in the distillate and their respective reference compounds.

In this way, 2- and 3-methyl-1-butanol (**11a,b**) and *cis*-whisky lactone (**29**) were identified as having the highest FD factor of 2048 (Table 2). Thereby, 2-methyl-1-butanol was present at a level of more than 99% in (*S*)-configuration. Additionally, the separation of 2- and 3-methyl-1-butanol was achieved showing a ratio of 2-methyl-1-butanol to 3-methyl-1-butanol of 23%–77%. 2-Phenylethanol (**27**), decanoic acid (**38**), and vanillin (**39**) all showed an FD factor of 1024. Further important odorants were methylpropanol (**8**), 2-methoxyphenol (**24**), 4-ethyl-2-methoxyphenol (**31**), 4-propyl-2-methoxyphenol (**34**), 1,1-diethoxyethane (**3**), ethyl 2-methylbutanoate (**5**), ethyl cyclohexanoate (**14**), sotolon (**37**), and (*E*)-*β*-damascenone (**23**; cooked apple-like with an FD factor of 64).

### Table 2. FD Factors of the Most Aroma-Active Compounds

| No.[a] | Compound[b] | Odor Quality[c] | RI[d] | | FD Factor[e] |
|--------|-------------|-----------------|-------|--------|--------------|
| | | | DB-FFAP | DB-5 | |
| 1a, b | 2- and 3-methyl-butanal[f] | malty | 930 | 658 | 16 |
| 2 | 2,3-butanedione | butter-like | 980 | 593 | 8 |
| 3 | 1,1-diethoxyethane | fruity | 1013 | 733 | 256 |
| 4 | ethyl butanoate | fruity | 1033 | 804 | 32 |
| 5 | ethyl 2-methylbutanoate | fruity | 1047 | 854 | 256 |
| 6 | ethyl 3-methylbutanoate | blueberry-like | 1067 | 861 | 128 |
| 7 | hexanal | grassy, green | 1080 | 800 | 32 |
| 8 | methylpropanol | malty | 1093 | 640 | 512 |
| 9 | 3-methylbutyl acetate | banana-like | 1115 | 879 | 32 |
| 10 | ethyl pentanoate | fruity | 1133 | 903 | 16 |
| 11a, b | 2- and 3-methyl-1-butanol[f] | malty | 1213 | 752 | 2048 |
| 12 | ethyl hexanoate | fruity | 1230 | 1001 | 32 |
| 13 | (Z)-3-hexenol | lettuce-like | 1392 | 858 | 16 |
| 14 | ethyl cyclohexanoate | fruity, sweet | 1409 | 1136 | 256 |
| 15 | acetic acid | vinegar-like | 1443 | 610 | 16 |
| 16 | 2-isobutyl-3-methoxypyrazine | bell pepper-like | 1517 | 1184 | 8 |
| 17 | (E)-2-nonenal | fatty, green | 1527 | 1160 | 8 |
| 18 | linalool | citrus-like, flowery | 1539 | 1102 | 16 |
| 19 | butanoic acid | sweaty | 1632 | 821 | 32 |

## Table 2. (Continued). FD Factors of the Most Aroma-Active Compounds

| No.[a] | Compound[b] | Odor Quality[c] | RI[d] DB-FFAP | RI[d] DB-5 | FD Factor[e] |
|---|---|---|---|---|---|
| 20a, b | 2- and 3-methyl-butanoic acid[f] | sweaty, fruity | 1668 | 874 | 64 |
| 21 | (E,E)-2,4-nonandienal | fatty, green | 1696 | 1216 | 16 |
| 22 | (E,E)-2,4-decadienal | fatty, deep-fried | 1800 | 1323 | 128 |
| 23 | (E)-β-damascenone | baked apple-like, grape juice-like | 1811 | 1389 | 64 |
| 24 | 2-methoxyphenol | smoky, sweet | 1860 | 1090 | 512 |
| 25 | ethyl 3-phenylpropanoate | flowery | 1875 | 1350 | 64 |
| 26 | trans-whisky lactone | coconut-like | 1890 | 1300 | 8 |
| 27 | 2-phenylethanol | flowery, honey-like | 1911 | 1116 | 1024 |
| 28 | β-ionone[g] | flowery | 1933 | 1492 | 8 |
| 29 | cis-whisky lactone | coconut-like | 1950 | 1331 | 2048 |
| 30 | γ-nonalactone | coconut-like | 2025 | 1360 | 16 |
| 31 | 4-ethyl-2-methoxyphenol | smoky, gammon-like | 2029 | 1284 | 512 |
| 32 | octanoic acid | fusty, carrot-like | 2052 | 1279 | 16 |
| 33 | 4-methylphenol | fecal, horse stable-like | 2094 | 1077 | 64 |
| 34 | 4-propyl-2-methoxyphenol | phenolic | 2106 | 1360 | 512 |
| 35 | 4-allyl-2-methoxyphenol | clove-like | 2165 | 1359 | 64 |
| 36 | 4-ethylphenol | phenolic | 2182 | 1169 | 32 |
| 37 | sotolon | seasoning-like | 2206 | 1108 | 256 |
| 38 | decanoic acid | soapy, musty | 2265 | 1373 | 1024 |
| 39 | vanillin | vanilla-like | 2573 | 1406 | 1024 |

[a] Numbering is identical to Figure 2.  [b] Odorant was identified by comparing the RIs on two columns with different polarities, the odor quality, the odor intensity perceived at the sniffing-port, and the mass spectra (EI, CI mode) with the data of reference compounds.  [c] Odor quality detected at the sniffing-port.  [d] Retention index determined using a homologous series of n-alkanes.  [e] FD factor determined by AEDA on capillary DB-FFAP.  [f] FD factors were determined in sum.  [g] No unequivocal mass spectra were obtained. Odorant was identified based on the remaining criteria mentioned in footnote b.

Furthermore, the determined enantiomeric ratio for 2-methylbutanal was 41% (R) and 59% (S) and ethyl 2-methylbutanote and 2-methylbutanoic acid were present at a level of more than 99% in the (S)-configuration. Simultaneously, the ratio of 2- to 3-methylbutanoic acid was determined to be 49% of 2-methylbutanoic acid to 51% of 3-methylbutanoic acid.

## Quantitation of Key Aroma Compounds and Calculation of OAVs

AEDA is an excellent method for screening the potential influence of the single odorants on the overall aroma to reduce the identification and quantitation experiments to a limited number of aroma-active compounds. However, the amount present in the air above a food, such as during consumption, is significantly influenced by the binding properties of the food matrix. Thus, the concentration of each aroma-active compound must be determined and correlated with the respective odor threshold. In total, 36 aroma compounds showing an FD factor of at least 32 were quantitated using SIDAs (Table 3). Concentrations of more than 1 mg//L were determined for ethanol (316 g/L), 3-methyl-1-butanol (824 mg/L), methylpropanol (452 mg/L), acetic acid (258 mg/L; determined via enzymatic assay), (S)-2-methyl-1-butanol (245 mg/L), 1,1-diethoxyethane (25.0 mg/L), 2-phenylethanol (17.1 mg/L), decanoic acid (5.41 mg/L), vanillin (2.35 mg/L), cis-whisky lactone (1.12 mg/L), and 3-methylbutylacetate (1.06 mg/L). Lower concentrations were found for the esters, for example, ethyl (S)-2-methylbutanoate (230 µg/L), ethyl pentanoate (51.5 µg/L), and ethyl 3-phenylpropanoate (3.54 µg/L), and the phenols, for example, 4-ethyl-2-methoxyphenol (272 µg/L), 4-ethylphenol (117 µg/L), and 4-propyl-2-methoxyphenol (19.3 µg/L).

### Table 3. Concentrations of the Most Important Aroma Compounds

| Compound | Concentration[a] [µg/L] |
| --- | --- |
| ethanol | 316000000 |
| 3-methyl-1-butanol | 824000 |
| methylpropanol | 452000 |
| acetic acid | 258000 |
| (S)-2-methyl-1-butanol | 245000 |
| 1,1-diethoxyethane | 25000 |
| 2-phenylethanol | 17100 |
| decanoic acid | 5410[b] |
| vanillin | 2350 |
| cis-whisky lactone | 1120 |
| 3-methylbutyl acetate | 1060 |
| 3-methylbutanoic acid | 890 |
| (S)-2-methylbutanoic acid | 865 |
| ethyl hexanoate | 844 |
| butanoic acid | 563 |
| ethyl butanoate | 370 |
| 4-ethyl-2-methoxyphenol | 272 |
| 3-methylbutanal | 233 |
| ethyl (S)-2-methylbutanoate | 230 |

**Table 3. (Continued). Concentrations of the Most Important Aroma Compounds**

| Compound | Concentration[a] [$\mu$g/L] |
|---|---|
| ethyl 3-methylbutanoate | 196 |
| hexanal | 128 |
| *trans*-whisky lactone | 124 |
| 4-ethylphenol | 117 |
| 4-allyl-2-methoxyphenol | 102[b] |
| ethyl pentanoate | 51.5 |
| 2-methoxyphenol | 47.0 |
| 2,3-butanedione | 43.7 |
| (*E*)-$\beta$-damascenone | 43.7 |
| (*S*)-2-methylbutanal | 27.4 |
| 4-propyl-2-methoxyphenol | 19.3 |
| (*R*)-2-methylbutanal | 18.9 |
| sotolon | 16.2[b] |
| ethyl 3-phenylpropanoate | 3.54 |
| 4-methylphenol | 3.41 |
| (*E,E*)-2,4-decadienal | 1.15 |
| ethyl cyclohexanoate | 1.03 |

[a] Mean values of triplicates, differing not more than ± 12%.　[b] Mean values of duplicates, differing not more than ± 10%.

Finally, to evaluate the contribution of each compound to the overall aroma, OAVs (ratios of concentration to odor threshold) were calculated. The odor thresholds were determined in sensory experiments in a mixture of ethanol/water (40/60 by vol.), except for the ethanol threshold, which was determined in water (Table 4).

Altogether, 29 of the 36 quantitated aroma compounds showed OAVs of at least 1 (Table 3). As expected, ethanol showed a high OAV (319), but ethyl (*S*)-2-methylbutanoate (1150) and (*E*)-$\beta$-damascenone (437) showed even higher OAVs. OAVs of at least 100 were also calculated for sotolon (147), ethyl 3-methylbutanoate (123), and vanillin (107). Lower OAVs were found for *cis*-whisky lactone (17), 2,3-butanedione (16), 3-methylbutanol (15), 4-allyl-2-methoxyphenol (14), and 3-methylbutanoic acid (11). In contrast, some compounds such as butanoic acid, *trans*-whisky lactone, and 4-methylphenol showed OAVs of less than 1 (Table 4). These components were overestimated by AEDA (based on odor thresholds in air) not representing the influence of the matrix (simulated by an aqueous solution containing 40% of pure ethanol (by vol.), which was also used for the determination of the odor thresholds, and thus, for the calculation of the respective OAVs considering matrix effects) and should not contribute to the overall rum aroma.

**Table 4. Orthonasal Odor Thresholds and OAVs of the Most Important Aroma Compounds**

| Compound | Odor Threshold[a] [µg/L] | OAV[b] |
|---|---|---|
| ethyl (S)-2-methylbutanoate | 0.2 | 1150 |
| (E)-β-damascenone | 0.1 | 437 |
| ethanol | 990000[c] | 319 |
| sotolon | 0.11[d] | 147 |
| ethyl 3-methylbutanoate | 1.6 | 123 |
| vanillin | 22 | 107 |
| 3-methylbutanal | 2.8 | 83 |
| (S)-2-methyl-1-butanol | 6100[d] | 40 |
| ethyl butanoate | 9.5 | 39 |
| 4-ethyl-2-methoxyphenol | 6.9 | 39 |
| 1,1-diethoxyethane | 719 | 35 |
| ethyl hexanoate | 30 | 28 |
| cis-whisky lactone | 67[e] | 17 |
| 2,3-butanedione | 2.8 | 16 |
| 3-methyl-1-butanol | 56100 | 15 |
| 4-allyl-2-methoxyphenol | 7.1 | 14 |
| 3-methylbutanoic acid | 78[f] | 11 |
| 2-phenylethanol | 2600 | 7 |
| 2-methoxyphenol | 9.2 | 5 |
| ethyl pentanoate | 11[d] | 5 |
| 4-propyl-2-methoxyphenol | 4.4[d] | 4 |
| 3-methylbutyl acetate | 245 | 4 |
| ethyl cyclohexanoate | 0.3[d] | 3 |
| methylpropanol | 160000[d] | 3 |
| decanoic acid | 2800[f] | 2 |
| acetic acid | 230000[d] | 1 |
| hexanal | 87.9[g] | 1 |
| (E,E)-2,4-decadienal | 1.1 | 1 |
| (S)-2-methylbutanal | 20[d] | 1 |
| butanoic acid | 6900[d] | < 1 |
| (S)-2-methylbutanoic acid | 3500[g] | < 1 |
| trans-whisky lactone | 790[e] | < 1 |

**Table 4. (Continued). Orthonasal Odor Thresholds and OAVs of the Most Important Aroma Compounds**

| Compound | Odor Threshold[a] [µg/L] | OAV[b] |
|---|---|---|
| 4-methylphenol | 89[g] | < 1 |
| ethyl 3-phenylpropanoate | 14[f] | < 1 |
| (R)-2-methylbutanal | 110[d] | < 1 |
| 4-ethylphenol | 173 | < 1 |

[a] Odor thresholds were determined in ethanol/water (40/60 by vol.) (25).   [b] Odor activity values were calculated by dividing the concentrations (see Table 3) by the respective odor thresholds.   [c] Odor threshold was determined in water.   [d] Odor threshold was newly determined in ethanol/water (40/60 by vol.).   [e] Odor threshold previously reported in ref. (45).   [f] Odor threshold previously reported in ref. (40).   [g] Odor threshold previously reported in ref. (4).

## Aroma Recombination Experiments

To verify if all key aroma compounds were correctly identified and quantitated, an aroma recombination experiment was performed. An odorless matrix was produced by extracting the rum with pentane and diethyl ether (three times each) and the resulting odorless aqueous phase was freeze-dried. All 29 aroma compounds showing OAVs of at least 1 were dissolved in ethanolic solutions (40% by vol.) and added in their determined naturally occurring concentrations to the odorless matrix. The recombinate was analyzed compared to the original rum by a trained sensory panel using APA. Therefore, the intensities of defined odor qualities had to be rated from 0 (not perceivable) to 3 (strongly perceivable). The aroma profiles of the rum and the recombinate showed a perfect similarity proving a correct identification and quantitation of all key odorants (Figure 3).

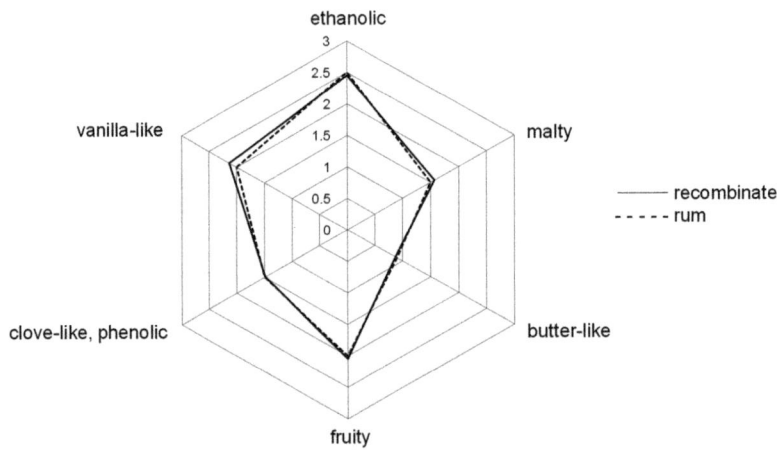

Figure 3. Aroma profile analyses of the original rum (dotted line) and of the corresponding aroma recombinate (solid line).

## Discussion of Formation Pathways of Key Odorants

The second highest OAV was determined for (*E*)-*β*-damascenone, which was already identified in sugar cane molasses by Masuda and Nishimura (*31*). Franitza et al. (*5*) studied the influence of the production process on the formation of key odorants in rum. The authors also found (*E*)-*β*-damascenone in molasses, which showed a clear increase in its concentration during distillation. This finding was also observed during the production of apple brandy (*32*), whisky (*33*), and Williams Christ pear brandy (*34*). This indicates that (*E*)-*β*-damascenone is released from precursors by heat treatment during distillation. Different precursors were previously discussed, where higher concentrations were determined during heat treatment under acidic conditions (*35–38*).

High concentrations and OAVs of at least 1 were determined for the following alcohols: 3-methyl-1-butanol, (*S*)-2-methyl-1-butanol, 2-phenylethanol, and methylpropanol. During fermentation, these alcohols are generated from the respective amino acids via enzymatic transamination to the corresponding α-keto acid, followed by enzymatic decarboxylation and reduction via the Ehrlich pathway (*39*). Besides the the alcohols, also the aldehydes can be formed as intermediates prior to the formation of the previously mentioned alcohols. The aldehydes (e.g., 2- or 3-methylbutanal) were described as key aroma compounds of alcoholic beverages in several studies (*33, 40, 41*). However, the aldehydes can also react to the respective acids via oxidation. An overview of the Ehrlich pathway is shown in Figure 4, exemplified by L-isoleucine and the corresponding products.

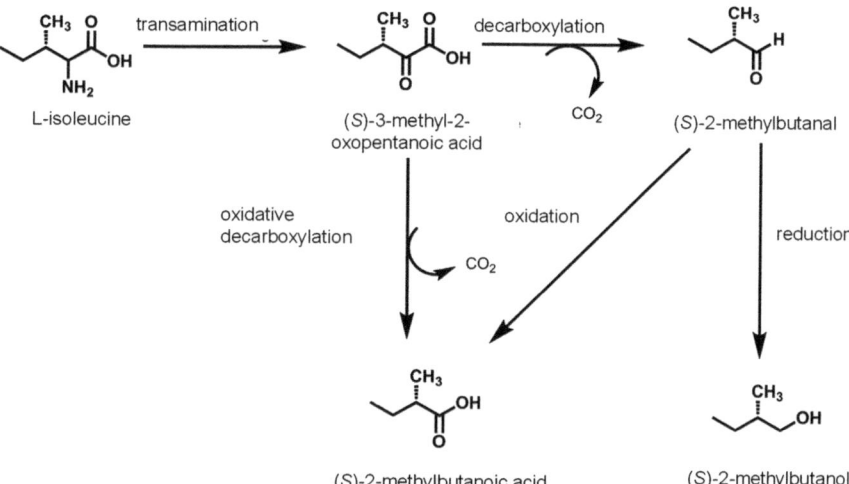

*Figure 4. Degradation of the amino acid L-isoleucine via the Ehrlich pathway.*

Matheis et al. (*29*) analyzed the enantiomeric ratios of the degradation products of L-isoleucine in different foods. In this analysis, 2-methyl-1-butanol was present in almost all samples with more than 99% of the (*S*)-enantiomer. In contrast, 2-methylbutanal was determined in different ratios, varying between 52% of the (*S*)-enantiomer in pretzels and 74% of the (*S*)-enantiomer in tequila. Further experiments showed that a racemization is not possible due to the food production process. This study indicated that the alcohol cannot be formed exclusively from the aldehyde. Also, in the actual study, 2-methyl-1-butanol showed more than 99% of the (*S*)-enantiomer, while 59% of (*S*)-2-methylbutanal was found. In comparison to the rums already analyzed by Franitza et al. (*5*), different concentrations of the degradation products of the amino acids were determined (e.g., the concentrations of 3-methyl-1-butanol ranged from 667 μg/L to 824 mg/L). Using different raw materials (and thereby different concentrations of amino acids) and different yeasts can lead to these different concentrations.

The amino acid-derived acids can further react with ethanol to the respective ethyl esters. In this study, ethyl 2-methylbutanoate showed greater than 99% of the (S)-enantiomer and was determined with the highest OAV.

2-Methoxyphenol, 4-allyl-2-methoxyphenol, and vanillin are characteristic aroma compounds for spirits matured in oak barrels. The origin and age of the wood used for the barrels, as well as the toasting level, have a significant influence on the spirits' concentrations. Decomposition products of lignin are formed during toasting of the barrel and can be extracted during the aging process of the spirit in the barrels (42, 43).

In addition, *cis*- and *trans*-whisky lactone were determined, which were firstly characterized by Masuda and Nishimura (44) in different oak barrels. The lactones can be extracted during maturation and their concentrations rise with increasing storage times (45). Although four different isomers are possible, until now, only (3S,4S)/(*cis*)- and (3S,4R)/(*trans*)-isomers were identified in oak wood (46) and, hence, also in the spirits held within (e.g., in cognac (47) or whisky) (33).

2,3-Butanedione is also formed as a yeast by-product during fermentation. As a result, (S)-2-hydroxy-2-methyl-3-oxobutanoic acid (α-acetolactate) can be synthesized from pyruvate, ending in 2,3-butanedione formed by oxidative decarboxylation (48, 49). The concentration of 2,3-butanedione is usually decreasing during storage due to its volatility. Consequently, in a cheap rum (i.e., not stored for a longer time), a clearly higher concentration was determined (621 μg/L) (4).

All in all, 2,3-butanedione, (E)-β-damascenone, 1,1-diethoxyethane, ethanol, ethyl butanoate, ethyl hexanoate, ethyl (S)-2-methylbutanoate, ethyl 3-methylbutanoate, 3-methylbutanal, and sotolon were determined both in the sugar cane rum in this study and in two molasses rums from a previous study (4) in different concentrations over their odor thresholds and, as a result, were characterized as key odorants. However, sensory experiments showed that these aroma compounds were not sufficient to mimic the overall aroma profile of rum. Depending on the type of rum, further key aroma compounds in different concentrations are characteristic and necessary for the unique aroma profile of the three rums.

## References

1. Bober, A.; Haddaway, L. W. Gas Chromatographic Identification of Alcoholic Beverages. *J. Chrom. Sci.* **1963**, *1*, 8–13.

2. Maarse, H.; Ten Noever de Brauw, M. C. Analysis of Volatile Components of Jamaica Rum. *J. Food Sci.* **1966**, *31*, 951–955.

3. Pino, J. A.; Tolle, S.; Gök, R.; Winterhalter, P. Characterisation of Odour-Active Compounds in Aged Rum. *Food Chem.* **2012**, *132*, 1436–1441.

4. Franitza, L.; Granvogl, M.; Schieberle, P. Characterization of the Key Aroma Compounds in Two Commercial Rums by Means of the Sensomics Approach. *J. Agric. Food Chem.* **2016**, *64*, 637–645.

5. Franitza, L.; Granvogl, M.; Schieberle, P. Influence of the Production Process on the Key Aroma Compounds of Rum: From Molasses to the Spirit. *J. Agric. Food. Chem.* **2016**, *64*, 9041–9053.

6.  Franitza, L.; Nicolotti, L.; Granvogl, M.; Schieberle, P. Differentiation of Rums Produced from Sugar Cane Juice (Rhum Agricole) from Rums Manufactured from Sugar Cane Molasses by a Metabolomics Approach. *J. Agric. Food. Chem.* **2018**, *66*, 3038–3045.

7.  Schieberle, P.; Hofmann, T. Mapping the Combinatorial Code of Food Flavors by Means of Molecular Sensory Science Approach. In *Chemical and Functional Properties of Food Components Series. Food Flavors. Chemical, Sensory and Technological Properties*; Jelen, H. , Ed.; CRC Press: Boca Raton, FL, 2012; pp 413–438.

8.  Schieberle, P.; Hofmann, T. Evaluation of the Character Impact Odorants in Fresh Strawberry Juice by Quantitative Measurements and Sensory Studies on Model Mixtures. *J. Agric. Food Chem.* **1997**, *45*, 227–232.

9.  Schieberle, P.; Gassenmeier, K.; Guth, H.; Sen, A.; Grosch, W. Character Impact Odour Compounds of Different Kinds of Butter. *Lebensm.-Wiss. Technol.* **1993**, *26*, 347–356.

10. Sen, A.; Laskawy, G.; Schieberle, P.; Grosch, W. Quantitative Determination of $\beta$-damascenone in Foods Using a Stable Isotope Dilution Assay. *J. Agric. Food Chem.* **1991**, *39*, 757–759.

11. Guth, H.; Grosch, W. Deterioration of Soya-Bean Oil: Quantification of Primary Flavor Compounds Using a Stable Isotope Dilution Assay. *Lebensm.-Wiss. Technol.* **1990**, *23*, 513–522.

12. Czerny, M.; Schieberle, P. Influence of the Polyethylene Packaging on the Adsorption of Odour-Active Compounds from UHT-Milk. *Eur. Food Res. Technol.* **2007**, *225*, 215–223.

13. Guth, H. Quantitation and Sensory Studies of Character Impact Odorants of Different White Wine Varieties. *J. Agric. Food Chem.* **1997**, *45*, 3027–3032.

14. Semmelroch, P.; Laskawy, G.; Blank, I.; Grosch, W. Determination of Potent Odourants in Roasted Coffee by Stable Isotope Dilution Assays. *Flavour Frag. J.* **1995**, *10*, 1–7.

15. Guth, H.; Grosch, W. Quantitation of Potent Odorants of Virgin Olive Oil by Stable-Isotope Dilution Assays. *J. Am. Oil Chem. Soc.* **1993**, *70*, 513–518.

16. Pfnür, P. *Studies on Chocolate Aroma (in German)*. Ph.D. Thesis, Technical University of Munich, Munich, Germany, 1998.

17. Blank, I.; Schieberle, P.; Grosch, W. Quantification of the Flavour Compounds 3-Hydroxy-4,5-Dimethyl-2(5H)-Furanone and 5-Ethyl-3-Hydroxy-4-Methyl-2(5H)-Furanone by a Stable Isotope Dilution Assay. In *Progress in Flavour Precursor Studies: Analysis, Generation, Biotechnology: Proceedings of the International Conference*; Schreier, P., Winterhalter, P. , Eds.; Allured Publishing Corporation: Carol Stream, IL, 1993; pp 103–109.

18. Cerny, C.; Grosch, W. Quantification of Character-Impact Odour Compounds of Roasted Beef. *Z. Lebensm.-Unters. Forsch.* **1993**, *196*, 417–422.

19. Schaller, T. *Characterization of the Key Aroma Compounds in Ginger (Zingiber officinale Roscoe) and Changes Due to Thermal Treatment (in German)*. Ph.D. Thesis, Technical University of Munich, Munich, Germany, 2013.

20. Schieberle, P.; Grosch, W. Changes in the Concentrations of Potent Crust Odourants During Storage of White Bread. *Flavour Frag. J.* **1992**, *7*, 213–218.

21. Guth, H.; Grosch, W. Identification of the Character Impact Odorants of Stewed Beef Juice by Instrumental Analyses and Sensory Studies. *J. Agric. Food Chem.* **1994**, *42*, 2862–2866.

22. Gassenmeier, K.; Schieberle, P. Potent Aromatic Compounds in the Crumb of Wheat Bread (French-Type) — Influence of Pre-Ferments and Studies on the Formation of Key Odorants During Dough Processing. *Z. Lebensm.-Unters. Forsch.* **1995**, *201*, 241–248.

23. Fuhrmann, E. *Study on the Aroma of Apples in Dependancy on the Variety and Research on Structure-Odor-Relationships of Involved Esters (in German)*. Ph.D. Thesis, Technical University of Munich, Munich, Germany, 1998.

24. Schmitt, R. *On the Role of Ingredients as Sources of Key Aroma Compounds in Crumb Chocolate (in German)*. Ph.D. Thesis, Technical University of Munich, Munich, Germany, 2005.

25. Poisson, L.; Schieberle, P. Characterization of the Key Aroma Compounds in an American Bourbon Whisky by Quantitative Measurements, Aroma Recombination, and Omission Studies. *J. Agric. Food Chem.* **2008**, *56*, 5820–5826.

26. Engel, W.; Bahr, W.; Schieberle, P. Solvent Assisted Flavour Evaporation — A New and Versatile Technique for Careful and Direct Isolation of Aroma Compounds from Complex Food Matrices. *Eur. Food Res. Technol.* **1999**, *209*, 237–241.

27. Bemelmans, J. M. H. Review of Isolation and Concentration Techniques. In *Progress in Flavour Research*; Land D. G., Nursten H. E., Eds.; Applied Science: London, UK, 1979; pp 79–98.

28. Schieberle, P. Primary Odorants of Pale Lager Beer. Differences to Other Beers and Changes During Storage. *Z. Lebensm.-Unters. Forsch.* **1991**, *193*, 558–565.

29. Matheis, K.; Granvogl, M.; Schieberle, P. Quantitation and Enantiomeric Ratios of Aroma Compounds Formed by an Ehrlich Degradation of L-Isoleucine in Fermented Foods. *J. Agric. Food Chem.* **2016**, *64*, 646–652.

30. Czerny, M.; Christlbauer, Ma.; Christlbauer, Mo.; Fischer, A.; Granvogl, M.; Hammer, M.; Hartl, C.; Hernandez, N.; Schieberle, P. Re-investigation on Odour Thresholds of Key Food Aroma Compounds and Development of an Aroma Language Based on Odour Qualities of Defined Aqueous Odorant Solutions. *Eur. Food Res. Technol.* **2008**, *228*, 265–273.

31. Masuda, M.; Nishimura, K. I. C. Occurence and Formation of Damascenone, trans-2,6,6-Trimethyl-1-crotonyl-cyclohexa-1,3-diene, in Alcoholic Beverages. *J. Food Sci.* **1980**, *45*, 396–397.

32. Schreier, P.; Drawert, F.; Steiger, G. Application HTST-Heating of the Mash and Its Influence on the Aroma Composition During the Production of Apple Brandy (in German). *Z. Lebensm.-Unters. Forsch.* **1978**, *167*, 16–22.

33. Vocke, M. *The influence of the Processing Steps on the Formation of Important Aroma Compounds in American Whiskey (in German)*. Ph.D. Thesis, Technical University of Munich, Munich, Germany, 2008.

34. Willner, B. *Characterization of Key Aroma Compounds in Williams Christ Pear Brandy — Analyses of the Influence of the Process Technology (in German)*. Ph.D. Thesis, Technical University of Munich, Munich, Germany, 2014.

35. Isoe, S.; Katsumura, S.; Sakan, T. The Synthesis of Damascenone and $\beta$-Damascone and the Possible Mechanism of their Formation from Carotenoids. *Helv. Chim. Acta.* **1973**, *56*, 1514–1516.

36. Ohloff, G.; Rautenstrauch, V.; Schulte-Elte, K. H. Model Reactions of the Biosynthesis of Compounds of the Damascone-Series and Their Preparative Use. *Helv. Chim. Acta.* **1973**, *56*, 1503–1513.

37. Skouroumounis, G. K.; Massy-Westropp, R. A.; Sefton, M. A.; Williams, P. J. Precursors of Damascenone in Fruit Juices. *Tetrahedron Lett.* **1992**, *33*, 3533–3536.

38. Skouroumounis, G. K.; Sefton, M. A. Acid-Catalyzed Hydrolysis of Alcohols and Their *β*-D-Glucopyranosides. *J. Agric. Food Chem.* **2000**, *48*, 2033–2039.

39. Ehrlich, F. The Chemical Processes Accompanying Yeast Fermentation (in German). *Ber. Dtsch. Chem. Ges.* **1907**, *40*, 1027–1047.

40. Willner, B.; Granvogl, M.; Schieberle, P. Characterization of the Key Aroma Compounds in Bartlett Pear Brandies by Means of the Sensomics Concept. *J. Agric. Food Chem.* **2013**, *61*, 9583–9593.

41. Frank, S.; Wollmann, N.; Schieberle, P.; Hofmann, T. Reconstitution of the Flavor Signature of Dornfelder Red Wine on the Basis of the Natural Concentrations of Its Key Aroma and Taste Compounds. *J. Agric. Food Chem.* **2011**, *59*, 8866–8874.

42. Baldwin, S.; Black, R. A.; Andreasen, A. A.; Adams, S. L. Aromatic Congener Formation in Maturation of Alcoholic Distillates. *J. Agric. Food Chem.* **1967**, *15*, 381–385.

43. Mosedale, J. R.; Puech, J. L. Wood Maturation of Distilled Beverages. *Trends Food Sci. Technol.* **1998**, *9*, 95–101.

44. Masuda, M.; Nishimura, K. Branched Nonalactones from Some *Quercus* Species. *Phytochemistry* **1971**, *10*, 1401–1402.

45. Otsuka, K.; Zenibayashi, Y.; Itoh, M.; Totsuka, A. Presence and Significance of Two Diastereomers of *β*-Methyl-*γ*-octalactone in Aged Distilled Liquors. *Agri. Biol. Chem.* **1974**, *38*, 485–490.

46. Günther, C.; Mosandl, A. Stereoisomeric Aroma. XV. Chirospecific Analysis of Natural Aroma Components: 3-Methyl-4-octanolide-"Quercus-, Whiskylactone" (in German). *Z. Lebensm. Unters. Forsch.* **1987**, *185*, 1–4.

47. Uselmann, V.; Schieberle, P. Decoding the Combinatorial Aroma Code of a Commercial Cognac by Application of the Sensomics Concept and First Insights into Differences from a German Brandy. *J. Agric. Food Chem.* **2015**, *63*, 1948–1956.

48. Lewis, K. F.; Weinhouse, S. Studies in Valine Biosynthesis. II. α-Acetolactate Formation in Microorganisms. *J. Am. Chem. Soc.* **1958**, *80*, 4913–4915.

49. Suomalainen, H.; Ronkainen, P. Mechanism of Diacetyl Formation in Yeast Fermentation. *Nature* **1968**, *220*, 792–793.

# Chapter 19

# Implementation of Stir Bar Sorptive Extraction (SBSE) for the Analysis of Volatile Compounds in Tequila

Miriam G. Rodríguez-Olvera,[1] Luisa I. Rodríguez-Rodríguez,[1] Michael C. Qian,[2] Yan Ping Qian,[2] and Pedro A. Vazquez-Landaverde*,[1]

[1]Centro de Investigación en Ciencia Aplicada y Tecnología Avanzada del Instituto Politécnico Nacional Unidad Queretaro, Cerro Blanco 141 Colinas del Cimatario, Queretaro, Queretaro, Mexico, 76090

[2]Department of Food Science and Technology, Oregon State University, 100 Wiegand Hall, Corvallis, Oregon 97331, United States

*E-mail: pavazquez@ipn.mx.

Tequila has a particular aroma and taste that distinguishes it from other alcoholic beverages, giving it international appeal. Tequila's unique sensory attributes are the result of a particular and complex combination of several chemical compounds. Therefore, a high sensitivity chemical analysis has been developed to evaluate the composition of tequila. The technique of solid phase extraction with sorption bar (SBSE) coupled with gas chromatography from mass spectrometry is highly sensitive and has the ability to identify trace compounds that could serve as quality markers and help with authentication. The objective of this work was to implement the SBSE methodology in the analysis of the volatile composition of tequila. Best analysis conditions for the polydimethylsiloxane (PDMS) bar were defined as follows: 15 mL of a sample with 59% dilution of tequila in water, extraction for 140 min at 1200 rpm, a desorption temperature of 264 °C, and a cryogenic temperature of -124 °C. For the ethylene glycol/silicone (EG/Silicone) bar, best conditions were defined as: 15 mL of a sample with a 70% dilution of tequila in water, extraction for 85 min at 835 rpm, a desorption temperature of 220 °C, and a cryogenic temperature of -130 °C. Several tequila samples were analyzed in their different types (silver, gold, aged, extra aged, ultra aged, and mixed), along with other distilled agave alcoholic drinks. A total of 590 chemical compounds were found, which is the greatest number reported so far for tequila. Principal component analysis indicated that SBSE methodology can differentiate between tequila (and different tequila types) and other distilled agave beverages. In the future, this methodology may be used to develop authentication tests, and thus guarantee the quality and authenticity of tequila around the world.

# Introduction

Tequila is a regional alcoholic beverage in Mexico that has been granted the Designation of Origin distinction. This beverage is made from distilling the fermented juice of *Agave tequilana weber* variety blue (*1*).

According to Mexican regulations, there are five types of tequila. All types can be either 100% agave (all fermented syrup comes from *A. weber* variety blue) or mixed (at least 51% of the fermented syrup comes from *A. weber* variety blue with the rest coming from any other sugar source). The five types of tequila are:

- Silver or white: Colorless transparent product, obtained by distillation. The alcohol content must be adjusted to 38–42% by adding water.
- Gold or young: Results of the mixture between silver tequila and aged, extra aged, or ultra aged tequila.
- Aged: Silver tequila matured for at least two months in oak barrels.
- Extra aged: Silver tequila matured for at least one year in oak barrels.
- Ultra aged: Product matured for at least three years in oak barrels.

The delicate taste of tequila is a combination of aroma and flavor, both determining consumer acceptance. Although the aroma (which is determined by volatile compounds) is considered to have the greatest impact on the perception of taste in all distilled beverages, these compounds combined with the non-volatile compounds constitute a complex mixture in a water-ethanol matrix (*2*).

The composition of tequila is similarly complex to other products that are distilled. For this reason, different compounds have been reported as responsible for the aroma and taste of tequila, such as alcohols, fatty acids, esters, aldehydes, terpenes, phenols, lactones, and sulfur compounds. (*3*).

In 1996, Benn and Peppard identified more than 175 compounds in a tequila extract with dichloromethane. Sixty of these compounds had aromatic properties. Among the components with higher concentrations are alcohols and esters that contribute to the sweet and floral smells in tequila and produce pleasant fruity flavors. However, the less volatile components also contribute to the overall development of flavor and aroma (*4*).

Solid phase microextraction methodology (SPME) along with gas chromatography coupled to mass spectrometry has been used to analyze tequila for the characterization of volatile compounds and the specific quantification of ethyl esters (*5*).

In another qualitative and quantitative study, differences between tequila types were evaluated by gas chromatography and olfactometry of extracts with dichloromethane obtained by liquid-liquid extraction (*6*).

A solid phase extraction procedure followed by high performance liquid chromatography (HPLC) analysis with UV-vis photodiode detector was proposed to simultaneously determine 11 markers in the tequila aging process. The proposed methodology was applied to a set of 15 samples of tequila grouped by the state of aging (silver, aged, and ultra aged) (*7*).

In another study reported by Martínez-López et al. (2011), the prism base surface plasmon resonance (SPR) technique is used to differentiate between three classes of tequilas (silver, extra aged, and ultra aged) (*8*).

Using Raman spectroscopy and principal component analysis (PCA), a method was developed to qualitatively study the ethanol content in tequila samples to find the differences between each of the classes (9).

Ceballos-Magaña et al. (2014) used SMPE-HS with gas chromatography coupled with mass spectrometry to obtain the differences between each tequila (silver, gold, extra aged, and ultra aged) according to certain chemical compounds. This method was used together with the Kruskal-Wallis test to highlight the significant differences between the tequila classes (10).

Barbosa-García et al. (2007) used UV-vis spectroscopy and PCA, with partial least squares analysis (PLS-DA), to differentiate between 60 samples of silver tequilas from the 100% agave and mixed categories (11).

To differentiate and authenticate the classes of tequila, a study was carried out by Contreras et al. (2010) where after verifying in a previous study the efficiency of the UV-vis spectroscopy and the PCA, the authors studied eight brands of silver and aging tequila focusing on identifying fake and adulterated tequilas. The described method combines UV-vis spectroscopy and chemometric techniques to confirm the originality of the product (12).

The stir bar sorptive extraction (SBSE) technique was first used in 1999 by Baltussen et al. for the extraction and enrichment of organic compounds from aqueous matrices. This method is based on solid extraction-desorption, which extracts the solutes into a polymer coating in the surface of a magnetic stirring bar. The SBSE technique presents a series of clear advantages compared to the rest of the extractive techniques. Firstly, the technique does not use solvents, which has several additional advantages, including the fact that the samples do not come into contact with any solvent and are less likely to be contaminated or are of artifact formation during the extraction process. Another advantage is its high degree of automation, which makes it a simple and quick technique to use. In addition, it does not require manipulation of the sample by the analyst, nor does not require prior sampling. This diminishes analytical error. Compared with SPME, the SBSE technique has a higher analytical sensitivity because it reaches very low detection and quantification limits in ppt levels (13).

The analysis of volatile compounds in wine using the SBSE technique has been studied in various ways. It has been applied to wines aged in oak barrels, in the study of the primary aromas of the wine, or to the study of the possible effect of the maturation of the grape in the aroma of the wine (14).

The study of the presence of 2,4,6-trichloroanisole using SBSE technique was carried out by Zalacain et al. (2004), who developed an analysis method that lacked a pre-concentration stage and had a relatively short analysis time (2 h) for the detection and quantification of 2,4,6-trichloroanisole, 2,3,4,5-tetrachloroanisole, 2,3,4,5,6-pentachloroanisole, and their respective phenols in samples of red and white wine (15).

Diez-Dominguez et al. (2004) developed a method of analysis using the SBSE technique for quantifying the volatile phenols 4-ethylphenol, 4-ethylguaiacol, 4-vinylphenol, and 4-vinylguaiacol, which are the cause of negative alterations produced in the aroma of the wine (14, 16).

Delgado et al. (2010) used the SBSE technique to analyze volatile compounds in sherry brandy. The optimization of the extraction procedure was carried out using a statistical approach based on a factorial design. The developed method was ultimately applied to different sherry brandies. The results obtained show that SBSE is a suitable technique for the reliable analysis of volatile compounds in brandy. Later in 2011, the study characterized and differentiated aromatic compounds in 48 samples of commercial sherry brandies. The results obtained showed that each category of sherry brandy can be differentiated by using SBSE to analyze the volatile profile (17, 18).

There are no reports showing the successful application of the SBSE technique in Tequila, so the objective of the present study was to implement SBSE in the analysis of chemical compounds in different tequila classes (silver, gold, aged, extra aged, ultra aged, and mixed).

## Materials and Methods

### Samples of Tequila and Agave Distillates

The most popular brands of tequila available in the market were chosen for sampling. These samples were obtained in different liquor stores in Mexico. Tequila samples were: 14 silver 100% agave, 1 silver mixed, 4 gold mixed, 16 aged 100% agave, 3 aged mixed, 10 extra aged 100% agave, 2 ultra aged 100% agave, and 2 ultra aged mixed. Other agave distillates were: one silver bacanora, one aged bacanora, one silver sotol, one extra aged sotol, one silver mezcal, one aged mezcal, and one extra aged mezcal.

### Extraction of Volatile Compounds

Extraction of volatile compounds was carried out using SBSE bars of 10 mm length (Gerstel GmbH & Co. KG, Germany) coated with two types of polymers: PDMS (Polydimethylsiloxane) and EG/Silicone (Ethylene glycol with silicone).

Samples were put in a 20 mL glass vial, and a magnetic stirrer was used for agitation. As an internal standard (Sigma Aldrich, Saint Louis, Missouri, USA), 25 µl of a solution of $p$-cymene were at a concentration of 0.174 mg/Kg. After extraction time, the SBSE bar was taken out the vial using a magnet, and excess liquid was whipped out using a lint-free tissue.

### Search for Best Extraction-Desorption Conditions

A $2^3$ central composite design (CCD with $\alpha = 1.68179$, K $= 3$) was used to evaluate extraction conditions. The values corresponding to the high and low points for the extraction conditions of the PDMS and EG/silicone bar are shown in Table 1.

**Table 1. Levels of the Factors for the Study of Extraction Conditions for PDMS and EG/Silicone Bars**

| | Levels | | | Points | |
| --- | --- | --- | --- | --- | --- |
| Factors | Low | High | Central | Axial ($-\alpha$) | Axial ($+\alpha$) |
| Stirring rate (rpm) | 500 | 1200 | 850 | 261.37 | 1438.63 |
| Extraction time (min) | 60 | 120 | 90 | 39.55 | 140.45 |
| Dilution (% tequila) | 25 | 75 | 50 | 7.96 | 92.04 |

To determine the desorption conditions of each of the bars, a $2^2$ central composite design (CCD with $\alpha = 1.41421$, K $= 2$) was used. Values corresponding to each factor are shown in Table 2.

The response surface methodology was used to obtain the best extraction conditions for tequila with the SBSE technique. Two answers were chosen (the number of peaks detected and the total area obtained) to determine the influence of the factors with their levels. Data obtained were analyzed with the statistical package Design Expert V7 (Stat-Ease, Minneapolis, USA).

**Table 2. Levels of the Factors for the Study of Desorption and Cryoconcentration Conditions for PDMS and EG/Silicone Bars**

| | | PDMS | | | | |
| --- | --- | --- | --- | --- | --- | --- |
| | *Levels* | | | | *Points* | |
| *Factors* | *Low* | *High* | *Central* | *Axial (-α)* | *Axial (+α)* | |
| **Desorption temperature (°C)** | 260 | 270 | 250 | 245 | 274 | |
| **Cryoconcentration temperature (°C)** | -50 | -130 | -90 | -33 | -146 | |
| | | *EG/Silicone* | | | | |
| | *Levels* | | | | *Points* | |
| *Factors* | *Low* | *High* | *Central* | *Axial (-α)* | *Axial (+α)* | |
| **Desorption temperature (°C)** | 200 | 220 | 210 | 195 | 224 | |
| **Cryoconcentration temperature (°C)** | -50 | -130 | -90 | -33 | -146 | |

## Instrumental Analysis

A MPS2XL autosampler equipped with the ALEX liner exchanger (Gerstel GmbH & Co. KG, Germany) was used to manipulate the stirring bars, which were thermally desorbed in a thermal desorption unit model TDU (Gerstel GmbH & Co. KG, Germany) connected to a programmable temperature injector model CIS-4 (Gerstel GmbH & Co. KG, Germany). The desorption temperature was programmed in the TDU from 40 °C to a desired temperature (according to Table 2) at a rate of 60 °C/min under helium flow (1 mL/min), which was maintained for 10 min. Desorbed analytes were cryoconcentrated in the CIS-4 system to a selected freezing temperature (according to Table 2) using liquid nitrogen. Finally, the CIS-4 was programmed from the freezing temperature to 280 °C at a speed of 10 °C/s and then kept for 5 min in splitless mode.

Both the desorption unit and programmable injector were installed in a 7890A gas chromatograph (Agilent Technologies, Inc., Santa Clara, CA, USA) equipped with a HP-5MS capillary column (Agilent Technologies, Inc., Santa Clara, CA, USA) that is 60 m long, 0.25 mm diameter, and 0.25 μm in film thickness. The carrier gas was helium at 1 mL/min. The oven program was at an initial temperature of 40 °C for 3 min, then increased to 60 °C at a rate of 3 °C/min and held for 3 min, increased to 80 °C at a rate of 3 °C/min and held for 3 min, increased to 110 °C at a rate of 3 °C/min and held for 3 min, increased to 130 °C at a rate of 3 °C/min and held for 3 min, increased to 150 °C at a rate of 3 °C/min and held for 3 min, increased to 170 °C at a rate of 3 °C/min and held for 3 min, increased to 200 °C at a rate of 3 °C/min, and finally increased to 280 °C at a rate of 3 °C/min and held for 5 min.

A transfer line at 280 °C conducted the compounds to a 5975C mass spectrometer with a single quadrupole (Agilent Technologies, Inc., Santa Clara, CA, USA) set in the electronic impact mode at 70 eV. The quadrupole and ionization source temperatures were 150 °C and 230 °C, respectively, with a mass range from 30 to 800 uma.

Volatile compound identification was performed by comparing their mass spectra with that of the database NIST/EPA/NIH Mass Spectra Library, version 1.7 (NIST Standard Reference Database, Gaithersburg, Maryland, USA). An 80% match was taken as a positive identification.

Data were analyzed by Principal Component Analysis using Minitab Statistical Software (State College, Pennsylvania, USA).

## Results and Discussion

### Exploration of Extraction Conditions

Given that the sample size was defined as a non-significant factor in previous research work, it was not evaluated in the present study (19).

The addition of salt into the matrix was not evaluated either because it is not necessary to add salt to obtain a greater response from the compounds extracted from the matrix as compared to SPME. The reason is that adding NaCl in the SBSE technique has been evaluated and does not have any significant effect (20). Furthermore, for SBSE there is a high risk of damaging the bar due to salt residues (17).

Ethanol peak area was eliminated for data analysis since it is present at high concentrations and can mislead the conclusions of the statistical analysis. Also, not all siloxane compounds were included in the analysis, since they are commonly artifacts coming from the SBSE bar.

For the PDMS bar extraction conditions, a quadratic model is adequate to explain the effects between extraction conditions using a correlation coefficient of $R^2 = 0.94$ for the total area and 0.92 for the number of peaks. This coefficient determines that a high percentage of the data is explained by the model.

The linear effect of the extraction time and the quadratic effect of the dilution were statistically significant for both experimental responses. Agitation speed did not have a significant effect on either response. It is known that intensive agitation shortens the extraction time (17). The linear effect occurs because a long extraction time results in the adsorption of a greater number of compounds.

A curvature in the quadratic effect for the dilution can be noticed. In Figure 1, it is observed that when using higher amounts of tequila, there is a lower amount of analytes extracted. This effect is due to the PDMS bar which is used for non-polar compounds, and tequila has a greater amount of polar compounds (like most alcohols).

Figure 2 shows that there is no significant effect on the stirring rate, suggesting that almost any value can be used.

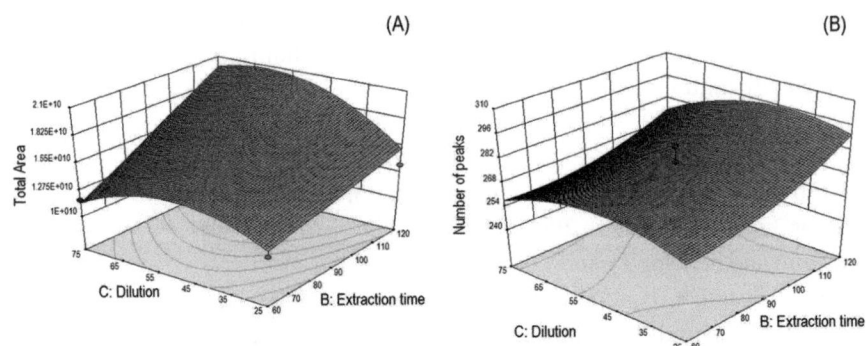

*Figure 1. Estimated response surface for the total chromatographic area (A) and the number of chromatographic peaks (B) when plotting the extraction time versus dilution for the PDMS bar.*

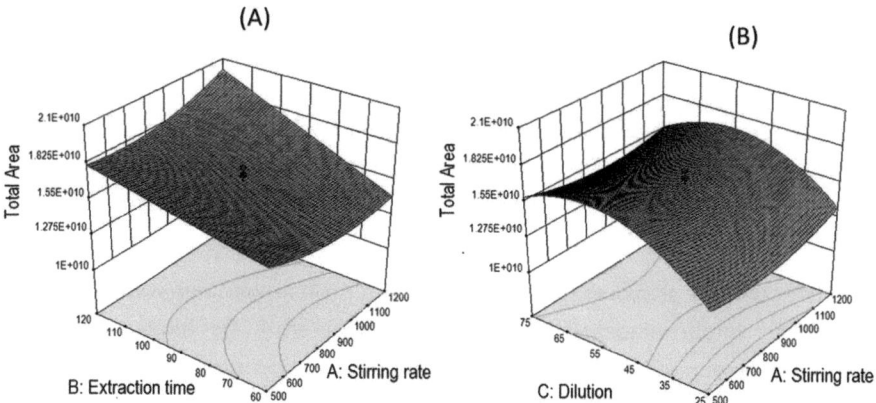

*Figure 2. Estimated response surface for the total chromatographic area of the stirring rate versus time (A) and stirring rate against dilution (B) for the PDMS bar.*

The best extraction conditions for the PDMS bar were a stirring rate of 1200 rpm, an extraction time of 140 min, and a 59% dilution of tequila in water.

For the EG/Silicone bar, the quadratic model was significant only for its total area, which makes the quadratic model unsuitable for measuring the number of peaks because several extracted compounds are in high concentrations, which mainly affects the total area. The linear effect of time can be considered significant, which shows that the number of peaks would be better explained with a linear model.

Figure 3 shows the response surface for the total area and number of peaks for the extraction conditions of tequila using the EG/Silicone bar where the quadratic effect can be observed for the dilution. This effect is caused by an increase in the percentage of the response attributed to the nature of this bar because it has a hydrophilic coating and more affinity for the alcohols present in tequila. A curvature is observed due to the increase of the tequila concentration lead to a negative effect observed by the diluent. The high concentration of ethanol in alcoholic beverages requires a dilution because ethanol is one of the main constituents that can compete with the chemical compounds in the extraction process. Some authors (20, 21) have found that an increase in the content of ethanol decreases the efficiency of chemical compound extraction present in samples such as alcoholic beverages.

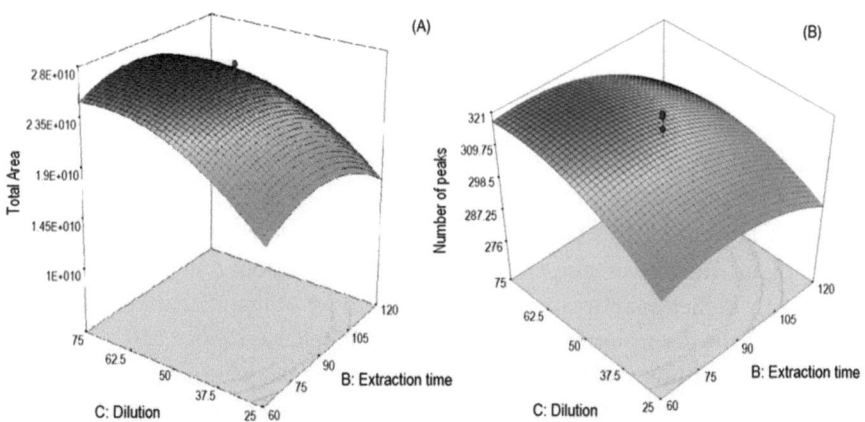

*Figure 3. Estimated response surface for the total chromatographic area (A) and the number of chromatographic peaks (B) obtained by plotting the extraction time versus the dilution for the EG/Silicone bar.*

The quadratic effect for time is significant, as it shows a maximum time where a greater number of polar compounds are adsorbed in the bar. If the bar is subjected to long extraction times, the amount of polar compounds decreases.

The best extraction conditions for the EG/Silicone bar were 70% dilution of tequila in water, a time of 85 min, and a stirring speed of 835 rpm.

**Exploration of Desorption Conditions**

The quadratic effects of desorption cryoconcentration temperatures were significant and the interaction was only significant for the total area, as seen in Figure 4. This interaction shows that a low cryoconcentration temperature produces the desorption of a greater number of chemical compounds when a high desorption temperature is used, which causes a greater number of compounds to be removed from the bar. It is for this reason that interaction is significant only for the total area because there is the same number of compounds. However, when undergoing a reconcentration process, there is an increase in the total area.

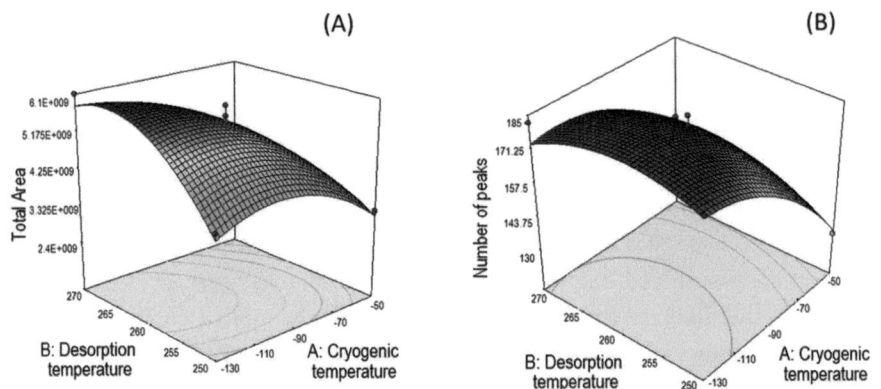

*Figure 4. Estimated response surface for the total chromatographic area (A) and the number of chromatographic peaks (B) obtained by plotting the desorption temperature versus the cryogenic temperature for the PDMS bar.*

The best desorption conditions for the PDMS bar were a cryogenic temperature of -124 °C and a desorption temperature of 264 °C.

For desorption conditions of the EG/Silicone bar, Figure 5 shows that a higher amount of compounds is obtained at a lower temperature of cryoconcentration. A high desorption temperature is required to have a higher quantity of compounds. Because of manufacturer recommendation for the EG/Silicone bar, the desorption temperature must not be raised above 220 °C.

The analytical desorption conditions for the EG/Silicone bar were a cryogenic temperature of -130 °C and a desorption temperature of 220 °C. Table 3 shows a compilation of the extraction and desorption conditions for both SBSE bars used in this work.

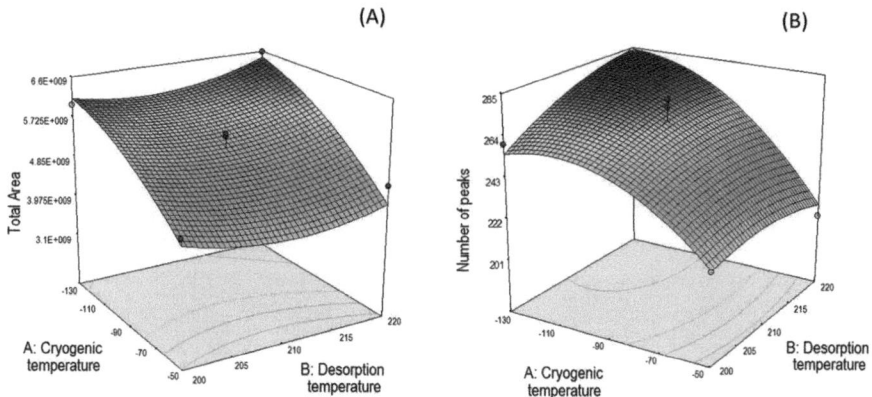

*Figure 5. Estimated response surface for the total chromatographic area (A) and the number of chromatographic peaks (B) obtained by plotting the desorption temperature versus the cryogenic temperature for the EG/Silicone bar.*

**Table 3. Extraction and Desorption Conditions for the PDMS and EG/Silicone Bars**

| *PDMS* | |
|---|---|
| **Extraction** | |
| **Stirring rate** | 1295 rpm |
| **Extraction time** | 140 min |
| **Dilution** | 59% of tequila and 41% of water |
| **Desorption** | |
| **Cryoconcentration temperature (°C)** | -123.71 °C |
| **Desorption temperature (°C)** | 264 °C |
| *EG/Silicone* | |
| **Extraction** | |
| **Stirring rate** | 835 rpm |
| **Extraction time** | 85 min |
| **Dilution** | 70% of tequila and 30% of water |
| **Desorption** | |
| **Cryoconcentration temperature (°C)** | -130 °C |
| **Desorption temperature (°C)** | 220 °C |

## Application of the SBSE Methodology for the Analysis of Chemical Compounds in Tequila and Other Agave Spirits

A total of 590 chemical compounds were identified using the PDMS bar and 560 compounds with the EG/Silicone bar (a complete list is not shown because of space limit). Sixty-five compounds were different between the identified lists for the two bars. This is by far the greatest number of chemical compounds reported for tequila, demonstrating that the SBSE technique is highly sensitive and provides better information that any previous methodology.

Principal component analyses of tequila samples and other agave spirits obtained with the PDMS bar (Figure 6) reveal the 59 samples of tequila and agave distillates analyzed can be arranged in eight different groups, according to their chemical profiles.

There is an overlap between the aged, silver, and gold tequila groups. Due to their elaboration processing, the aging time for the aged tequilas (2 months) is not enough to clearly establish the formation of compounds that mark the separation from silver and gold tequilas.

Among the extra aged tequilas, there are some samples that are within the group of the aged tequilas, which suggests that these extra aged tequilas may not have the necessary time to age.

The mixed tequilas group can be differentiated from the 100% agave tequilas group; this can be explained by the lower content of syrup from the agave which causes different compounds to be produced during fermentation. An important observation came for sample AE1, which is an ultra aged tequila supposedly made of 100% agave, but its chemical profile falls into the ultra aged mixed tequilas group, suggesting that it is not a 100% agave tequila.

Mezcal, sotol, and bacanora spirits can also be differentiated from the tequila samples. This differentiation is based on the raw material from which the samples are made as different species of agave are used for each, thus generating different chemical compounds.

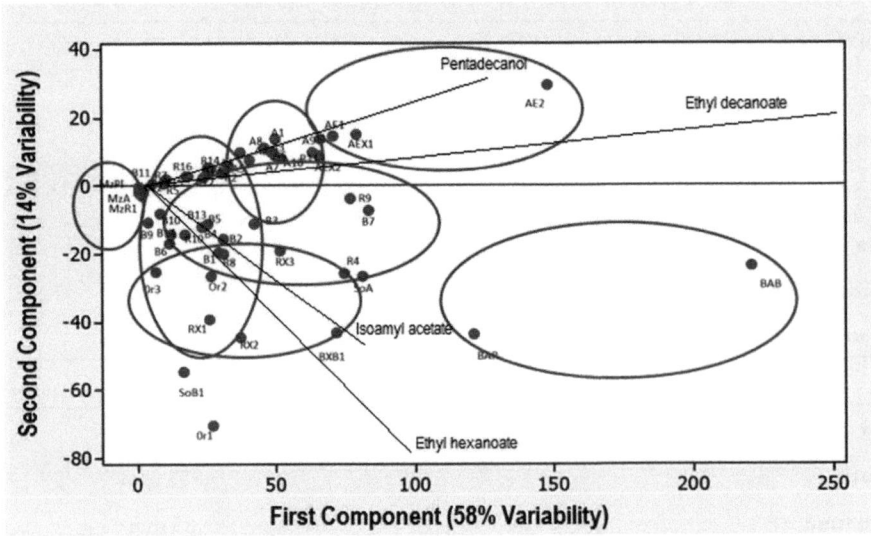

*Figure 6. Principal component analysis for the chemical profile of Tequila and agave distillates obtained by SBSE with the PDMS bar. Silver tequilas 100% agave: B, B1, B2, B3, B4, B5, B6, B7, B8, B9, B10, B11, B12, B13. Aged tequilas 100% agave: R1, R2, R3, R4, R5, R6, R7, R8, R9, R10, R11, R12, R13, R14, R15, R16. Extra aged tequilas 100% agave: A, A1, A2, A3, A4, A5, A6, A7, A8, A9, A10. Ultra aged tequilas 100% agave: AE, AE1, AE2, AE3. Silver mixed tequila: BX, BX1. Gold mixed tequila: Or, Or1, Or2, Or3, Or4. Aged mixed tequila: RX, RX1, RX2, RX3. Ultra aged mixed tequila: AEX, AEX1, AEX2. Silver bacanora: BAB. Aged bacanora: BAR. Silver sotol: SoB1. Aged sotol: SoA2. Silver mezcal: MzPl. Aged mezcal: MzR1. Extra aged mezcal: MzA.*

Figure 6 shows that esters were the compounds that most differentiated the samples analyzed. Although aging can contribute to the production of esters, the fermentation stage is the main one that contributes to the production of these compounds from silver tequila. Ethyl esters are present in other beverages such as whiskey, cognac, and rum due to the yeast metabolism during fermentation; these compounds are associated with pleasant fruit aromas. For example, ethyl decanoate has sweet notes of apple, wood, and butter; ethyl hexanoate also has sweet notes (22).

The extra aged and ultra aged tequilas differ when the ethyl decanoate occurs in greater concentration. It is reported that most esters can be the result of yeast metabolism in fermentation; however, they are also formed during the aging process by esterifying fatty acids in the presence of ethanol at high concentrations (4).

Pentadecanol has alcohol notes, having a projection towards samples of mezcal and aged tequilas.

Isoamyl acetate has green and sweet notes and was mainly found in silver, mixed, and gold tequilas.

Along with ethanol, yeast strain is the most important factor that influences the amount of higher alcohols produced because the yeasts metabolize carbohydrates, amino acids, and fatty acids, which transform them into higher alcohols, aldehydes, and ketones that are important fermentation products (23, 24).

Principal component analysis of tequila samples and other agave spirits obtained with the EG/Silicone bar (Figure 7) identified a similar grouping to what was obtained with the PDMS bar.

Sample AE1 follows the same behavior as in the PDMS bar, which suggests that it is not a 100% agave tequila because it is grouped closely with ultra aged mixed tequilas.

As with the PDMS bar, esters established the main trend toward extra aged tequilas; this behavior was also previously observed by Vallejo-Córdoba et al. (2004) (5).

Even though the EG/Silicone bar has a more polar nature, it did not show a difference in terms of the differentiation of the samples of tequila and agave distillates when compared to the PDMS bar. This behavior is must likely explained by the same compounds in higher concentration that are extracted by both bars.

The extraction with EG/Silicone bar shows ethyl octanoate as the major ester with aroma descriptors like wine, sweet, and butter. This compound occurs in higher concentration during the aging process (25).

Ethyl dodecanoate is a compound that marks the difference in the principal component analysis for the silver and aged tequilas. It has floral and fruit notes, which is somewhat expected because the agave has notes that are present in this compound (25).

Ethyl decanoate has sweet apple and woody notes; ethyl hexanoate has sweet notes as well as isoamyl acetate which has green notes. The differences between silver, gold, and mixed tequilas are categorized based on the concentration of isoamyl acetate.

According to Figure 7, mezcal and sotol differ from the rest of the samples mainly due to the existence of ethyl octanoate and ethyl hexanoate. This finding is in agreement with a De León-Rodríguez et.al. (2006) study that quantified these compounds in mezcal samples that were present in higher concentration.

Agave distillates were clearly separated from the rest of the tequila samples; this is due to the greater number of chemical compounds present in mezcal, sotol, and bacanora compared to the raw material. For example, mezcal has more than 10 varieties of agave with the most used being the *Agave salmiana* and the *Agave angustifolia*. The most used for the sotol is *Agave dasylirion* and the most used for bacanora is *Agave angustifolia haw*. A great variety of agaves are allowed for elaboration, and the specifications are not as clear as those for tequila.

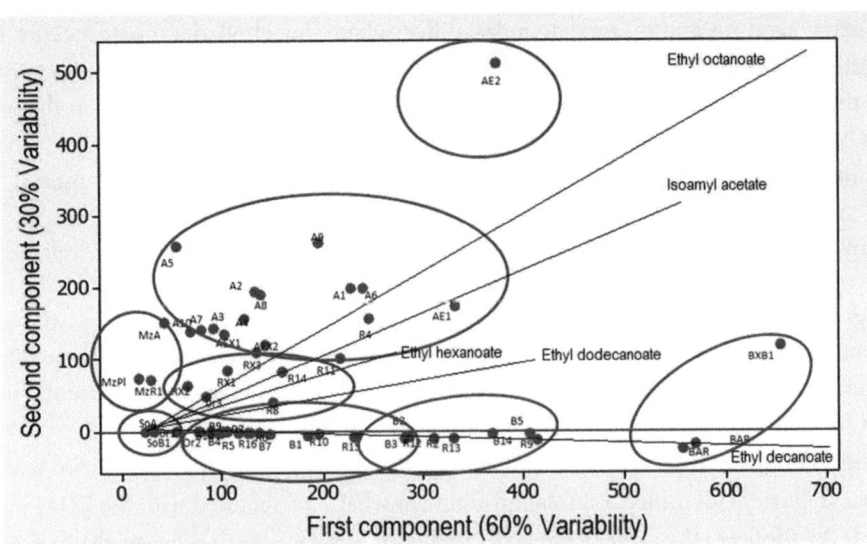

*Figure 7. Principal component analysis for the chemical profile of Tequila and agave distillates obtained by SBSE with the EG/Silicone bar. Silver tequilas 100% agave: B, B1, B2, B3, B4, B5, B6, B7, B8, B9, B10, B11, B12, B13. Aged tequilas 100% agave: R1, R2, R3, R4, R5, R6, R7, R8, R9, R10, R11, R12, R13, R14, R15, R16. Extra aged tequilas 100% agave: A, A1, A2, A3, A4, A5, A6, A7, A8, A9, A10. Ultra aged tequilas 100% agave: AE, AE1, AE2, AE3. Silver mixed tequila: BX, BX1. Gold mixed tequila: Or, Or1, Or2, Or3, Or4. Aged mixed tequila: RX, RX1, RX2, RX3. Ultra aged mixed tequila: AEX, AEX1, AEX2. Silver bacanora: BAB. Aged bacanora: BAR. Silver sotol: SoB1. Aged sotol: SoA2. Silver mezcal: MzPl. Aged mezcal: MzR1. Extra aged mezcal: MzA.*

## Conclusions

The SBSE technique was highly sensitive for the analysis of chemical compounds in tequila and identified 590 compounds using the PDMS bar and 560 compounds using the EG/Silicone bar. This is the highest amount of chemical compounds reported for tequila.

The most influential chemical compounds for differentiating between tequila classes were pentadecanol, isoamyl acetate, ethyl octanoate, ethyl hexanoate, ethyl decanoate, and ethyl dodecanoate. It was also possible to differentiate tequila when compared to other agave distillates.

The SBSE technique implemented in this work could help develop new methodologies based on specific compounds that could serve as markers of the aging process, recognize complex patterns for authenticating tequila origin, and help to avoid fraud thus guaranteeing the best quality for consumers.

## References

1.  Diario Oficial de la Federacion. NOM-006-SCFI-2012. *Bebidas alcohólicas-Tequila-Especificaciones.* http://www.dof.gob.mx/nota_detalle.php?codigo=5282165&fecha=20/10/2018 (accessed Oct 20, 2018).

2.  MacNamara, K.; Hoffmann, A. Gas Chromatographic Technology in Analysis of Distilled Spirits. *Dev. Food Sci.* **1998**, *39*, 303–346.

3. Peña-Alvarez, A.; Capella, S.; Juárez, R.; Labastida, C. Determination of Terpenes in Tequila by Solid Phase Microextraction-Gas Chromatography–mass Spectrometry. *J. Chromatogr. A.* **2006**, *1134*, 291–297.

4. Benn, S. M.; Peppard, T. L. Characterization of Tequila Flavor by Instrumental and Sensory Analysis. *J. Agric. Food Chem.* **1996**, *44*, 557–566.

5. Vallejo-Cordoba, B.; González-Córdova, A. F.; del Carmen Estrada-Montoya, M. Tequila Volatile Characterization and Ethyl Ester Determination by Solid Phase Microextraction Gas Chromatography/Mass Spectrometry Analysis. *J. Agric. Food Chem.* **2004**, *52*, 5567–5571.

6. López, M. G.; Dufour, J. P. Tequilas: Charm Analysis of Blanco, Reposado, and Anejo Tequilas. In *Gas Chromatography-Olfactometry*; Leland, J. V.; Schieberle, P.; Buettner, A.; Acree, T. E., Eds.; ACS Symposium Series 782; American Chemical Society: Washington, DC, 2001; pp 6–62.

7. Muñoz-Muñoz, A. C.; Grenier, A. C.; Gutiérrez-Pulido, H.; Cervantes-Martínez, J. Development and Validation of a High Performance Liquid Chromatography-Diode Array Detection Method for the Determination of Aging Markers in Tequila. *J. Chromatogr. A* **2008**, *1213*, 218–223.

8. Martínez-López, G.; Luna-Moreno, D.; Monzón-Hernández, D.; Valdivia-Hernández, R. Optical Method to Differentiate Tequilas Based on Angular Modulation Surface Plasmon Resonance. *Opt. Lasers Eng.* **2011**, *49*, 675–679.

9. Frausto-Reyes, C.; Medina-Gutiérrez, C.; Sato-Berrú, R.; Sahagún, L. R. Qualitative Study of Ethanol Content in Tequilas by Raman Spectroscopy and Principal Component Analysis. *Spectrochim. Acta Part A Mol. Biomol. Spectrosc.* **2005**, *61*, 2657–2662.

10. Ceballos-Magaña, S. G.; de Pablos, F.; Jurado, J. M.; Martín, M. J.; Alcázar, Á.; Muñiz-Valencia, R.; Gonzalo-Lumbreras, R.; Izquierdo-Hornillos, R. Characterisation of Tequila According to Their Major Volatile Composition Using Multilayer Perceptron Neural Networks. *Food Chem.* **2013**, *136*, 1309–1315.

11. Barbosa-García, O.; Ramos-Ortíz, G.; Maldonado, J. L.; Pichardo-Molina, J. L.; Meneses-Nava, M. A.; Landgrave, J. E. A.; Cervantes-Martínez, J. UV–vis Absorption Spectroscopy and Multivariate Analysis as a Method to Discriminate Tequila. *Spectrochim. Acta Part A Mol. Biomol. Spectrosc.* **2007**, *66*, 129–134.

12. Contreras, U.; Barbosa-García, O.; Pichardo-Molina, J. L.; Ramos-Ortíz, G.; Maldonado, J. L.; Meneses-Nava, M. A.; Ornelas-Soto, N. E.; López-de-Alba, P. L. Screening Method for Identification of Adulterate and Fake Tequilas by Using UV–VIS Spectroscopy and Chemometrics. *Food Res. Int.* **2010**, *43*, 2356–2362.

13. Baltussen, E.; Sandra, P.; David, F.; Cramers, C. Stir Bar Sorptive Extraction (SBSE), a Novel Extraction Technique for Aqueous Samples: Theory and Principles. *J. Microcolumn Sep.* **1999**, *11*, 737–747.

14. Alves, R. F.; Nascimento, A. M. D.; Nogueira, J. M. F. Characterization of the Aroma Profile of Madeira Wine by Sorptive Extraction Techniques. *Anal. Chim. Acta* **2005**, *546*, 11–21.

15. Zalacain, A.; Alonso, G. L.; Lorenzo, C.; Iñiguez, M.; Salinas, M. R. Stir Bar Sorptive Extraction for the Analysis of Wine Cork Taint. *J. Chromatogr. A* **2004**, *1033*, 173–178.

16. Díez, J.; Domínguez, C.; Guillén, D. A.; Veas, R.; Barroso, C. G. Optimisation of Stir Bar Sorptive Extraction for the Analysis of Volatile Phenols in Wines. *J. Chromatogr. A* **2004**, *1025*, 263–267.

17. Delgado, R.; Durán, E.; Castro, R.; Natera, R.; Barroso, C. G. Development of a Stir Bar Sorptive Extraction Method Coupled to Gas Chromatography-Mass Spectrometry for the Analysis of Volatile Compounds in Sherry Brandy. *Anal. Chim. Acta* **2010**, *672*, 130–136.

18. Durán Guerrero, E.; Cejudo Bastante, M. J.; Castro Mejías, R.; Natera Marín, R.; García Barroso, C. Characterization and Differentiation of Sherry Brandies Using Their Aromatic Profile. *J. Agric. Food Chem.* **2011**, *59*, 2410–2415.

19. Camino-Sánchez, F. J.; Rodríguez-Gómez, R.; Zafra-Gómez, A.; Santos-Fandila, A.; Vílchez, J. L. Stir Bar Sorptive Extraction: Recent Applications, Limitations and Future Trends. *Talanta* **2014**, *130*, 388–399.

20. Marín, J.; Zalacain, A.; De Miguel, C.; Alonso, G. L.; Salinas, M. R. Stir Bar Sorptive Extraction for the Determination of Volatile Compounds in Oak-Aged Wines. *J. Chromatogr. A* **2005**, *1098*, 1–6.

21. Ebeler, S. E.; Terrien, M. B.; Butzke, C. E. Analysis of Brandy Aroma by Solid-Phase Microextraction and Liquid–Liquid Extraction. *J. Sci. Food Agric.* **2000**, *80*, 625–630.

22. Mestres, M.; Sala, C.; Martı, M.; Busto, O.; Guasch, J. Headspace Solid-Phase Microextraction of Sulphides and Disulphides Using Carboxen–Polydimethylsiloxane Fibers in the Analysis of Wine Aroma. *J. Chromatogr. A* **1999**, *835*, 137–144.

23. Christoph, N.; Bauer-Christoph, C. Flavour of Spirit Drinks: Raw Materials, Fermentation, Distillation, and Ageing BT. In *Flavours and Fragrances: Chemistry, Bioprocessing and Sustainability*; Berger, R. G., Ed.; Springer Berlin Heidelberg: Berlin, 2007; pp 219–239.

24. Pinal, L.; Cedeño, M.; Gutièrrez, H.; Alvarez-Jacobs, J. Fermentation Parameters Influencing Higher Alcohol Production in the Tequila Process. *Biotechnol. Lett.* **1997**, *19*, 45–47.

25. De León-Rodríguez, A.; González-Hernández, L.; Barba de la Rosa, A. P.; Escalante-Minakata, P.; López, M. G. Characterization of Volatile Compounds of Mezcal, an Ethnic Alcoholic Beverage Obtained from Agave Salmiana. *J. Agric. Food Chem.* **2006**, *54*, 1337–1341.

# Editors' Biographies

## Brian Guthrie

Brian Guthrie currently performs research to understand the chemical and physical origins of human sensations during the oral processing of food. He has worked in the food and ingredients industries with responsibilities spanning from knowledge building to utilizing fundamental science to formulation and product development. Brian has also worked extensively in food sensory science, from studies on the peripheral cellular events of olfaction and gustatory signal transduction to developing the understanding of consumer preference and choice. He has also been involved in exploring the use of sensory technology in food and ingredient applications.

## Jonathan D. Beauchamp

Jonathan Beauchamp holds a physics degree from University College London, UK and a PhD in environmental physics from the University of Innsbruck, Austria. He currently works as a research associate at the Fraunhofer Institute for Process Engineering and Packaging IVV in Freising, Germany, where he is group manager for volatile emissions and deputy head of the department of sensory analytics. Jonathan's primary research focus encompasses the characterization of emissions of volatile organic compounds (VOCs) in the field of food and flavor, in non-food applications, and from the human volatilome, primarily via exhaled breath. His further scientific interests and activities relate to the detection and perception of odor-active compounds in human olfaction. Jonathan's technical expertise revolve around the use of chemical ionization mass spectrometry (CIMS), in particular proton transfer reaction mass spectrometry (PTR-MS) for analyzing volatile emissions in real time.

Jonathan has published over 50 papers and book chapters that have been cited over 1000 times, and he has coedited a previous book within this American Chemical Society (ACS) Symposium Series. He has co-organized and chaired several ACS symposia and has been on the organizing committee of other international conferences, notably the PTR-MS Conference series and the International Association of Breath Research (IABR) Breath Summit series. He has been an active member of ACS and the Agricultural and Food Chemistry Division (AGFD) since 2013 and additionally is a member of the IABR board. Jonathan is affiliated with several journals, including being a current associate editor of *Journal of Breath Research*, and he sits on the editorial boards of *Heliyon* and *Food Packaging and Shelf Life*.

## Andrea Buettner

Andrea Buettner is a professor of flavor science (a food chemist with a PhD in aroma research) and currently holds the Chair of Aroma and Smell Research at Friedrich-Alexander-Universität Erlangen-Nürnberg (FAU) in Erlangen, Germany. She is also deputy director of the Fraunhofer Institute of Process Engineering and Packaging IVV in Freising, Germany and head of the department of sensory analytics. She recently coinitiated the Campus of the Senses initiative, which she also currently heads, that combines interdisciplinary basic research with application-oriented research on the subject of digital sensory transition.

Andrea's work has demonstrated the importance of the combined effects of the food and matrix composition, saliva, mucosa, mastication, and swallowing on flavor release and perception. She has identified new odorous compounds with a special focus on citrus and other food materials and has worked on structure-odor relationships. Recent work additionally addresses packaged foods and articles of everyday use in relation to their chemosensorially active constituents. She specializes in chemical trace analysis, monitoring of physiological processes, psychophysical measurements of sensory data, and determination of chemosensorially active and volatile compounds and their metabolites in vivo. Recently, Andrea has broadened her research interests to include the digitization of the human senses, monitoring volatiles and chemosensorially active compounds in the physiological context, and exploration of the physiological impact of such derivatives in humans. Her work has contributed to our understanding of psychological and behavioral aspects in human nutrition. Andrea has been awarded with several accolades including the ACS Fellow Award from the Agricultural and Food Chemistry Division (AGFD), the 2013 Nutricia Science Award, the 2012 Danone Innovation Prize, the ACS Young Investigator Award (AGFD), and the 2010 Kurt-Täufel Award for Young Scientists.

## Stephen Toth

Stephen Toth is a senior research investigator in the research analytical services department at International Flavors and Fragrances (IFF). He has spent his lengthy career at IFF working in fragrance delivery, chromatography, and ultimately the structure elucidation department. He is active in the Agricultural and Food Chemistry Division of the American Chemical Society and currently serves as treasurer of the division. He has organized several analytical and flavor chemistry symposia at the American Chemical Society's national meetings. He holds a B.S. and M.S. in chemistry from Seton Hall University and a PhD in food science from Rutgers University.

## Michael C. Qian

Michael C. Qian received his BS in chemistry from Wuhan University and a PhD from University of Minnesota. He is a full professor of flavor chemistry at Oregon State University. Professor Qian's research at Oregon State University involves aroma and flavor chemical and biochemical generation in food and beverage systems, with a focus on wine and wine grapes. He has made significant contributions to the basic understanding of viticultural practices on volatile and volatile precursor formation in wine grapes and the implication on wine quality. He applied the flavor chemistry theory and principle and pioneered the flavor chemistry research in Chinese liquor (baijiu). Dr. Qian has published extensively in *J. of Agricultural and Food Chemistry, Food Chemistry, J. Chromatography A*, and other leading journals. His publications have been cited 3800 times by international researchers, with an h-index of 36 and i10-index of 64. Due to his outstanding contribution to the area of flavor chemistry, he was elected as the Fellow of Agricultural and Food Chemistry Division of the American Chemical Society. Dr. Qian is a leader in his profession. He has organized over ten flavor chemistry symposia at the American Chemical Society's national meetings as well as the Pacifichem conference. He coedited five ACS Symposium Series books on flavor chemistry. He is the initiator and the chair of the International Flavor and Fragrance Conference series. He has held various officer positions in the Agricultural and Food Chemistry Division of the American Chemical Society, including the chair position of the division in 2014. He is the recipient of the Distinguished Service Award from the Agricultural and Food Chemistry Division (AGFD) of the American Chemical Society of 2018 due to his outstanding service contributions to the field of agricultural and food chemistry to improve the safety, supply, or quality of food.

# Indexes

# Author Index

# Subject Index